AF892719

The Uses of Cyclotrons in Chemistry, Metallurgy and Biology

The Uses of Cyclotrons in Chemistry, Metallurgy and Biology

Proceedings of a Conference held at
St. Catherine's College, Oxford,
22-23 September 1969

Edited by
C. B. Amphlett, Ph.D., D.Sc.,
Chemistry Division, A.E.R.E., Harwell, Berks.

Published And Distributed In The United States By
DANIEL DAVEY & CO., INC.
Hartford, Connecticut

LONDON
BUTTERWORTHS

ENGLAND:	BUTTERWORTH & CO. (PUBLISHERS) LTD. LONDON: 88 Kingsway, WC2B 6 AB
AUSTRALIA:	BUTTERWORTH & CO. (AUSTRALIA) LTD. SYDNEY: 20 Loftus Street MELBOURNE: 343 Little Collins Street BRISBANE: 240 Queen Street
CANADA:	BUTTERWORTH & CO. (CANADA) LTD. TORONTO: 14 Curity Avenue, 374
NEW ZEALAND:	BUTTERWORTH & CO. (NEW ZEALAND) LTD. WELLINGTON: 49/51 Ballance Street AUCKLAND: 35 High Street
SOUTH AFRICA:	BUTTERWORTH & CO. (SOUTH AFRICA) (Pty) LTD DURBAN: 33/35 Beach Grove

© U.K.A.E.A. 1970

ISBN: 0 408 70029 7

Typeset by Peridon Ltd

Printed in Great Britain by
J. W. Arrowsmith Ltd., Bristol

CONTENTS

SESSION 6

SESSION 7

SESSION 8

ROUND-TABLE DISCUSSION

FOREWORD

The development of the variable energy cyclotron and the production of 'compact' cyclotrons which may be purchased to order have greatly increased the scope of this type of accelerator for research in fields other than nuclear physics. It was therefore felt appropriate that the 5th International Cyclotron Conference should be followed by a meeting in which cyclotron users from many fields could describe and discuss their work and problems. The level of interest shown in what were widely diverse topics is indicated by the high proportion of the total number of delegates who were present at any given session; furthermore, an appreciable overlap in attendance between the two conferences suggests the possibility of useful cross-fertilisation between designers, operators and users.
At the end of the meeting it was suggested that a similar meeting should be held in a few years' time, possibly in connection with the 6th Cyclotron Conference. It is important that there should be other meetings similar to the present one, since only in this way will the maximum effect be extracted from so important and costly a field of work.
In conclusion, I would like to thank all those who participated in the Conference, not least of all those whose untiring work behind the scenes made it possible.

C. B. Amphlett

A.E.R.E.,
Harwell

SESSION 1

Nuclear Chemistry

Chairman: Mr. G. N. Walton

Nuclear charge dispersion in the fission of heavy nuclei by medium-energy protons[1]

L. Yaffe

Radiochemistry Laboratory, Department of Chemistry, McGill University, Montreal, Quebec, Canada

ABSTRACT

Fission products from the fission of ^{233}U, ^{235}U, ^{238}U, ^{232}Th, and ^{239}Pu have been studied with a view to elucidating the distribution of nuclear charge in fission caused by protons of energies 20–85 MeV.

The importance of the neutron-to-proton ratio of the fissioning species is apparent. Excitation functions are shown for fission products with mass numbers $A = 130-140$. The energy at which the excitation function of these individual fission products reaches its maximum shows the dependence on nuclear constitution.

Charge distribution data show the role of the neutron-to-proton ratio of the fissioning system. In general, with increasing bombarding energy, the distribution curves become broader, and the most probable charge, Z_p, moves closer to stability. The experimental values of Z_p are compared with those obtained from calculations assuming various charge distribution postulates.

Some comments are made about the numbers of neutrons emitted from the various fissioning species at different excitation energies.

INTRODUCTION

Nuclear fission was discovered just over thirty years ago. It is the basis of a multi-billion dollar industry and, in the opinion of some people, is one of the major means of solving the problem of raising the standard of living in some of the industrially-poorer nations. It is somewhat surprising to find that no theory of fission exists which can explain the experimental facts available, much less predict results. It is even more surprising to find that many of the experimental data are still missing.

With this in mind our laboratory is investigating three aspects of nuclear

[1] This research was supported by the National Research Council of Canada

fission: (a) a study of the nuclear charge distribution as a function of bombarding energy. This is particularly important because, particularly at higher energies, this information is essential for determining so fundamental a parameter as the mass distribution of the fission products. Such information can throw a good deal of light on the mechanistic aspects of fission. (b) Studies of isomer ratios of fission products as a function of bombarding energies, which give information about angular momentum in the fission process. (c) Recoil studies of fission products, which give information about the deposition energy involved in the fission process.

This paper deals with the distribution of nuclear charge in the fission of ^{233}U, ^{235}U, ^{238}U, ^{232}Th, and ^{239}Pu as a function of energy by protons in the energy region 20–85 MeV. This energy region is intrinsically very interesting because it is generally believed that a transition gradually takes place from the low-energy compound-nucleus type of reaction to the direct interaction type.

At thermal energies most of the definitive work on charge distribution is due to Wahl and his co-workers[1-3] and deals only with ^{235}U. Wahl[4] also investigated charge distribution in the fission of ^{235}U with 14-MeV neutrons. Pate, Foster, and Yaffe[5] were the first to initiate a systematic study of nuclear charge distribution in fission as a function of energy, by bombarding ^{232}Th with protons of energy 8–87 MeV. They showed that the full-width at half-maximum of the charge distribution curves remained constant up to 25-MeV bombarding energy, and then the curves widened as the bombarding energy increased. This broadening was accompanied by a shift of the most probable charge, Z_p, towards Z_A, the nuclear charge associated with stability for mass A. The constancy of the full-width at half-maximum (FWHM) was interpreted as possibly indicating a single fissioning species at these energies. Pate[6] attempted to analyse these data in terms of the two charge distribution postulates current at that time (and still current):

(a) Unchanged Charge Distribution (UCD)[7], i.e. the fission products will have the same neutron-to-proton ratio (N/Z) as the parent fissioning nucleus. According to this prescription, the most probable charge, Z_p, of an isobaric chain of fission products f_p of mass A from a target nucleus with mass A_f and nuclear charge Z_f can be calculated from the following expression:

$$(Z_p)_{f_p} = (A_{f_p} + \nu_f)\frac{Z_f}{A_f - \nu_{pf}} \tag{1}$$

where ν_f and ν_{pf} are the average number of neutrons evaporated by the fission fragment after fission and by the target nucleus prior to fission respectively. For thermal neutron fission this obviously simplifies to

$$(Z_p)_{f_p} = (A_{f_p} + \nu_f)\frac{Z_f}{A_f} \tag{2}$$

(b) Equal Charge Displacement (ECD)[8]. This postulates that the two fragments adopt such a charge distribution that $(Z_A - Z_p)_l = (Z_A - Z_p)_h$ where l and h refer to the light and heavy fragments respectively.

Much of the thermal neutron data can be correlated with the ECD hypothesis. Pate[6] found that the thorium data[5] were 'not inconsistent' with the ECD postulate.

Friedlander et al[9] made a comprehensive study of the charge dispersion of various caesium isotopes formed in the fission of ^{238}U by protons of energies from 0·1 to 6·2 GeV by radiometric and mass spectrometric measurements. The shift of Z_p towards Z_A and the broadening of the charge dispersion curves as the bombarding energy was increased are clearly evident from this work. Consideration of the recoil behaviour of some of the fission products showed that the neutron-deficient products were formed in high-deposition-energy processes. The data were best interpreted by the UCD mechanism. These and other high-energy data are extremely well summarised by Friedlander,[10] and no further reference to energies higher than 100 MeV will be made in this paper unless the work can elucidate the lower-energy processes. This paper summarises and correlates results obtained in our laboratory by a large number of research workers. Pertinent details are to be found in references 11–16, and much of the data presented here has also recently been given as paper SM-122/3 at the Symposium on the Physics and Chemistry of Fission (International Atomic Energy Agency).

EXPERIMENTAL

(a) Irradiations: All irradiations were made in the circulating beam of the McGill 82 in proton synchrocyclotron. In general the target foils were irradiated in stacks of three to compensate for fission product recoil, and the middle foil was used for radiochemical separation. The beam was monitored by either aluminium or copper foils, depending on the energy range, and the monitor reactions used were ^{27}Al (p, 3pn) ^{24}Na, ^{63}Cu (p, n) ^{63}Zn, and ^{65}Cu (p, pn) ^{64}Cu in the usual way. ^{233}U was irradiated as a powder (U_3O_8) in an aluminium tube and the monitor intimately mixed with it. ^{239}Pu was irradiated as the Pu-Al alloy. The beam current was of the order of 1 μA and the irradiation time varied from 2 to 15 min depending on the half-life of the nuclides being investigated.

(b) Chemical Separations: Standard radiochemical techniques were used to separate the radioactive nuclides.[11-16] For nuclides which were shielded from formation by β-decay by the presence of either stable or long-lived neighbouring isobars only one chemical separation and further purification was necessary. In other cases where the half-lives of the precursors were relatively short, either one irradiation and several timed separations were made, or in some cases two separate irradiations with varying times of separation. The method is described elsewhere[9] in detail. In this way it was possible to determine the *independent* yields of the fission products as distinct from the chain or *cumulative* yields.

(c) Activity Measurements: After chemical separation and purification the activities of the nuclides were measured by end-window β^-, $4\pi\beta^-$, β^+ coincidence, NaI and Ge (Li) counters. The decay curves were analysed by the Cumming CLSQ programme.[17]

(d) Treatment of data: The counting rates were transformed to disintegration rates by correcting for counter efficiencies, branching ratios, and chemical yields. The rates were then converted into formation cross sections, based on the monitor cross sections, using standard equations.[9]

RESULTS AND DISCUSSION

(a) Excitation Functions: Friedlander et al[9] have shown that in the fission of ^{238}U the probability of the independent production of neutron-deficient fission products increases as the bombarding energy is increased (Figs. 1 and 2). Fig. 1 combines results of Davies and Yaffe[11] and of Friedlander et al[9] and shows the behaviour of the excitation function for the *independent* formation of a typical neutron-excess nuclide in this mass region, viz. ^{136}Cs in the fission of ^{235}U. The excitation function reaches a maximum at ~ 60 MeV, i.e. at this energy ^{136}Cs is a highly probable product. Fig. 2 shows the behaviour of the excitation

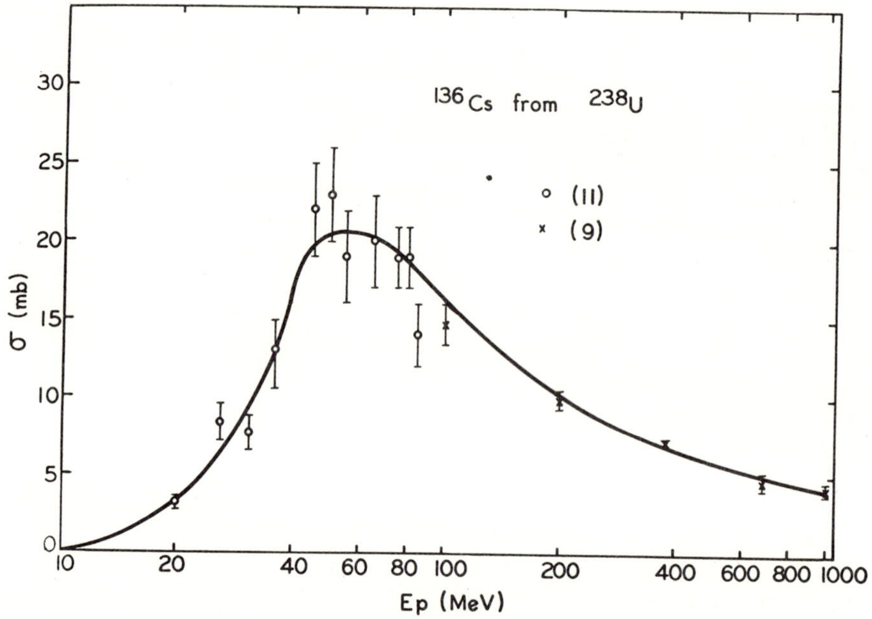

Fig. 1. Excitation Function of ^{136}Cs *from the Fission of* ^{238}U

functions of the independent formation of neutron-deficient isotopes of caesium. As the energy of bombardment increases, the formation of the more neutron-deficient isotopes becomes more probable. Friedlander et al[9] showed (Fig. 3), that for the fission of ^{238}U there is a correlation between the bombardment energy at which the excitation function for the independent formation of fission product will reach a maximum and the neutron-to-proton ratio (N/Z) of the *product*. This work has been extended in our laboratory by studying the fission of other heavy nuclei. In some cases the studies have not been taken to high enough energies to get complete excitation functions of the most neutron-deficient isotopes of caesium. However, the same relation between the bombardment energy at which the excitation function reaches its maximum and the value of N/Z of the product holds. This is clearly evident in Fig. 4. There is of course a fair amount of uncertainty in determining the maxima of these peaks, and thus the slopes of the lines have a large uncertainty and need more data for a clearer definition. However, the trends are perfectly obvious. The lines seem to converge at a value of $N/Z \approx 1{\cdot}55$. This would mean that ^{140}Cs, for example, would be formed as a fission product with very high probability at very low energies for all targets, and this has indeed been observed in thermal and spontaneous fission.

What is probably even more important is the correlation between the neutron-to-proton ratio of the target and the probability of formation of the

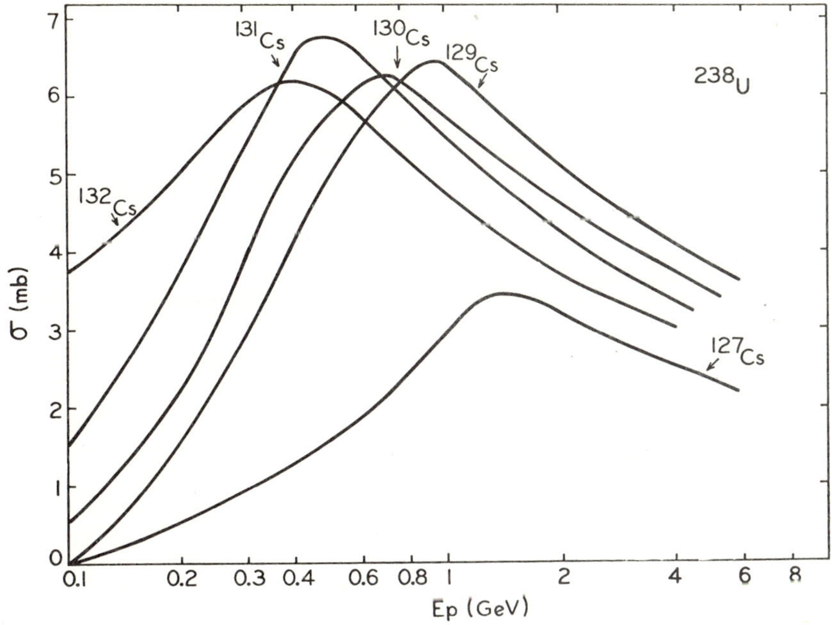

Fig. 2. Excitation Functions of Neutron-Deficient Isotopes of Caesium from the Fission of ^{238}U

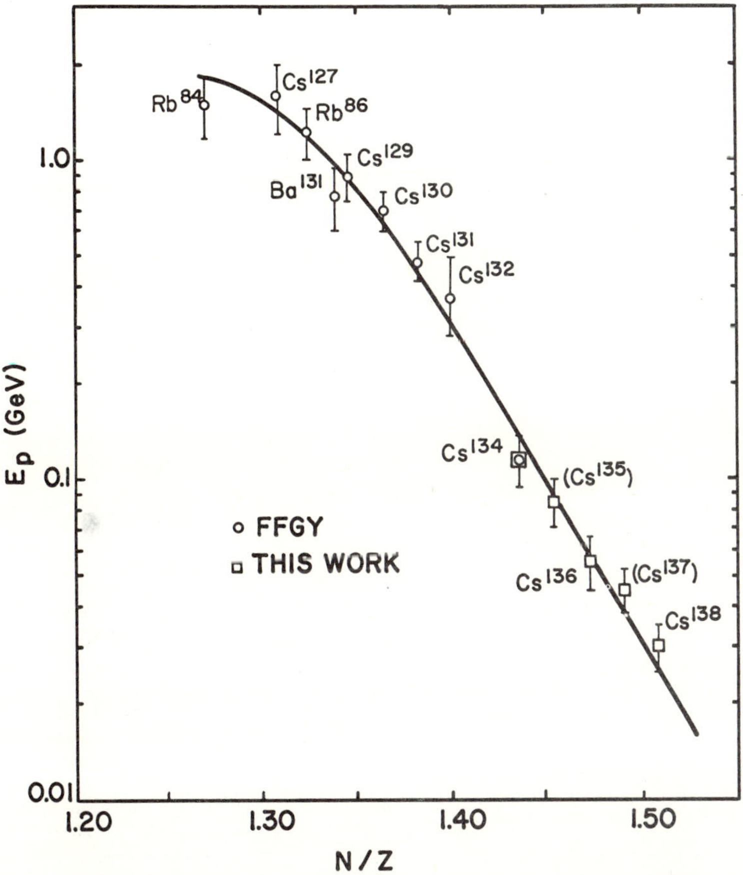

Fig. 3. Energies at which the Excitation Functions of various Cs *isotopes from the fission of* ^{238}U *reach their maxima plotted against the neutron-to-proton ratio of the fission product*

fission product as a function of its N/Z ratio. Fig. 5 shows data for ^{136}Cs and ^{130}Cs from both ^{238}U and ^{233}U targets. It is readily seen that the excitation function of ^{136}Cs reaches a maximum at a lower energy when the target is ^{233}U (N/Z = 1·533) than for ^{238}U (N/Z = 1·587). Even more striking is the fact that the cross-sections in the rising portion of the excitation function are a factor of 5 higher for formation from ^{233}U than from ^{238}U at the same bombardment energy. The same is seen to be true for the formation of ^{130}Cs. For example, the independent formation cross-section of ^{129}Cs in the fission of ^{238}U was barely measurable (<0·05 mb) by Friedlander et al[9] at a bombarding energy of 100 MeV. This nuclide is formed with a sizable independent formation cross-section (2·8 mb) at 85 MeV[14] in the fission of ^{233}U.

This shows clearly the importance of the neutron-to-proton ratio in the target

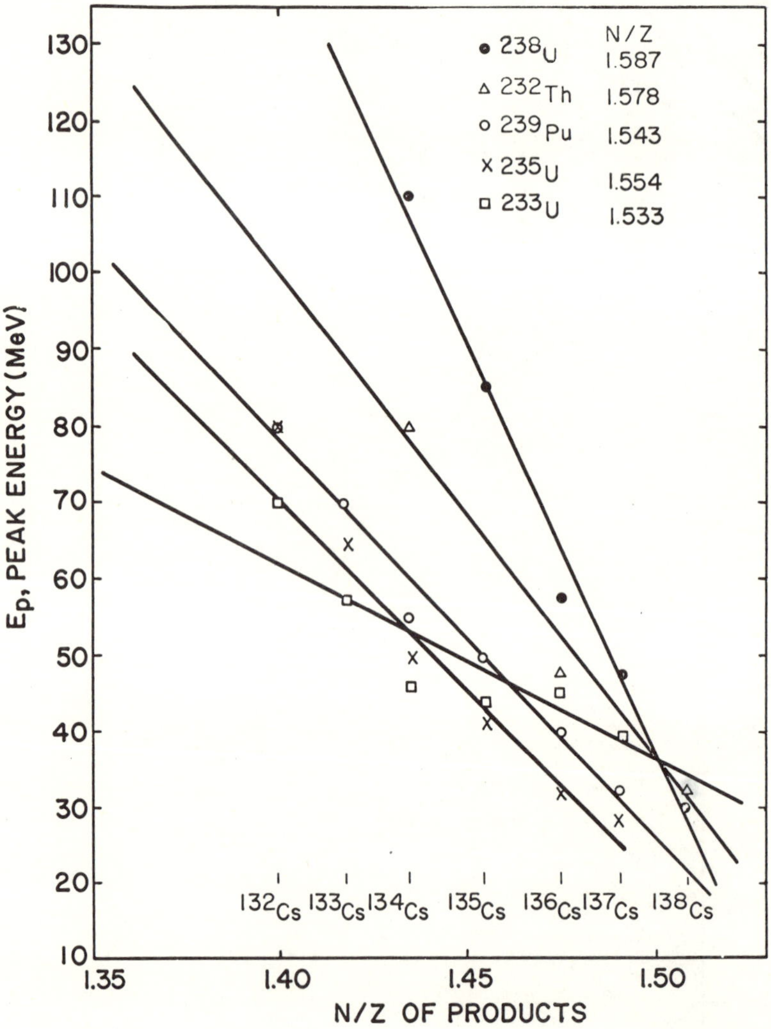

Fig. 4. Energies at which the Excitation Functions of various caesium isotopes from the fission of ^{233}U, ^{235}U, ^{238}U, ^{232}Th, *and* ^{239}Pu *reach their maxima plotted against the neutron-to-proton ratio of the fission product*

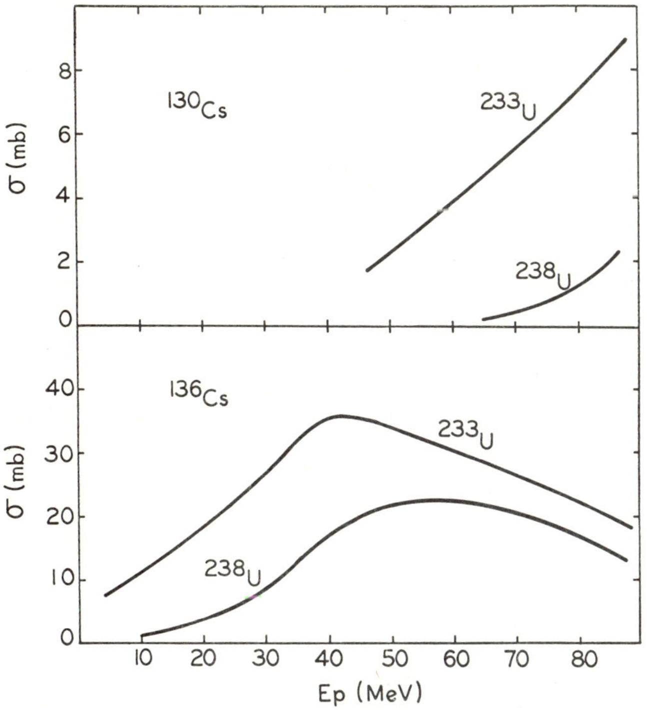

Fig. 5. Excitation Functions of ^{130}Cs *and* ^{136}Cs *from the Fission of* ^{233}U *and* ^{238}U

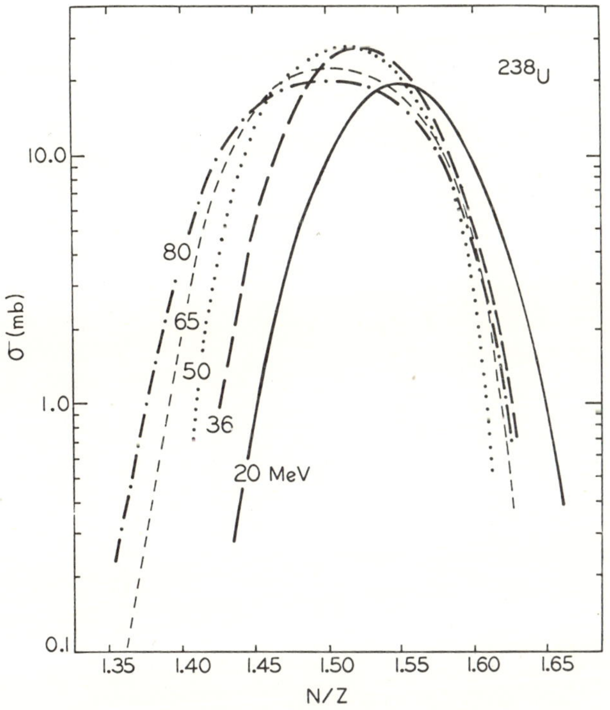

Fig. 6. Charge Dispersion Curves from the Fission of ^{238}U

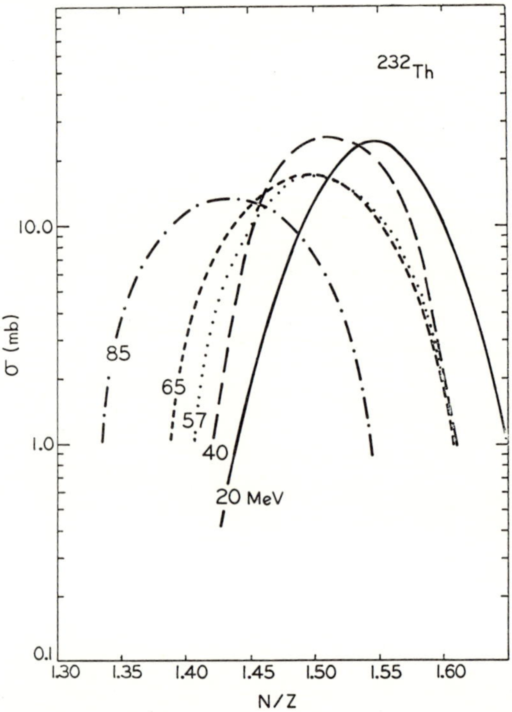

Fig. 7. Charge Dispersion Curves from the Fission of ^{235}U

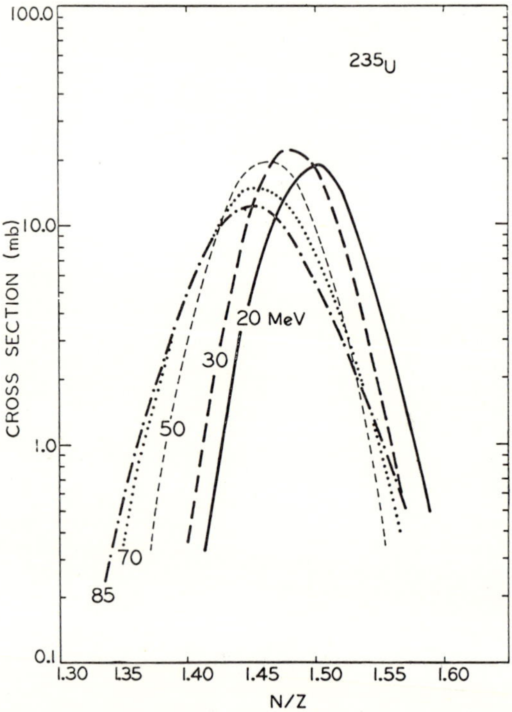

Fig. 8. Charge Dispersion Curves from the Fission of ^{233}U

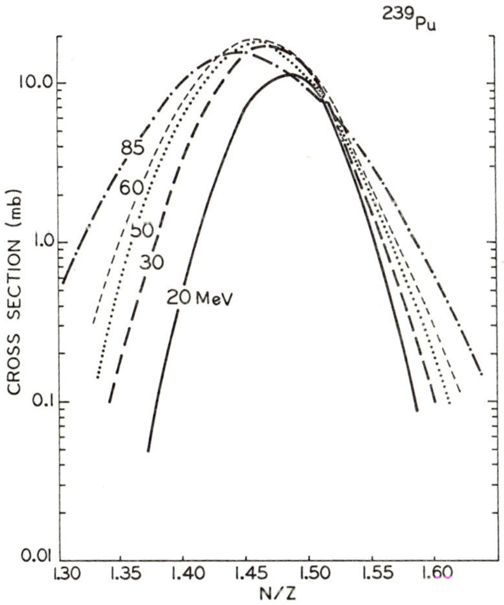

Fig. 9. Charge Dispersion Curves from the Fission of ^{232}Th

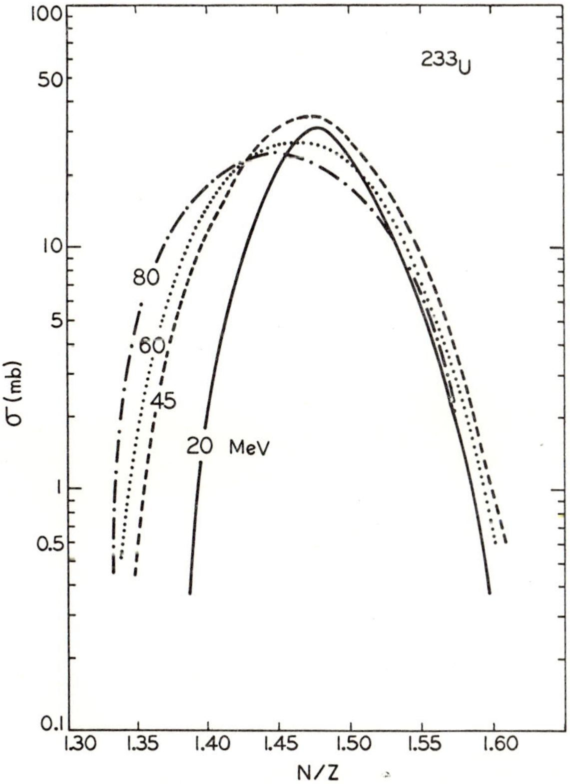

Fig. 10. Charge Dispersion Curves from the Fission of ^{239}Pu

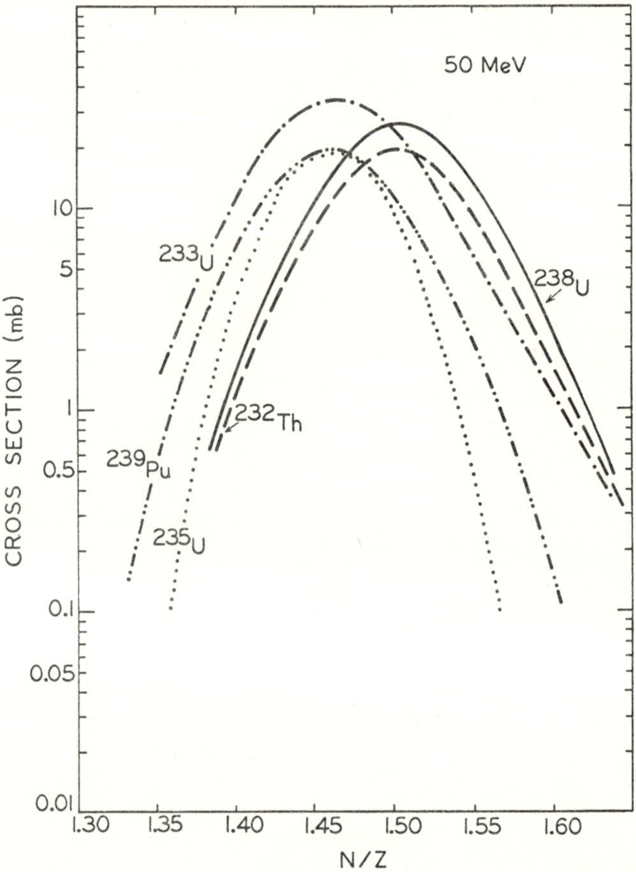

Fig. 11. Charge Dispersion Curves at 50-MeV Bombarding Energy for ^{238}U, ^{232}Th, ^{235}U, ^{239}Pu, *and* ^{233}U

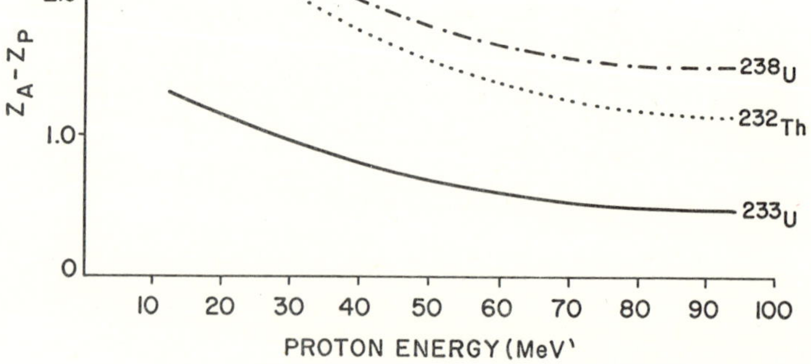

Fig. 12. Displacement of Most Probable Charge, Z_p, *from Stability* Z_A *for* ^{238}U, ^{232}Th, *and* ^{233}U *as a function of Bombardment Energy*

in determining the nature of the fission process and argues strongly for the Unchanged Charge Distribution Hypothesis.

(b) Charge dispersion curves: If we assume an essentially flat mass distribution in the narrow mass region studied (A = 129–138), the cross-section data may be transformed into charge dispersion curves. Such a series of curves is shown in Figs. 6–10 for each of ^{238}U, ^{232}Th, ^{235}U, ^{239}Pu, and ^{233}U.
The cross-sections were plotted against N/Z for the product, rather than the conventional $Z - Z_A$ used at lower energies. This is done to minimise the effect of shell-edge discontinuities, as suggested by Friedlander et al[9], but also because of the importance of the N/Z parameter as indicated in this work. The right-hand portion of the curve is formed in such a manner that the sum of the yields, read from the curves for the various isobars, must agree with the isobaric yield determined experimentally. This virtually defines the curve uniquely, unless one assumes some curious discontinuity. With the more neutron-deficient targets, e.g. ^{233}U, ^{235}U, and ^{239}Pu, the more neutron-excessive fission products fall on the right-hand portion of the curve. The combination of the two does define the curve exactly in these cases and lends even greater credence to this method for the other targets.

The behaviour of the charge dispersion curves with increasing energy is virtually the same—an increase in full-width at half-maximum and a decrease in $Z_A - Z_p$, i.e. a shift towards β-stability and increased production of neutron-deficient nuclides with increase in bombardment energy. Fig. 11 shows the curves for the various targets at a bombardment energy of 50 MeV and their relative displacement from Z_A.

Soon after this work commenced, the importance of the N/Z value of the target became obvious. From the information available for ^{238}U, ^{232}Th, and ^{233}U the values of $Z_A - Z_p$ were plotted as a function of bombardment energy (Fig. 12). The values of $Z_A - Z_p$ at each energy were then plotted against the

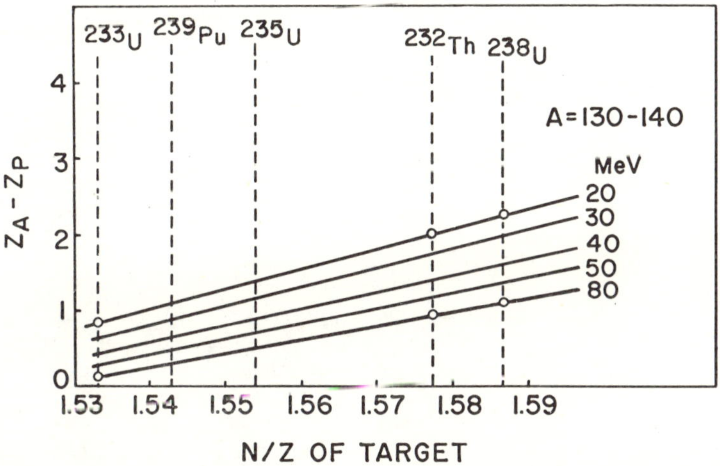

Fig. 13. Displacement of Most Probable Charge from Stability as a function of the neutron-to-proton ratio of the target for various Bombarding Energies

N/Z value of the *target* (Fig. 13) and predictions read from this for ^{239}Pu and ^{235}U, which were then being investigated in our laboratory. The fit for ^{239}Pu (Fig. 14) was very good. These data were added to those in Fig. 12 and the curve re-defined to predict the results for ^{235}U (Fig. 15). Fig. 16 thus contains the

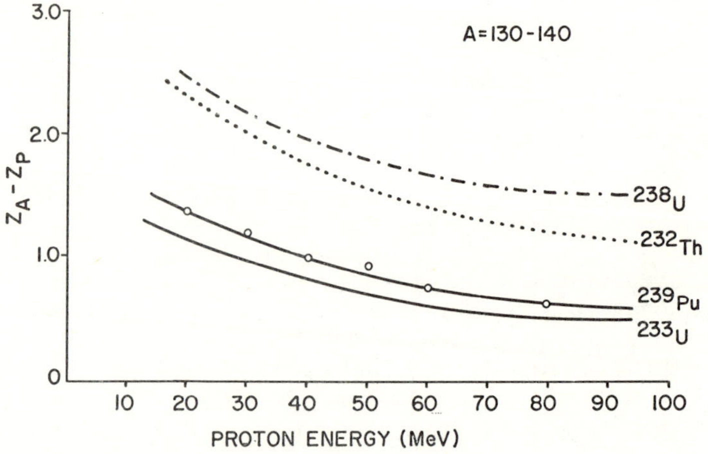

Fig. 14. Displacement of Most Probable Charge from Stability as a function of Bombardment Energy, showing predicted and experimental values for ^{239}Pu

——— *predicted*

—o—o— *experimental*

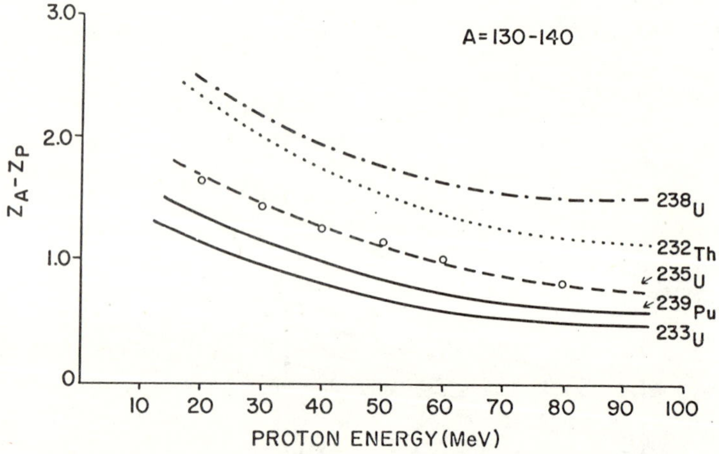

Fig. 15. Displacement of Most Probable Charge from Stability as a function of Bombardment Energy, showing predicted and experimental values for ^{235}U

——— *predicted*

—o—o— *experimental*

information for all 5 targets of different N/Z, and is now being used as an empirical method to predict $Z_A - Z_p$ values, and thus the most probable charge for a variety of fissioning systems now being investigated in our laboratory. A word of caution is however in order. All these data are for the very narrow mass range $A = 129-138$. The behaviour in the light mass region is very much different[18] and it is certainly conceivable that different behaviour will be observed in other parts of the mass yield curve, e.g. symmetrical fission.

From Fig. 16 we see that at 85 MeV bombarding energy ^{238}U has the same value of $Z_A - Z_p$ as ^{233}U at 20 MeV and ^{235}U at 40 MeV. Thus there is a strong suggestion that the average fission parent (after neutron emission) is the same in each case.

(c) Z_p calculations: It is of interest to compare the experimental (and predicted) values of Z_p with those obtained by calculation assuming each of the two postulates—Equal Charge Displacement (ECD) and Unchanged Charge Distribution (UCD). We have calculated Z_p for $A = 136$ at bombarding energies of 30, 40, and 50 MeV where one could reasonably assume that most of the reactions occurred by the compound-nucleus mechanism. The steps in the calculation were as follows:

(1) the compound nucleus $^{239}_{93}Np$ was assumed to be able to undergo fission and/or neutron evaporation. Its excitation energy was calculated from a knowledge of the kinetic energy of the incident proton in the centre-of-mass system and the Q-value of the reaction;

(2) after each reaction evaporation step the excitation energy of the residual nucleus was calculated as

$U'' = U' - B_n - 2T$ where

U' and U'' = excitation energies of the nuclei of masses A and A-1 respectively,

B_n = the neutron binding energy[19]

$2T$ = kinetic energy of the neutron

T = nuclear temperature, calculated from

$U'' = aT^2 - 4T$ [20] where

a = level-density parameter, 10·5 MeV^{-1} [21] with the odd-even parameter ∂, taken to be 1·1 MeV

(3) the fission branching ratio $G = \Gamma_f/(\Gamma_f + \Gamma_n)$ was calculated for each parent nucleus;[20]

(4) several fragment pairs were selected randomly in the mass range $A = 136-141$ (with $Z = 55$, since charged-particle emission will be rare due to the neutron-excess characteristics of the target nuclei);

(5) the total excitation energy of each fragment was calculated and the average kinetic energy of the fragments, as calculated by Milton and Fraser,[22] was subtracted from the total energy released in a particular fission mode;

(6) the excitation energy was then assumed to be divided between the two complementary fragments in proportion to their masses;

(7) neutron evaporation from the fragments was then considered as discussed above from the parents (proton emission was again neglected for the reason given above);

(8) only those fragments giving rise to $A = 136$ were then sorted;

(9) Z_p values were then calculated for both the ECD and UCD postulates.

The values for fission of ^{239}Pu and ^{235}U, which are typical of the systems studied, are shown in Table 1. It can be seen that the agreement with the UCD postulate is quite good. However, we should not lose sight of some of the assumptions involved in such a calculation. For example, the excitation energy

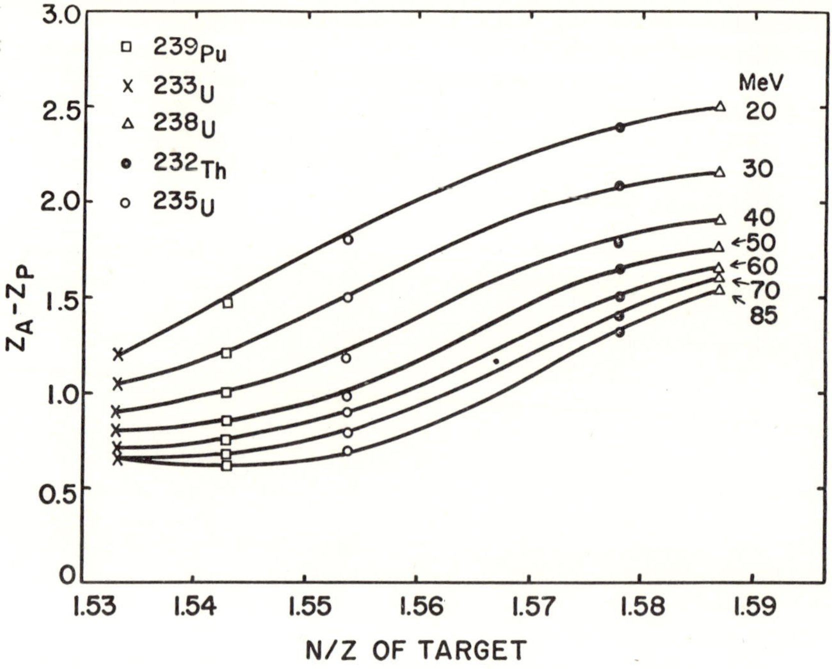

Fig. 16. Displacement of Most Probable Charge from Stability as a function of the neutron-to-proton ratio of the target for a variety of bombarding energies

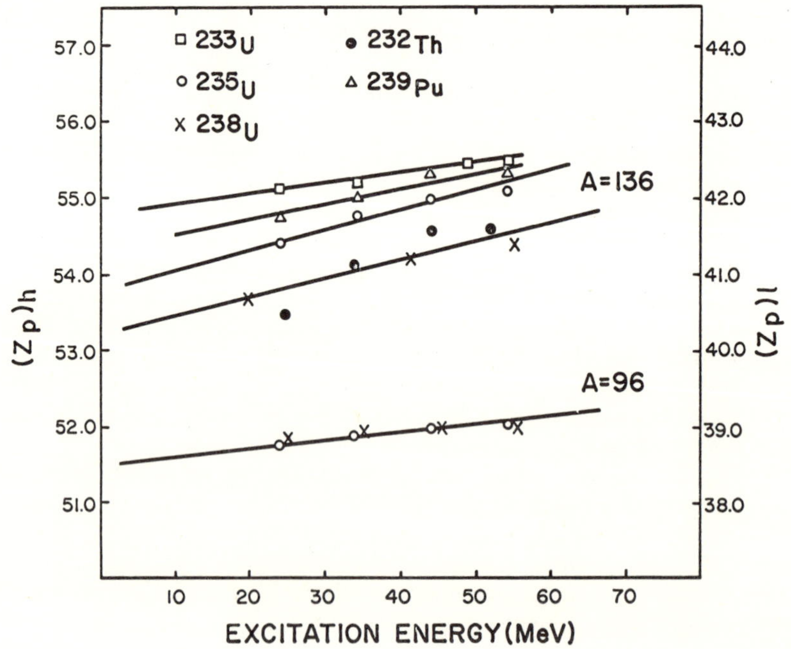

Fig. 17. The Most Probable Charge of heavy and light fragments as a function of Excitation Energy for various Fissioning Systems

was assumed to be shared between the fragments in proportion to their masses, since this was assumed to be the most reasonable assumption to make. If one makes the gross assumption that the heavy product received only one-half the energy determined in this way, the calculated Z_p values decrease by 0·6 charge units, but the agreement with UCD is still rather good. If one makes the inverse

Table 1 COMPARISON OF EXPERIMENTAL AND CALCULATED Z_p VALUE

Proton Energy (MeV)	Z_p					
	^{239}Pu			^{235}U		
	ECD	*UCD*	*Expt.*	*ECD*	*UCD*	*Expt.*
30	54·1	55·1	55·0	53·9	54·8	54·8
40	54·3	55·3	55·3	54·2	55·0	55·0
50	54·6	55·6	55·3	54·6	55·4	55·2

gross assumption that the heavy fragment receives 1·5 times this energy, the calculated Z_p values increase by 1·6 charge units and agreement is good with neither mechanism–but closer to ECD. These assumptions, however, seem so gross as to fall outside the realm of plausibility, and the weight of evidence does indeed point to the UCD mechanism.

(d) Total number of neutrons emitted: McHugh and Michel[23] have shown that one can calculate the total number of neutrons, ν_T, emitted in fission as follows:

$$(\partial Z_p/\partial E)_A = -(\partial Z_p/\partial A)_E\,(\partial A/\partial E)_{Z_p}$$

and

$$(\partial A/\partial E)_{Z_p} = -(\partial \nu/\partial E)_{Z_p}$$

then

$$(\partial \nu/\partial E)_{Z_p} = (\partial Z_p/\partial E)_A/(\partial Z_p/\partial A)_E$$

At any given excitation energy, the variation of Z_p with mass number can be approximated by the variation of Z_A with mass number. Coryell has shown that $(\partial Z_A/\partial A)_E = 0{\cdot}38$

$$\therefore\ \mathrm{d}\nu/\mathrm{d}E = (\mathrm{d}Z_p/\mathrm{d}E)/0{\cdot}38.$$

The value for $\mathrm{d}Z_p/\mathrm{d}E$ for the heavy fragment in this mass region is obtained for the various fissioning systems from Fig. 17. However, very few such data exist for the light fragment. Khan et al[18] have recently completed such a study for light-mass products in the fission of ^{235}U and ^{238}U and these data are also shown in Fig. 17 for $A = 96$. This enables one to obtain values for both $\mathrm{d}\nu_h/\mathrm{d}E$, and $\mathrm{d}\nu_l/\mathrm{d}E$, when h and l stand for heavy and light fragments respectively, and from this $\mathrm{d}\nu_T/\mathrm{d}E$, where ν_T = total neutron emission, since $\mathrm{d}\nu_T/\mathrm{d}E = \mathrm{d}\nu_h/\mathrm{d}E + \mathrm{d}\nu_l/\mathrm{d}E$. These data are shown in Table 2 for $A = 136$ and $A = 96$, the heavy and light fragment respectively. It must be borne in mind that they may not be complementary fragments. However, the reasonable assumption is made that $\mathrm{d}\nu_h/\mathrm{d}E$ and $\mathrm{d}\nu_l/\mathrm{d}E$ are constant over this very narrow mass range.

Fig. 18 shows the values of ν_T versus energy. The other fissioning systems are included, even though no data for the light fragment are yet available for these systems, the assumption being made that, like ^{235}U and ^{238}U, $\nu_h = 2\nu_l$ for ^{233}U, ^{232}Th, and ^{239}Pu.

Table 2

	^{235}U	^{238}U
$(dZ_p/dE)_h$	0·0274 MeV^{-1}	0·0250 MeV^{-1}
$(dZ_p/dE)_l$	0·0111 MeV^{-1}	0·0111 MeV^{-1}
$d\nu_h/dE$	0·072 MeV^{-1}	0·066 MeV^{-1}
$d\nu_l/dE$	0·029 MeV^{-1}	0·029 MeV^{-1}
$d\nu_T/dE$	0·101 MeV^{-1}	0·095 MeV^{-1}

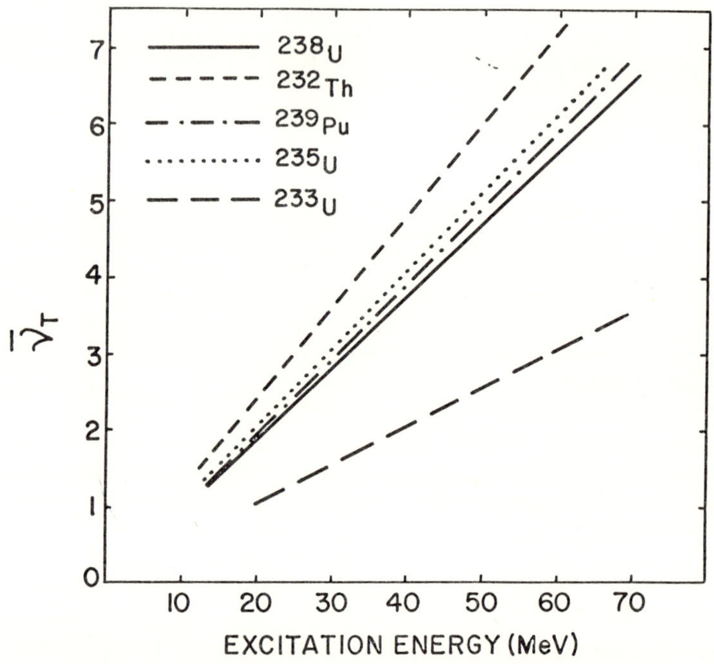

Fig. 18. Total numbers of neutrons emitted from various Fissioning Systems as a function of Excitation Energy

REFERENCES

1. Wahl, A. C., *J. Inorg. Nucl. Chem.* **6**, 263, (1958).
2. Wahl, A. C., Ferguson, R. L., Nethaway, D. R., Troutner, D. E. and Wolfsberg, K., *Phys. Rev.* **126**, 1112, (1962).
3. Norris, A. E. and Wahl, A. C., *Phys. Rev.* **146**, 926, (1966).
4. Wahl, A. C., *Phys. Rev.* **99**, 730, (1955).
5. Pate, B. D., Foster, J. S. and Yaffe, L., *Can. J. Chem.* **36**, 1691, (1958).
6. Pate, B. D., *Can. J. Chem* **36**, 1707, (1958).
7. Goeckerman, R. H. and Perlman, I., *Phys. Rev.* **76**, 628, (1949).
8. Glendenin, L. E., Coryell, C. D. and Edwards, R. R., *Phys. Rev.* **75**, 337, (1949).
9. Friedlander, G., Friedman, L., Gordon, B. and Yaffe, L., *Phys. Rev.* **129**, 1809, (1963).
10. Friedlander, G., Symposium on the Physics and Chemistry of Fission, International Atomic Energy Agency, Vienna (1965), Vol. II p. 265.
11. Davies, J. H. and Yaffe, L., *Can. J. Phys.* **41**, 762, (1963).
12. Benjamin, P. P., Marsden, D. A., Porile, N. T. and Yaffe, L., *Can. J. Chem.* **47**, 301, (1969).
13. Parikh, S. S., Marsden, D. A., Porile, N. T. and Yaffe, L., *Can. J. Chem.* **45**, 1863, (1967).
14. Tomita, I. and Yaffe, L., *Can. J. Chem.* **47**, 2921, (1969).
15. Saha, G. B. and Yaffe, L. To be published.
16. Saha, G. B., Tomita, I. and Yaffe, L. To be published.
17. Cumming, J. B., U.S. Atomic Energy Commission NAS-NS-3107, p. 25 (1962).
18. Khan, A. H., Saha, G. B. and Yaffe, L., To be published.
19. Myers, W. D. and Swiatecki, W. J., University of California Radiation Laboratory Report UCRL-11980, (1965).
20. Vonach, H. K., Vandenbosch, R. and Huizenga, J. R., *Nucl. Phys.* **60**, 70, (1964).
21. Huizenga, J. R. and Vandenbosch, R., 'Nuclear Reactions', edited by P. M. Endt and P. B. Smith, Vol. II, North-Holland Publishing Co., Amsterdam (1962).
22. Milton, J. C. D. and Fraser, J. S., *Can. J. Phys.* **40**, 1626, (1962).
23. McHugh, J. A. and Michel, M. C., *Phys. Rev.* **172**, 1160, (1968).

DISCUSSION

G. N. Walton (A.E.R.E., Harwell): In your paper the neutron emission is estimated from the split in excitation energy between the two fragments in proportion to their mass. In any split where momentum is conserved the energy is divided unevenly with respect to mass, i.e. the lighter fragment receives the greater energy. Would you anticipate that excitation energy may also be divided in this way?

L. Yaffe: No. We are dealing with excitation energies where the reaction is of the compound nucleus type. Equipartition of energy takes place and when the fission occurs it would be expected that the energy per nucleon be the same whether the latter found itself in the light or the heavy fragment. The reasonable assumption is to divide the excitation energy according to mass. In the paper we have however divided the excitation energy in other ways; these were purposely made so gross as to be outside expected reality. Any reasonable split in excitation energy would not be expected to affect our results greatly.

J. G. Cuninghame (A.E.R.E.): Can you say how your data compare with those of other workers on prompt neutron emission?

L. Yaffe: Other data are sparse, and our data refer only to neutrons emitted *before* fission and by the fission products of masses $A = 130-140$. Our data compare with Leachman's figure (0·12 MeV^{-1}, if my memory is correct) for 14 MeV neutron fission of ^{235}U, but we should not of course extrapolate them to the thermal energies. This would give values below the average of 2·5 neutrons/fission, a value which is also very sensitive to the mass of the fission fragments.

Studies of the decay of the compound nucleus ^{68}Ge

S. Barkan, J. M. D'Auria, Brian D. Pate, Francis H. Ruddy, K. Vaughan
Department of Chemistry, Simon Fraser University, Burnaby 2, B.C., Canada

and E. W. Vogt
Department of Physics, University of British Columbia, Vancouver, B.C., Canada

ABSTRACT

The decay of the compound nucleus ^{68}Ge, formed via the reactions ^{4}He + ^{64}Zn, ^{12}C + ^{56}Fe, and ^{16}O + ^{52}Cr, has been studied through measurements on product activation cross-sections, recoil nucleus ranges, and gamma-ray emission (the latter via in-beam experiments). The experimental results for cross-sections are compared with those from compound statistical calculations featuring explicit inclusion of angular momentum effects. Good agreement is observed when literature values of binding energy and other parameters are employed.

Data on the emission of gamma-rays are expected to test the predictions of the calculations in considerably more detail, particularly as to the distribution in energy and angular momentum of the nuclei produced as a result of particle evaporation. The preliminary results presented are interpreted in a qualitative way as to the origin of the spectrum characteristics discussed.

INTRODUCTION

The postulate[1] that the decay of compound nuclei is independent of their means of formation is known to be inaccurate in that angular momentum conservation plays an important role in the de-excitation process.[2-7] α-particles and heavy-ion beams currently available from cyclotrons and other accelerators make it possible to study such nuclear reactions over a wide range of angular momentum as well as energy. Modern computers make possible also the intricate calculations which must be performed if the results are to be interpreted in terms, for example, of the compound statistical and optical models[8] without the approximations and simplifications previously necessary.

Previous studies[9-11] have dealt with the compound nucleus ^{68}Ge produced by bombarding ^{64}Zn with ^{4}He ions, but without experimental confirmation of the reaction mechanism by product-nucleus range measurement. In the present work, range measurements were made to determine the bombarding energy interval within which compound nucleus formation was dominant; product formation

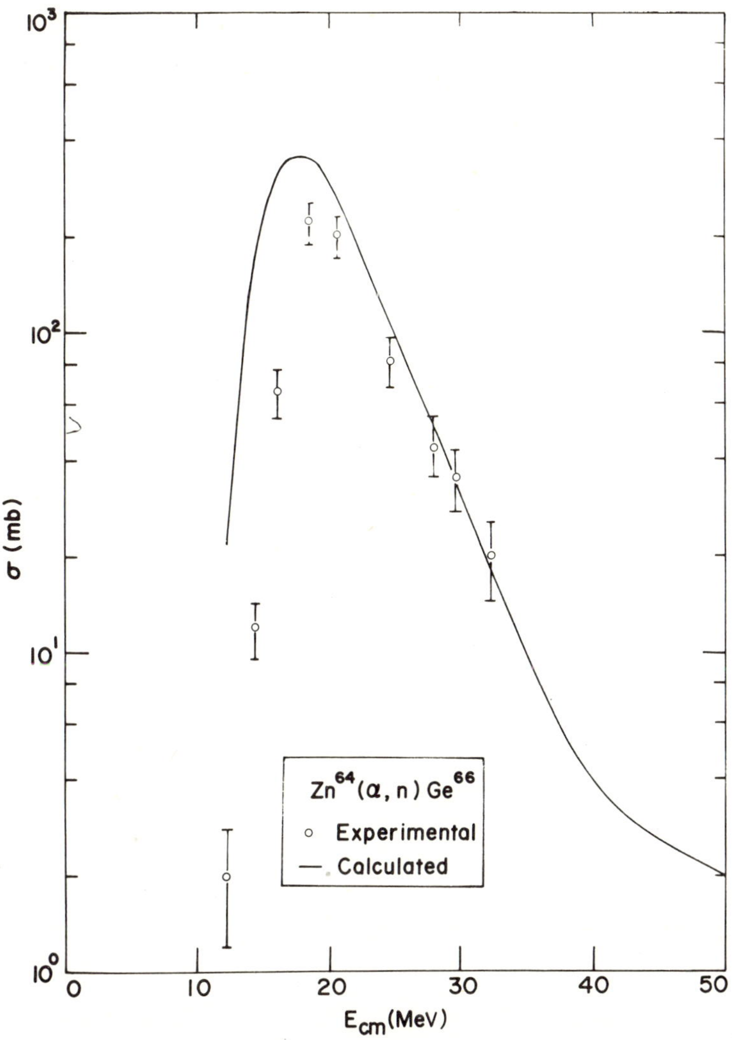

Fig. 4. Comparison of calculated excitation function with experimental data for the $^{64}Zn(\alpha, n)\ ^{67}Ge$ *reaction*

and angular momentum of the nuclei produced by the emission of α-particles, protons p_1 and p_2, and neutrons n_1 and n_2. The solid line in each case at constant E indicates the compound nucleus energy, while the dotted line indicates the latter energy minus the emitted particle binding energy. The contours represent calculated yields, successive contour lines separated by factors of ten in yield. The shaded area on each figure is the region where no nuclear states exist, that is where the energy calculated for each J-value to be taken up in the form of rotation exceeds the total available excitation energy.

Comparison with experimental total formation cross-sections is achieved by summation of the yield under the contours calculated for each bombarding energy. As representative examples, Fig. 4 compares the calculated excitation function for the reaction $^{64}Zn(\alpha, n)^{67}Ge$ with the experimental data and Fig. 5 makes the comparison for $^{64}Zn(\alpha, 2n)^{66}Ga$. It should be reiterated that no calculation normalisation or parameter adjustment has taken place, and the

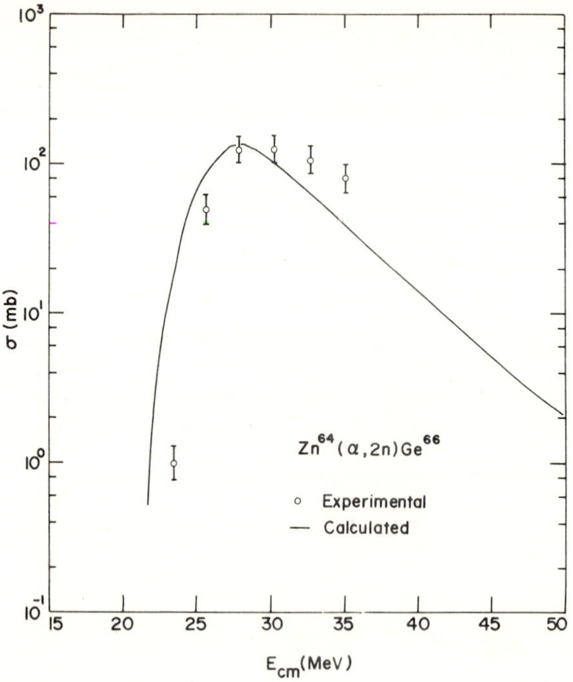

Fig. 5. Comparison of calculated excitation function with experimental data for the ^{64}Zn(α, 2n) ^{66}Ge *reaction*

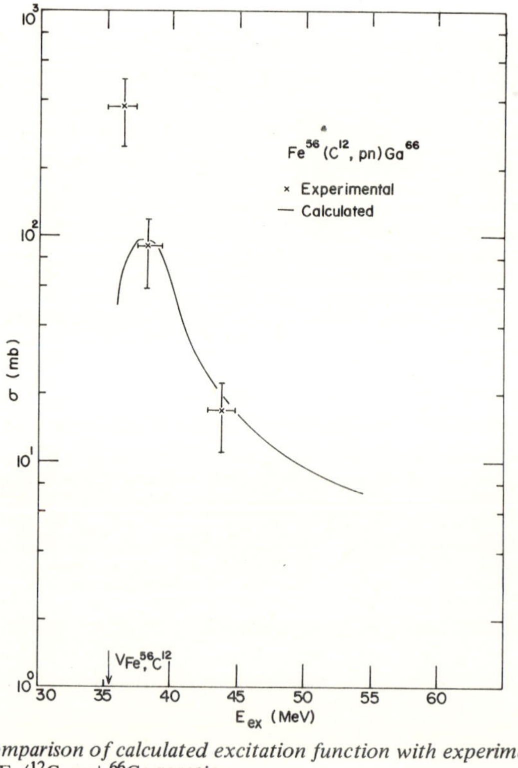

Fig. 6. Comparison of calculated excitation function with experimental data for the ^{56}Fe(^{12}C, pn) ^{66}Ga *reaction*

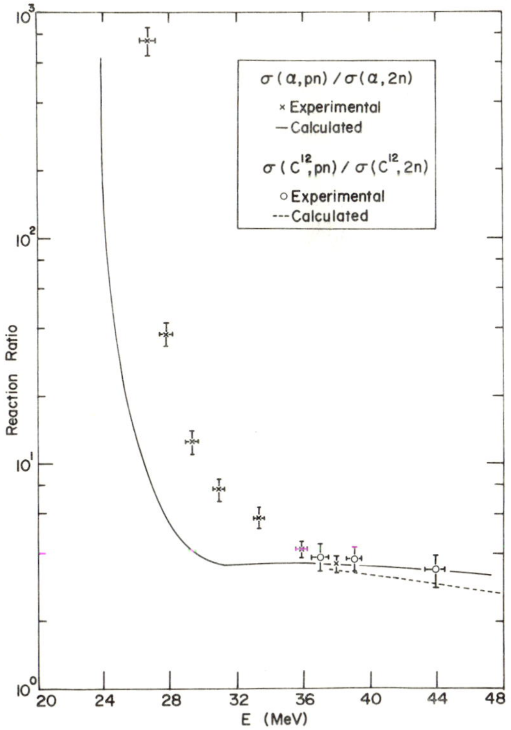

Fig. 7. Comparison of calculated and experimental cross-section ratios

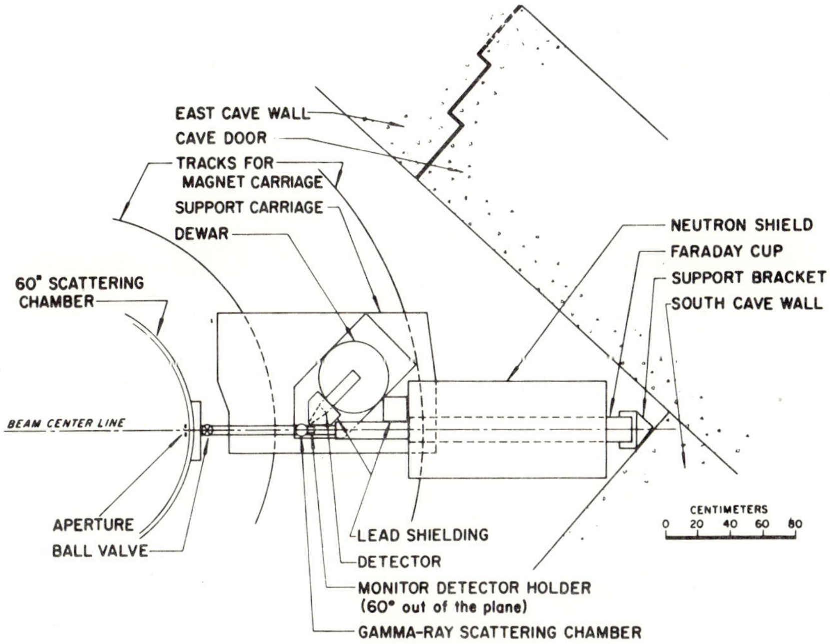

Fig. 8. Experimental arrangement for γ-ray emission measurements

suggestion is strong that some binding energies or nuclear radius values are in error. However, the yield magnitudes and their variation with energy are in all cases examined reproduced better than in many previous calculations.

Fig. 6 makes the comparison for $^{56}Fe(^{12}C, pn)^{66}Ga$, and agreement is seen to be good for energies above those close to the coulomb barrier value.

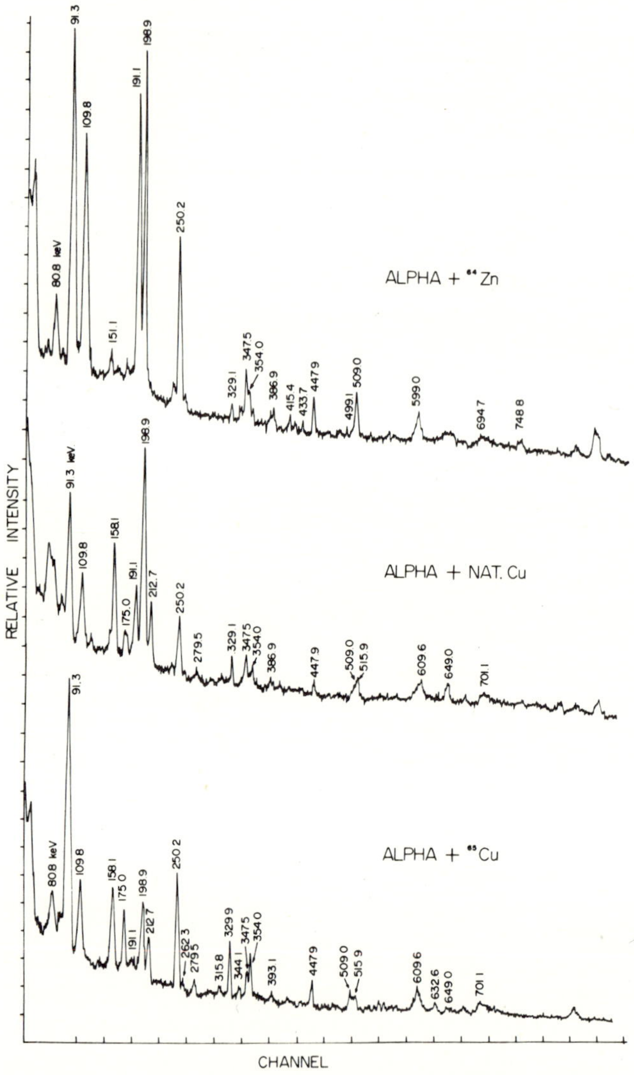

Fig. 9. Low energy γ-spectra from 41 MeV α-particles incident on ^{64}Zn, ^{65}Cu *and natural copper. Peak energies are in keV*

The calculated yield at the lowest energies is however seen to be too small, reflecting evidently too high a coulomb barrier energy calculated with the optical model parameters employed.

Fig. 7 compares experimental and calculated yield ratios for the reactions (α, pn)/(α, 2n) and (^{12}C, pn)/(^{12}C, 2n). Beyond the low-energy region where discrepancies are attributed to binding energy errors, good agreement is observed between calculation and experiment. Furthermore, the data for α- and carbon-induced reaction yield ratios are seen to merge near 35 MeV, where the calculated values of the average angular momenta of the respective compound nuclei are very close. Thus these data provide evidence for validity of the 'independence hypotheses' under these conditions.

Satisfaction with the reproduction of experimental total cross-sections by the calculation is tempered by the fact that the test is a relatively crude one, and that many detailed features of the distributions such as those in Fig. 3 are lost when the yields differential in energy and angular momentum are summed. It was thus decided to attempt experimental measurement of γ-rays emitted in the de-excitation process, most of which would arise after particle evaporation and hence from the decay of the nuclei in distributions such as those represented in Fig. 3.

PRELIMINARY EXPERIMENTS ON GAMMA-EMISSION

EXPERIMENTAL

Due to the courtesy of Dr. I. Halpern and Mr. S. Ferguson of the University of Washington, Seattle, the first experiments have been performed with an existing experimental arrangement, shown in Fig. 8. The external beam from the University of Washington cyclotron passed through a reaction chamber of about 10 cm diam., at the centre of which a metal target foil was mounted. γ-rays from the target emerged from the chamber through a window about 1 mm thick and impinged on a Ge(Li) detector of 40 cm^3 volume. The detector was heavily shielded with lead against radiations incident from directions other than from the target, and the detector assembly could be rotated about the target foil axis to permit angular distribution data to be obtained. The beam currents of about 1 nA were monitored via the intensity of α-particles scattered by the target from the beam, and via a Faraday cup. Backgrounds with the target removed were found to be negligible.

Measurements were made first with 41 MeV α-particles from the cyclotron without energy reduction (for reasons of experimental convenience) and with foils of enriched ^{64}Zn and ^{65}Cu, and with a foil of the natural ^{63}Cu + ^{65}Cu mixture. Some results are shown in Fig. 9; γ-rays in appreciable intensity were observed with energies up to many MeV and the figure shows only the first MeV of the spectrum.

The spectra contain two components, a continuum and, superimposed upon it, individually-resolved lines. The latter are expected to be identified with particular product nuclei (once the effects of oxygen contamination are subtracted) by cross-bombardment techniques—shown here—and by variation of the incident energy.

DISCUSSION

The spectrum continua are considered[30,31] to arise largely from statistical emission of γ-rays during the initial de-excitation steps after evaporation; one would expect to observe many γ-rays overlapping in energy and emitted isotropically. Such emissions would for the most part be expected to carry off relatively little angular momentum, so that nuclei in the distributions such as those of Fig. 3 would arrive in the region adjacent to the area of no states (shaded) with angular momentum values close to those they possessed at the end of evaporation. Thereafter rotation de-excitation is dominant, with individual lines being resolved and a rotational angular distribution being observed.

In addition to information of spectroscopic interest being available on the nuclei involved, it seems likely that the relative total intensities of continua and lines, and relative line intensities, should be calculable from distributions such as those of Fig. 3. This, combined, for example, with heavy-ion experiments in which substantially different angular momentum distributions are encountered, should constitute a more direct test of the validity of statistical model calculations.

ACKNOWLEDGEMENTS

It is a pleasure to acknowledge the courtesy of C. P. Baker and the operating crew of the Brookhaven National Laboratory cyclotron, of R. Vandenbosch, J. Heagney, the late T. Morgan and the operating crews of the University of Washington cyclotron, of J. Hood and the operating crew of the Washington University cyclotron and of I. Preiss, H. Bakhru and the operating crew of the Yale University heavy-ion accelerator for cooperative scheduling of bombardments.

It is also a pleasure to acknowledge the assistance of M. Blann and J. M. Miller in helpful discussions. Our gratitude is also due to T. W. Donnelly for use of his optical-model program and to the University of British Columbia Computing Centre for carrying out calculations. We are most grateful to H. Marusyk, E. K. Miller and the staff of the Simon Fraser University Data Processing Centre for assistance in carrying out calculations at the Simon Fraser University Computing Centre, which were supported in part by National Research Council Grant Number E-997. The general support of this work by the National Research Council under Grant Number A-2510 is gratefully acknowledged.

REFERENCES

1. Ghoshal, S. N., *Phys. Rev.* **80**, 939, (1959).
2. Williams, D. C. and Thomas, T. D., *Nucl. Phys.* **A92**, 1, (1967).
3. Blann, M., *Phys. Rev.* **157**, 860, (1967).
4. Grover, J. R. and Gilat, J., *Phys. Rev.* **157**, 802, (1967).
5. Grover, J. R. and Gilat, J., *Phys. Rev.* **157**, 814, (1967).
6. Grover, J. R. and Gilat, J., *Phys. Rev.* **157**, 823, (1967).
7. Alexander, J. M. and Simonoff, G. N., *Phys. Rev.* **133**, B93, (1964).
8. Vogt, E., in 'Advances of Nuclear Physics', Vol. I, Plenum Press, New York (1967).
9. Porile, N. T., *Phys. Rev.* **115**, 939, (1959).
10. Stelson, P. H. and McGowan, F. K., *Phys. Rev.* **133**, B911, (1964).
11. Cogneau, M. and Gilly, L., *Nucl. Phys.* **73**, 122, (1965).
12. O'Hanlon, J. F. and Haering, R. R., personal communication;
    O'Hanlon, J. F. Ph.D. thesis, Simon Fraser University (1967) unpublished.
13. Holland, L., 'Vacuum deposition of thin films', Chapman and Hall, London (1961).
14. Blann, M., Lanzafame, F. M., and Piscitelli, R. A., *Phys. Rev.* **133**, B700, (1964).
15. Cumming, J. B., personal communication (1967).
16. Houck, F. S. and Miller, J. M., *Phys. Rev.* **123**, 231, (1961).
17. Atkinson, J. H. Jr. and Willis, B. H. *UCRL2426 Rev.*, Vol. 2 35–36, (June 1957).
18. Northcliffe, I. C., *Phys. Rev.* **120**, 1744, (1960).
19. Northcliffe, I. C., *Ann. Rev. Nucl. Sci.* **13**, 67, (1963).

20. Marinsky, J. A., 'Nuclear Sciences Series' NAS-NS 3043, National Academy of Sciences–National Research Council, Washington 25, D.C. (1961).
21. Lewis, J. E., NAS-NS 3032 (1961).
22. Heath, R. L., 'Atomic Energy Commission Research and Development Report' IDO-16408 (1957).
23. Cumming, J. B., NAS-NS 3107 (1962).
24. Alexander, J. M., in 'Nuclear Chemistry', ed. by Yaffe, L., Academic Press, New York, Chapt. IV, 273, (1968).
25. Hazan, J. P. and Blann, M., *Phys. Rev.* **137**, B1202, (1965).
26. Blann, M. and Ewart, A., *Phys. Rev.* **134**, B783, (1964).
27. Lindhard, J., Scharff, M., and Schiøtt, H. E., *Mat. Fys. Medd. Dan. Vid. Selsk.* **33**, No. 14 (1963).
28. Ruddy, F. H. and Pate, B. D., *Nucl. Phys.* **A127**, 305, (1969).
29. Ruddy, F. H., Pate, B. D., and Vogt, E. W., *Nucl. Phys.* **A127**, 323, (1969).
30. Williamson, C. F., Ferguson, S. M., Shepherd, B. J., and Halpern, I., *Nucl. Phys.* (to be published) 1969.
31. Mills, S. J. and Rautenbach, W. L., *Nucl. Phys.* **A124**, 597, (1969).

DISCUSSION

J. L. Need (New England Nuclear Corporation): Is any correction made for pairing in the level density expression?

B. D. Pate: Cameron's pairing energy values were used. We also used superconductor theory to reduce the moment of inertia below the rigid body value for excitation energies less than a neutron binding energy above the ground state.

H. A. C. McKay (A.E.R.E.): How would your data look if they were plotted by the method of Ghoshal?

B. D. Pate: Very similar, for the same bombarding energy. My figure comparing cross-section ratios makes this point; when the average angular momenta are equal, the compound nuclei made via different target-projectile combinations display identical decay behaviour.

Search for new isotopes and new superheavy elements produced by heavy-ion reactions

Marc Lefort

Laboratoire de Chimie Nucléaire
Institut de Physique Nucléaire, Orsay, France

ABSTRACT

The production of new isotopes far from the line of stability by heavy-ion reactions is examined. Cross-sections for complete fusion reactions are given. It is shown that the projectile should be as heavy as possible. In order to produce transuranium elements with good yields, the bombardment of targets must also be done with very heavy ions, because the compound nucleus has to be produced as 'cold' as possible. This can be achieved if the Q-value of the nuclear reaction is very negative. Different reactions are proposed which will be obtained in the near future at Orsay. A brief account is also given of the identification method for the reaction products.

INTRODUCTION

It has been many times emphasised that only a small fraction of the bound nuclei are known.[1] In particular, only isotopes close to the stability line have been studied on the neutron-deficient side, and a few attempts have been made to search for delayed proton emitters. Although most of the new nuclides that can be discovered have short half-lives and therefore may not be of great practical interest, it is useful to obtain information on the validity of mass formulae by checking their predictions in regions far from stability. Furthermore, the question of the stability of peculiar nuclei such as $^{100}_{50}Sn$ (double magic shell) is very interesting in connection with nuclear models.

The third and perhaps the most interesting point is the search of elements heavier than $Z = 110$, because of the possibility of the existence of an island of stability connected with nuclear magic numbers for superheavy elements. There have appeared in the literature[2,3] several theoretical evaluations of where the next shells for neutrons, N_s, and for protons, Z_s, might be. The numbers have ranged from 114 to 126 for Z_s and from 164 to 194 for N_s. The calculations show that the shell effect both increases the fission barrier, B_f, and decreases the α-decay energy. Therefore both spontaneous fission and α-decay half-lives are enhanced. The production of very heavy nuclides may open a large field of new research not only in nuclear spectroscopy but also in chemistry, if the lifetimes are long enough. Astrophysicists are also interested to know whether or not such superheavy nuclei can occur in stars, supernovae, etc.

We believe that heavy-ion beams accelerated in cyclotrons are a good tool for the production of many of these nuclides.

COMPLETE FUSION BETWEEN HEAVY IONS AND TARGET NUCLEI

When a complex projectile of charge Z_i approaches a target nucleus of charge Z_t with an energy in the centre-of-mass system higher than the Coulomb barrier, a fusion nucleus may be formed and it will show the characteristics of an excited compound system. However, the trajectory may also be largely scattered according to the Rutherford laws and the nuclear interaction will be

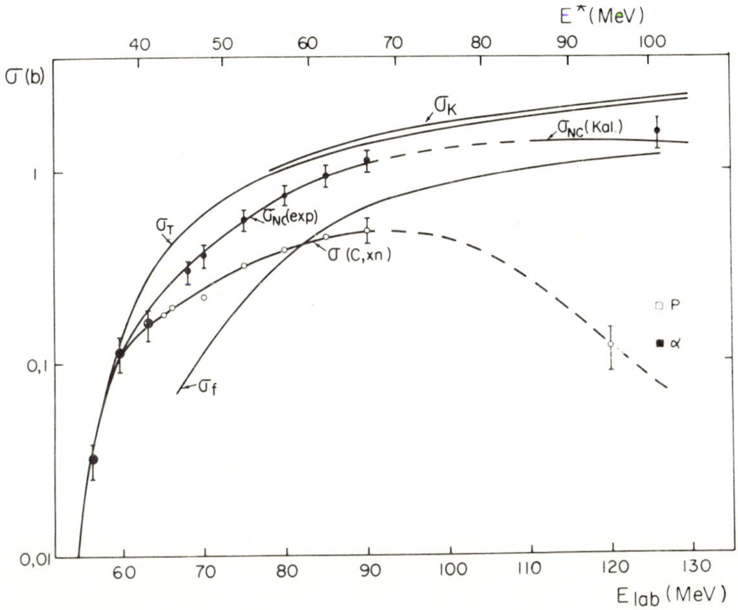

Fig. 1. Total reaction cross-section (σ_T) and compound nucleus cross-section $\sigma_{NC}(exp)$ as a function of heavy ion energy for the reaction $^{12}C + ^{197}Au$. σ_T is the calculated total cross-section by Thomas, $\sigma_{NC}(Kal)$ the compound nucleus cross-section calculated by Kalinkin

restricted to the transfer process of one or several nucleons. It has been shown in many experiments that at large impact parameters, or in other words at high angular momenta, compound nucleus formation is very improbable. For a given target, the ratio σ_{CF}/σ_R of the complete fusion cross-section to the total reaction cross-section decreases when the energy increases and also when the mass of the projectile increases. Both experimental results and calculations made on the assumption of a critical angular momentum number l_c, reach the conclusion that σ_{CF}/σ_R is very close to 1 just above the Coulomb barrier and decreases continuously for higher energies. This is illustrated in Fig. 1, taken from Bimbot *et al*,[4] for astatine compound nuclei. Experiments have also shown that complete fusion still takes place to a large extent with ions as heavy as ^{40}Ar. Sikkeland[5] has proposed an empirical formula $\sigma_{CF}/\sigma_R = (1 + 0{\cdot}03\,A_i)^{-1}$ which is available over an energy range of around 30 MeV above the Coulomb barrier. Very roughly, we may expect that a third of the total reaction cross-section will still correspond to a complete fusion mechanism in the bombardment of a heavy target such as Au or Th by a beam of bromine or krypton ions.

Calculations of the total reaction cross-section have been made with a Wood–Saxon potential and an optical model. As a result, we have given in Fig. 2 the energy to which a given ion should be accelerated in order to obtain a reaction cross-section of 50 mb on a thorium target. It is compared to the possibilities of acceleration of the Orsay machine for heavy ions (ALICE).

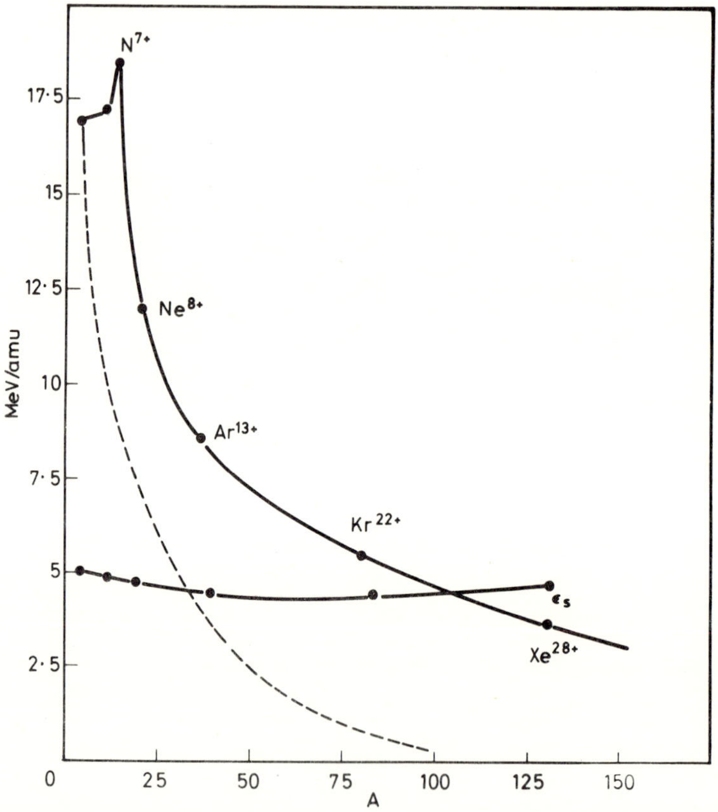

Fig. 2. Threshold energies for heavy-ion reactions on thorium as a function of the mass number of the projectile. Energies are given in MeV per nucleon. Also given are the kinetic energies that will be available for several ions (N, Ne, Ar, Kr, *and* Xe) *on the accelerator ALICE at Orsay*

DE-EXCITATION OF THE COMPOUND NUCLEUS

When a compound nucleus is formed, the excitation energy is usually dissipated by the emission of nucleons, preferentially uncharged neutrons, although high angular momenta will favour the emission of particles of heavier mass such as α-particles, or photon emission. If Q is the mass balance of the reaction, the excitation energy available for the emission of particles can be written

$$E^* = \bar{\epsilon}_i - E_{\mathrm{rot.}} + Q$$

where $\bar{\epsilon}_i$ is the energy of the incident ion in the centre-of-mass system and $E_{\mathrm{rot.}}$ is the rotational energy depending on the transfer of angular momentum. It should be noted that Q usually has a negative value when the projectile is heavy enough. We may take advantage of this negative value in order to compensate $\bar{\epsilon}_i$ when it is too large. An illustration is given in the production of ^{100}Sn. A possible reaction is

$$^{16}_{8}\mathrm{O} + ^{92}_{42}\mathrm{Mo} \rightarrow ^{108}_{50}\mathrm{Sn} \rightarrow ^{100}_{50}\mathrm{Sn} + 8\mathrm{n}$$

^{108}Sn must be produced at an excitation energy of around 117 MeV in order to evaporate eight neutrons (binding energies of eight neutrons + average kinetic energies). Mass tables give $Q = -11$ MeV, and even if we neglect the rotational energy the kinetic energy that should be necessary in the centre-of-mass will be 128 MeV. The cross-section for the production of $^{108}_{50}$Sn can be estimated at only 400 mb, because a large fraction of non-compound process will occur. Furthermore, the evaporation of a large number of neutrons cannot be achieved without a large competition with proton and α-particle emission. As a result the cross-section at the end of the chain will go down to less than 10 mb. Another reaction could be:

$$^{40}_{20}\mathrm{Ca} + ^{64}_{30}\mathrm{Zn} \rightarrow ^{104}_{50}\mathrm{Sn} \rightarrow ^{100}_{50}\mathrm{Sn} + 4\mathrm{n}$$

where the number of neutrons to be emitted is smaller. The Q-value is -32 MeV for the formation of the compound nucleus and the excitation energy for the evaporation of four neutrons is 60 MeV. The necessary energy in the centre-of-mass would be of the order of 92 MeV, for which σ_{CF}/σ_R is larger than at 128 MeV, and the charged particle evaporation will be much smaller. A cross-section of the order of 100 mb may be expected. Another advantage is that the recoiling nucleus will have 56 MeV instead of 22 MeV, and it will be easier to collect it outside the target.

Many other reactions can be suggested for the production of interesting new nuclides. Among them we may quote the following:

^{69}Ga (^{12}C, 6n) ^{76}Rb, ^{64}Zn (^{14}N, 3n) ^{75}Rb, ^{106}Cd (^{14}N, 6n) ^{114}Cs, ^{203}Tl (^{12}C, Sn) ^{207}Fr

which have been proposed by N'Guyen Long Den[6] in order to produce light alkali elements and to make a separation with a mass spectrometer on-line.

PRODUCTION OF SUPERHEAVY ELEMENTS: FISSION WIDTH FOR THE DE-EXCITATION OF HEAVY NUCLEI

It is expected that after a region of very great instability for $106 < Z < 112$, due to very low fission barriers, an island of stability may be observed.

Moreover, some calculations predict the possibility of superheavy elements surviving in nature.[7] The schemes we shall consider for the artificial production of isotopes of these elements involve reactions that which are characterised by a complete fusion of complex nuclei followed by neutron evaporation in competition with fission and γ de-excitation. Other types of reactions such as nuclear explosive or transfer reactions do not appear hopeful in making superheavy elements.

In the preceding pages, we have shown that a large part of the total cross-section can be attributed to the complete fusion cross-section σ_{CF}. Nevertheless, when the compound nucleus is heavy, the main de-excitation process is *fission,* and the production cross-section of a residual nucleus after the emission of x neutrons becomes very low. A semi-empirical formula for the cross-section is

$$\sigma_x = \sigma_{CF} \prod_{i=1}^{x} [\Gamma_n/(\Gamma_n + \Gamma_f)]_i$$

where Γ_n and Γ_f are the level widths for neutron emission and fission, respectively. The ratio Γ_n/Γ_f depends very much on the difference $B_f - B_n$ between the fission barrier and the neutron binding energy. It is also sensitive to the rotational energy of the system. Without dealing in detail with the estimation of this ratio, it is clear that its value is very small when B_f becomes lower than B_n, and this is the case in many isotopes for $Z > 92$. Theoretical predictions[8] show that for $Z > 112$ we may again expect to find $B_n > B_f$. But, even under these better conditions, it will always be desirable to have the product

$$\prod_{i=1}^{x} [\Gamma_n/(\Gamma_n + \Gamma_f)]_i$$

as close to 1 as possible and therefore to limit the number of emitted neutrons to 1 or, even better, to zero. This means that the compound nucleus should be formed with an excitation energy lower than B_n in such a way that the only de-excitation process would be γ-emission. We can achieve this if the mass balance of the reaction is very negative and if its absolute value is close to the kinetic energy that is needed to overcome the Coulomb barrier. To take an example, the mass balance for the reaction $^{80}Kr + ^{230}Th \rightarrow ^{310}126$ is estimated at −308 MeV from the Myers and Swiatecki mass formula.[8] The kinetic energy in the centre-of-mass system which is needed to make a fusion nucleus with a cross-section of the order of 50 mb is around 310 MeV. Therefore the excitation energy will be very small, and the fusion nucleus will in principle decay only by γ-emission. Its recoil energy will be of the order of 107 MeV.

We have considered a number of target-heavy projectile pairs and have calculated the excitation energies of the fusion nuclei. If the magic numbers are to be $Z = 126$ and $N = 184$, we propose reactions such as $^{80}Kr + ^{230}Th$ and $^{80}Kr + ^{232}Th$ followed by the emission of two neutrons. For $Z = 124$ and $N = 184$, $^{82}Kr + ^{226}Ra \rightarrow ^{308}124$.

If the magic numbers are $Z = 114$ and $N = 184$, it is impossible to produce a fusion nucleus with an excess of neutrons comparable with the N/Z ratio in any target-projectile pair. There is, however, some hope of reaching isotopes for which the instability by both spontaneous fission and α-decay will not be too large. One proposal is to produce a heavy nucleus such as ($^{84}_{36}$Kr + $^{232}_{90}$Th → ^{316}126) which will decay rapidly by α-emission into the more stable ^{308}122. According to Nilsson's estimation, the spontaneous fission half-life would still be of the order of one year and the α-decay half-life ~10^{-3} s. Also, reactions on $^{198}_{82}$Pt, $^{208}_{82}$Pb may produce element 114 if the closed shell for neutrons appears at 170 instead of 184, as has sometimes been suggested:

$$^{208}\text{Pb} + {}^{86}\text{Kr} \rightarrow {}^{292}118 + 2\text{n}$$

^{292}118 will undergo a fast α-decay into ^{288}116, which should have a certain degree of stability.

Another possibility, the cross-section for which is unknown, would be a transfer reaction with a very large number of nucleons. We have observed (unpublished work) that bombardment of a medium target nucleus by $^{40}_{18}$Ar ions gives yields around 1 mb of $^{20}_{10}$Ne and we expect that the complementary part of 8 protons and 20 neutrons is added to the target. It is therefore believable that $^{86}_{36}$Kr might be stripped off into the very stable $^{32}_{16}$S and that the following reaction may occur:

$$^{244}_{94}\text{Pu} + {}^{86}_{36}\text{Kr} \rightarrow {}^{298}114 + {}^{32}_{16}\text{S}$$

$^{28}_{14}$Si can also be obtained as follows:

$$^{238}_{92}\text{U} + {}^{86}_{36}\text{Kr} \rightarrow {}^{296}114 + {}^{28}_{14}\text{Si}$$

In both cases the reaction product would have a ratio N/Z in the vicinity of 184/114, where the largest stability is expected. However, the cross-section of these processes might only be of the order of a tenth of a mb.

IDENTIFICATION

Our project[9] consists at present of an identification of the mass of the recoiling nucleus with the help of a magnetic spectrometer together with a velocity measurement. A rough illustration of the apparatus is shown in Fig. 3. At a bombarding energy of the order of 425 MeV, the compound nucleus should recoil with an energy of 110 MeV and a velocity of $8{\cdot}2 \times 10^8$ cm s^{-1}, and will enter a double-focusing magnetic spectrometer (radius 85 cm) which has been described elsewhere.[10] The kinetic energy will be measured by a solid state detector located at the focus. It is well known that simultaneous measurements of the kinetic energy ϵ and the momentum $mv = Ze \times BR$ give a unique value of the ratio Z^2/m, where Z is the effective charge of the ion and m its mass

$$\epsilon = \left(\frac{eBR}{2}\right)^2 \frac{Z^2}{m}$$

For a given ion at a given energy, there are several possible effective charges. From extrapolation of Northcliffe's data[11] the most probable charges for mass 310 at 110 MeV are around $Z = 33$ and the corresponding BR-value is of the order of 8000 gauss-m. We expect to obtain events for several different values of BR corresponding to the various values of Z. Because of the unknown

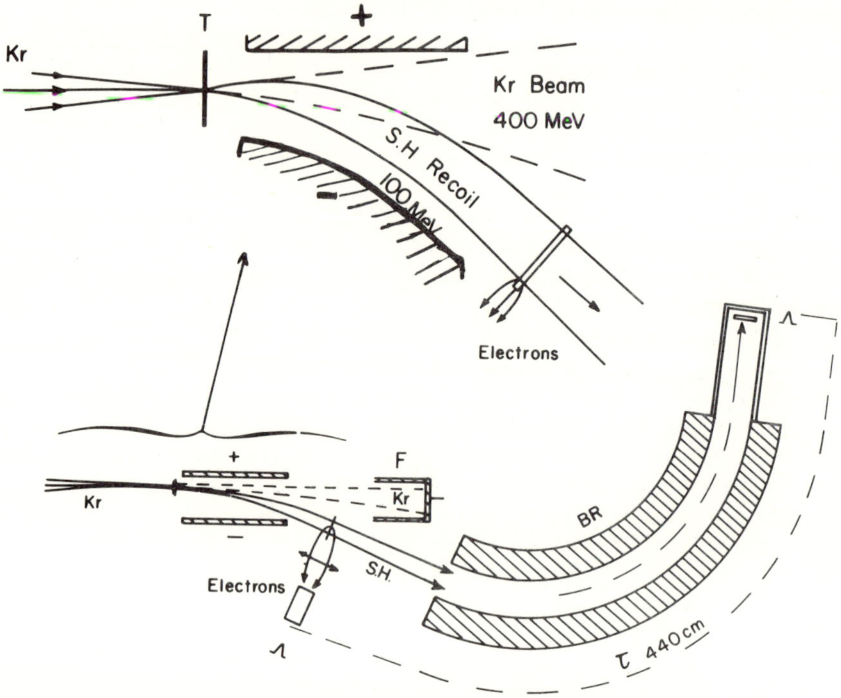

Fig. 3. Schematic diagram of the apparatus used for identification. Superheavy recoils (SH) *are deflected into the magnetic spectrometer* (BR). *A time of flight is measured over 440 cm between an electron multiplier and the energy detector*

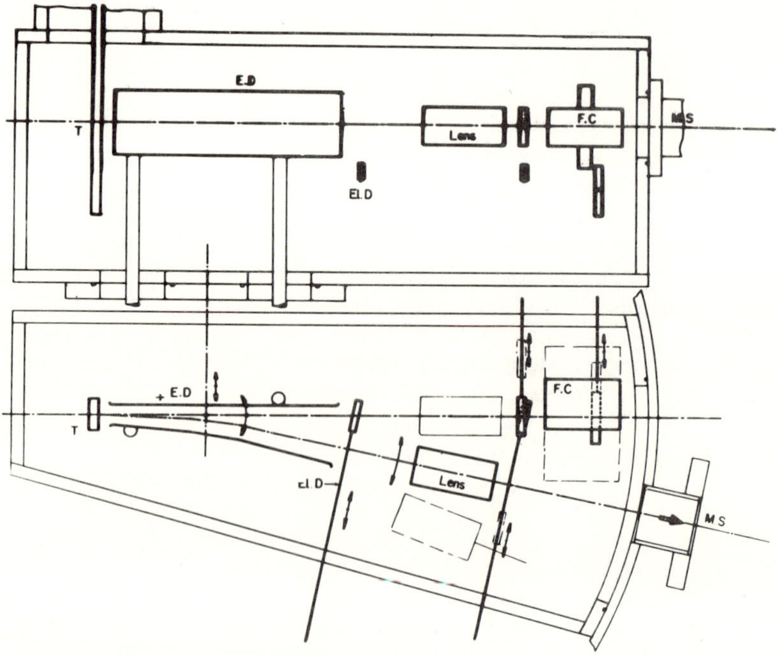

Fig. 4. Reaction chamber and electrostatic deflection of the recoiling nuclei

correction for the ionisation defect in the energy measurement,[12] we have assumed that the energy calibration of the detector might be subject to an error of 1%, which will appear in the resolution of the identification spectrum. Therefore the result on the mass assignment will be given with an accuracy within about three units.

The main difficulty is to get rid of the krypton beam itself. If krypton ions enter the aperture of the magnet, they will be deflected and will cause a large background on the energy measurement. Our proposal is to build behind the target an electrostatic septum that will deflect the 400 MeV krypton beam very slightly but will, on the other hand, change the direction of the recoiling superheavy ions.

At the same time, this scattering device (shown on a larger scale in Fig. 4) will give us the possibility of putting a thin foil on the trajectory of recoiling nuclei only. Electrons ejected from this foil will be used to initiate a pulse for a time-of-flight measurement, τ, between that point and the time of the energy measurement at the focus of the spectrometer. The distance is 440 cm, and it will take 5×10^{-7} s for mass 310 to travel this far, a much longer time than for any other ion that might also reach the energy detector. For example, fission fragments, which may behave like the recoiling nuclei ($\epsilon = 100$ MeV and $Z^2/m \simeq 3$), will travel in a shorter time (Fig. 5). We believe that such a time-of-flight measurement will allow the certain identification of a superheavy recoil.

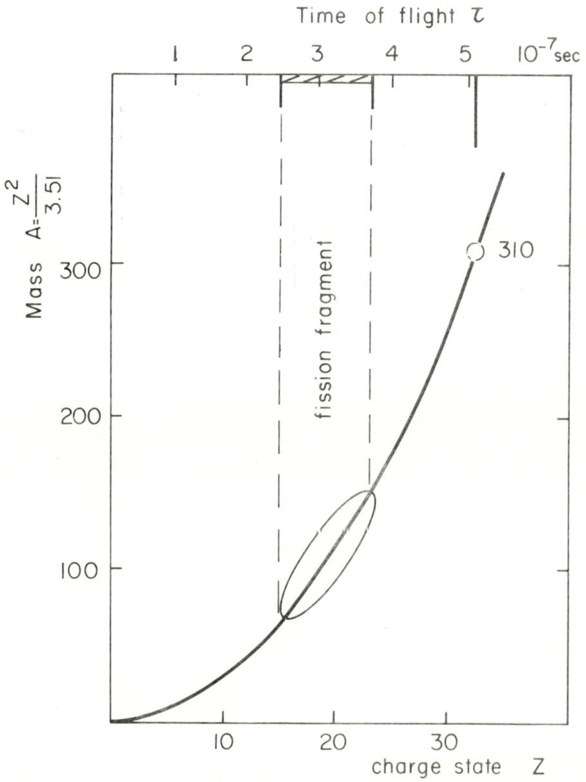

Fig. 5. Mass versus time of flight for particles of the same identification parameter $Z^2/A = 3{\cdot}51$

CONCLUSION

From the examples that are given, it will be clearly understood that the production of many new nuclides might be possible if superheavy ion beams were available. It is very often interesting to have the projectile nearly as heavy as the target in order to produce a compound nucleus that is either very neutron-deficient or as cold as possible. The problem of acceleration of medium and heavy nuclides is closely related to the charge state that can be reached in the centre of the cyclotron, since the energy is proportional to the square of that charge. One way to produce high charge states is to strip off ions already moving with high velocity through a thin foil. This process is being applied at Orsay, where heavy projectiles such as Ar^{4+} or Kr^{8+} will be accelerated at 1 MeV per nucleon in a Linac and injected into the cyclotron, at the centre of which they will be stripped to Ar^{13+} or Kr^{22+} (Project ALICE[13]). Another way is to improve the performances of multicharged ion sources, as is done in Dubna.[14]

Whatever the solution, a large amount of work might be opened up for nuclear chemists when various types of projectiles from Li and B to Kr or even Xe ions will be available.

REFERENCES

1. Conference on Nuclides far from Stability, Lysekil (1966), *Ark. Fys.* **36,** 1, (1967).
2. Myers, W. D. and Swiatecki, W. J., *Nucl. Phys.* **81,** 1, (1966).
3. Nilsson, S. G., Nix, J. R., and Sobiczewski, A., *Nucl. Phys.* **A115,** 545, (1968); *Nucl. Phys.* **A131,** 1, (1969).
4. Bimbot, R., Lefort, M., and Simon, A., *J. Phys.* **29,** 563, (1968).
5. Sikkeland, T., *Ark. Fys.* **36,** 539, (1967).
6. N'Guyen Long Den, personal communication.
7. Nilsson, S. G., Thomson, S. G., and Tsang, C. F., *Phys. Letters* **28B,** 458, (1969).
8. Myers, W. D. and Swiatecki, W. J., UCRL 11980 (1965).
9. Lefort, M., *Bull. Soc. Chim.* 1815 (1969).
   Lefort, M., Jacmart, J. C., Péter, J., Riou, M., Stéphan, C., and Tarrago, X., Heidelberg Conference on Nuclear Reactions induced by Heavy Ions, July 15-18 (1969).
10. Jacmart, J. C., Liu, M., Mazloum, F., and Riou, M., International Conference on Heavy Ions, Dubna (1966).
11. Northcliffe, L. C., *Ann. Rev. Nucl. Sci.* **13,** 67, (1963).
12. Kaufman, S. B., Wilkins, B. D., Steinberg, E. P., and Fluss, M. J., Symposium on Heavy Ion Nuclear Reactions, New York (September 1969).
13. Lefort, M. and Cabrespine, A., A.L.I.C.E. Project for Acceleration of Heavy Ions at Orsay, Heidelberg Conference, July 15-18 (1969).
14. Donetz, E. D., Ilioutchenko, H., and Alpert, D., Dubna Report (1969).

DISCUSSION

J. G. Cuninghame (A.E.R.E.): Various laboratories plan to bombard very heavy nuclei with very heavy projectiles, e.g. $^{238}U + {}^{238}U$, with the idea that asymmetric fission of the resulting compound nucleus will produce products in the stability region around $Z = 114$. Do you intend to try to do this, and do you think that this approach to the production of superheavy nuclei has a better chance of success than your nuclear transfer method?

M. Lefort: We are not ready to accelerate ions heavier than mass 100, because our ion source will not give high charge states, and the linear accelerator in front of the cyclotron will not be able to take a low charge/mass ratio. The fusion–fission process via the compound nucleus (U + U) has a rather small chance of success. It has been shown that the cross-section for complete fission decreases as the mass of the bombarding particle is increased, and we believe that it will be roughly half the total cross-section for Kr ions. It might become very small in the case of uranium. Moreover, the Coulomb induced fission that has been predicted will make the two uranium nuclei undergo fission before fusion. I therefore believe that the chance of producing superheavy elements by this method is very small.

H. Münzel (Karlsruhe): You have shown experimental results for many-neutron transfer reactions increasing the mass number of light nuclei. Do you expect that as many neutrons may be transferred to heavy nuclei also?

M. Lefort: Although the results for 4- and 6-neutron transfers concern transfer from heavy targets to light projectiles, it has been found that for 2-neutron transfers the yield is about the same in both ways, i.e. for pickup as well as for stripping. This is not so for proton transfers.

B. D. Pate (Simon Fraser University): I understand that your initial experiments–soon to be performed–rely on mass measurement for identification of the superheavy elements produced. Do you have any plans for chemical identification in later experiments?

M. Lefort: Yes. If we obtain evidence for masses in the range ~300, we have a programme for on-line mass spectrometry. Some work has been done by N'Guyen Long Den and Boussières on chemical separation in the ion source of the mass spectrometer. We also have plans for element 118, which should behave like radon and could be oxidised, as has recently been shown for radon at the Argonne National Laboratory.

G. N. Walton (A.E.R.E.): You point out that reactions with low or negative values of Q are likely to be the most promising for the formation of heavy compound nuclei. On the other hand, such reactions will have low cross-sections. Is there any way of optimising the conflicting requirements of high stability and high probability of formation?

M. Lefort: We must compromise on the choice of bombarding energy, i.e. as low as possible to get the lower excitation energy, but high enough to give a reaction cross-section of ~50 mb. Calculations have been made to determine the energy necessary for such a cross-section, and this was found to be ~5 MeV per nucleon whatever the bombarding ion.

L. Yaffe (McGill University): The last question dealt with the probability of *formation* of the very high Z elements. What probability would you assign to their production with half-lives greater than, say, 0·1 s?

M. Lefort: The answer lies with the theoreticians. If we take the predictions of Nilsson and others for $^{298}114$, the half-life is $\sim 10^{10}$y, but the uncertainty might be of the order of 10^6. Moreover, the chance of producing that particular nucleus is very small, because any type of bombardment will give a less neutron-rich product. This uncertainty has forced us to use an identification method that will not be restricted by too short or too long a half-life. Any nucleus with decay time $> 5 \times 10^{-7}$ s should be detected, and we still hope that there is a reasonable chance of discovering an element in the range $Z = 110$–126.

Nuclear chemistry with the Oak Ridge isochronous cyclotron *

R. J. Silva, R. L. Hahn, M. L. Mallory, C. E. Bemis, P. F. Dittner, O. L. Keller, J. R. Stokely, and K. S. Toth

Oak Ridge National Laboratory, Oak Ridge, Tennessee

ABSTRACT

The α-decay at an energy of 7·335 ± 0·003 MeV has been tentatively identified as a new californium isotope (^{241}Cf) having a half-life of 3·1 ± 1·2 min. The energy of the α-decay of ^{242}Cf was found to be 7·385 ± 0·004 MeV in agreement with a previous value obtained at Berkeley. A 7·351 ± 0·006 MeV α-peak that decays to the first excited state of ^{238}Cm has been resolved. Reduction potential measurements of ^{256}Md have been performed and the results indicate that the potential is less than +0·3 volts versus the normal hydrogen electrode. Preliminary experiments to produce element 104 have given results consistent with those obtained at Berkeley.

INTRODUCTION

The recently-developed carbon beam of the Oak Ridge Isochronous Cyclotron (ORIC), with 118 MeV energy and an intensity of 5 electrical microamperes (eμA) of +6 charge, is currently being used in discovery experiments of ^{241}Cf and ^{240}Cf. Bombardments of highly-enriched targets of ^{235}U, ^{234}U, and ^{233}U have produced a new α activity of 7·335 MeV whose excitation function behaves in the manner expected for ^{241}Cf. This activity is not produced in bombardments of ^{238}U. A new α activity has also been located at ~7·58 MeV in the ^{233}U bombardments. This activity is thought to be that of ^{240}Cf. Bombardments of ^{235}U have allowed us to confirm Berkeley's value of 7·385 MeV measured for the α particle energy of ^{242}Cf. We have also resolved the 7·351 MeV α particle that decays to the first excited state of the daughter ^{238}Cm. A cross bombardment of ^{231}Pa with nitrogen as well as chemical confirmation of the ^{241}Cf discovery are planned.

Studies of mendelevium chemistry are being performed using 1·5-h ^{256}Md produced by bombarding ^{253}Es (produced in the ORNL High Flux Isotope

*Research sponsored by the U.S. Atomic Energy Commission under contract with the Union Carbide Corporation

Reactor) with 30 MeV α particles. The objective of the current experiments is to measure more accurately the III-II reduction potential of ^{256}Md.

Preliminary experiments to produce element 104 (mass 257) by bombarding ^{249}Cf with the ^{12}C ions have been initiated. The objective is to carry out fast gas-phase chemistry on this 4·5 s, α-emitting isotope.

1. CALIFORNIUM

EXPERIMENTAL

Beam Characteristics. A portion of the experimental time was devoted to developing the ^{12}C beam from ORIC. Initial bombardments were carried out with ~1 eμA of carbon ions. For later bombardments, as the ORIC ion source was improved,[1] the current on the target increased to as much as 5 eμA, at which the Be degrading foils, which were uncooled, were burned through. All subsequent runs were limited to 3 eμA. At present the cyclotron is capable of delivering to the target as much as 10 eμA over a $^3/_8$ in diameter spot.

The energy of the beam was 118 MeV, and in order to change the bombarding energy rapidly a series of Be degrading foils, mounted on air pistons, could be remotely inserted into the beam path. Bombarding energies of 71, 76, 78, 81, 85, 90, 93 MeV and higher were employed.

Target Preparation. Targets approximately 1 mg/cm^2 thick were prepared from enriched uranium samples of the following isotopic purities: ^{238}U–99·99+%, ^{235}U–93%, ^{234}U–97%, and ^{233}U–99%. The uranium was chemically purified by selective adsorption onto and elution from an anion exchange resin column using 8 M HCl and 0·1 M HCl, respectively, followed by extraction from 8 M HNO_3 into diethyl ether. The dried nitrate was dissolved in ~1 ml of a solution of 60% ethyl alcohol and 40% acetone, and was then molecularly plated onto a 1 mil thick Be foil over a 1 cm^2 area using a glass chimney. The plating was carried out for about 1 hour with a voltage of ~300 volts and a current of 2 milliamperes applied between the Be foil cathode and a platinum anode stirring device.

Gas Jet and Rabbit. Another portion of the experimental time was devoted to developing a gas jet-rabbit system. The uranium target served as the entrance window for the beam to a cylindrical chamber of diameter 5 cm and length 7·5 cm, which contained helium at a pressure of ~250 mm Hg. In the middle of the chamber and at 90° to the beam direction was an entrance tube for the helium gas. Opposite this entrance tube was a 0·5 mm orifice through which the helium gas, containing the recoil products ejected from the target and slowed down by the gas, was pumped out by a 300 ft^3/min Kinney mechanical pump forming a gas jet.[2] The beam passed through a Be window and was then monitored in a Faraday cup. On the vacuum side of the gas jet the pressure was 500 μ. At a distance of 0·5 cm from the orifice was a platinum disc onto which the helium gas impinged, depositing the recoil atoms. The disc was attached by Eastman 910 cement to a polyethylene rabbit, which was contained in an aluminium cylinder mounted on an air piston. This arrangement allowed remote positioning of the rabbit from in front of the gas jet orifice and into the rabbit pneumatic tube. The rabbit could be transferred pneumatically through a $^5/_8$ in I.D. polyethylene tube from the experimental vault to a chemistry laboratory in

5 seconds. In the latter the rabbit was received in a device that allowed the rabbit to be rotated in front of a silicon surface-barrier α-particle detector. In addition, this device was designed so that another rabbit would be in position to be transferred to the experimental vault, hence allowing almost 100% utilisation of the cyclotron time.

Data Acquisition System. α particles were detected using an ORTEC 200 mm^2 Si(Au) surface barrier detector with a 100 μ depletion depth in conjunction with a low noise charge-sensitive preamplifier (Tennelec Model TC-130). Pulses from the preamplifier were amplified using a Tennelec Model TC-200 main amplifier and a biased post amplifier (Tennelec Model TC-250). The measured α-particle resolution for the 5·499 MeV α group in the decay of ^{238}Pu was ~18 keV FWHM.

α pulses in an energy range of 6·00 to 8·50 MeV were recorded with a 1600-channel pulse-height analyser (Victoreen Model 1600) which was divided into eight 200-channel or four 400-channel sections in order to follow the decay of various α groups through preset time intervals. In addition, the α pulses occurring during the total counting interval were also recorded using a 1024-channel pulse-height analyser (Nuclear Data Model 2200). The spectra recorded with the 1024-channel analyser provided a continuous sum spectrum at higher channel resolution so that more accurate energy values could be obtained.

In addition, the α groups of ^{213}Ra (6·623, 6·730), from reactions of ^{12}C on Pb impurities in the targets, as well as ^{217}Rn (7·735) and ^{214}Po (7·680) members of the decay chains of ^{225}Th and ^{226}Th produced in the bombardments, were used as internal energy calibration points.

RESULTS

^{242}Cf Experiment. Figure 1 depicts typical α-particle energy spectra obtained in the bombardment of ^{235}U, ^{234}U, and ^{233}U with various carbon energies. Only those parts of the spectra containing the α peaks at energies of 7·385, 7·351 and 7·335 MeV are shown in order to readily see the energy separations. Table 1 contains the energy and half-life of ^{242}Cf as measured by groups at Berkeley[3] and Argonne,[4] and from the results of this experiment. The energy obtained in this experiment for ^{242}Cf is 7·385 ± 0·004 MeV, in excellent agreement with the value obtained by Berkeley. The half-life obtained in this experiment is in good agreement with that found by both groups. Analysis of the data shows another α peak at 7·351 ± 0·006 MeV, the half-life of which is equal to the half-life of the ^{242}Cf peak at 7·385 MeV. The intensity of the 7·351 MeV peak relative to the 7·385 MeV peak is ~20%. The energy difference of ~40 keV and the relative intensities are consistent with the systematics of first excited states of even-even transuranium isotopes. Hence, from the above data we assign the 7·351 ± 0·006 MeV energy to the α decay of ^{242}Cf to the first excited state of ^{238}Cm.

^{241}Cf Experiment. A total of 250-300 irradiations of eight to twelve minutes duration were carried out, producing a total of 230 events at a new energy of 7·335 ± 0·003 MeV. The half-life of this peak was determined to be 3·1 ± 1·2 min (Fig. 2). The excitation functions for production of ^{242}Cf and the α activity at an energy of 7·335 MeV are depicted in Fig. 3 for ^{234}U and ^{233}U.

Theoretical calculations indicate that in general, for ^{12}C ions of bombarding

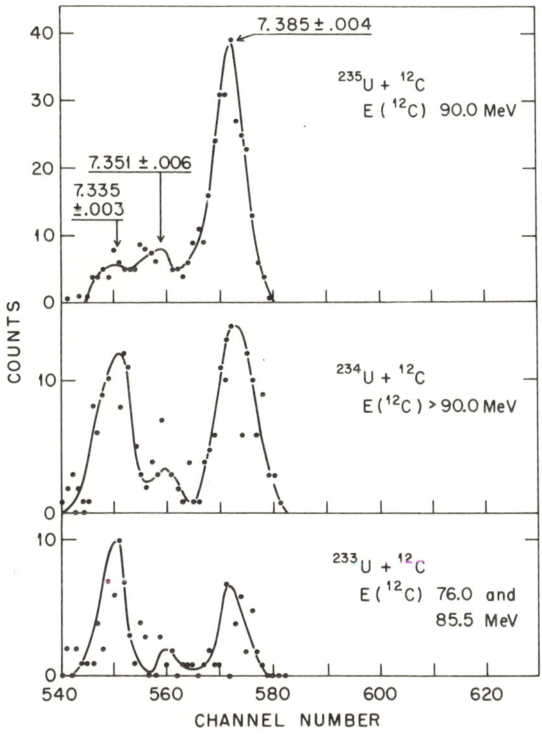

Fig. 1. A section of the α-energy spectrum obtained in the bombardment of three different isotopes of uranium with carbon. The relative intensity and energy shift of the 7·335 MeV a peak with respect to the known ^{242}Cf *peak (7·385 MeV) for the various uranium isotopes leads tentatively to the assignment of* ^{241}Cf *to the 7·335 MeV peak*

Table 1. NUCLEAR PROPERTIES OF $^{242-241}$Cf

Laboratory	*Energy (MeV)*	*Half-life (min)*	*Nuclide*	*Excitation Peak for* ^{233}*U (MeV)*
Argonne	7·35 ± 0·01	3·2 ± 0·5	^{242}Cf	
Berkeley	7·39 ± 0·02	3·4 ± 0·2	^{242}Cf	
Oak Ridge	7·385 ± 0·004	3·3 ± 0·4	^{242}Cf	78·5
Oak Ridge	7·351 ± 0·006	~3·3	^{242}Cf*	
Oak Ridge	7·335 ± 0·003	3·1 ± 1·2	^{241}Cf	81·5

*Excited state

energies sufficient to exceed the Coulomb barrier for transuranium elements, the resulting compound nucleus contains sufficient energy to evaporate most probably 4 to 5 neutrons, with decreasing probability for emission of fewer neutrons. The 3n emission probability can sometimes compete favourably with the 4n emission, particularly as one goes to the very neutron-deficient nuclei where the neutron binding energies become increasingly larger, thus shifting the maximum in the excitation functions to higher energies. There is, however, fission competition at each neutron evaporation step, so that the cross-sections decrease with the number of neutrons evaporated above 4–5 neutrons.

Experiments on transuranium elements do indeed show that ^{12}C, 4n and 5n reactions usually have the largest cross-sections.

Hence, in the bombardment of the various uranium isotopes with carbon ions, one would expect the following qualitative behaviour: (1) the cross-section of the ^{235}U(^{12}C, 5n)^{242}Cf reaction would be greater than that for the ^{235}U(^{12}C, 6n)^{241}Cf

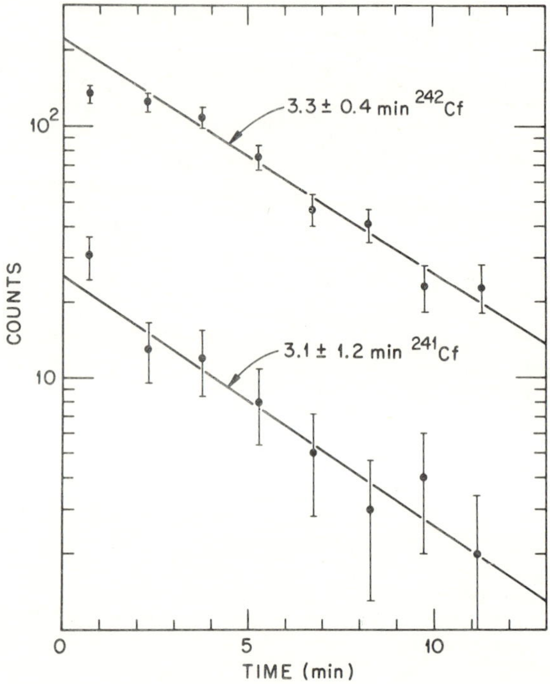

Fig. 2. The decay rates for ^{242}Cf *(top curve) and the* ^{241}Cf *a peak (bottom curve) give half-lives of 3·3 ± 0·4 and 3·1 ± 1·2 min respectively*

reaction; (2) the cross-section of the ^{234}U(^{12}C, 4n)^{242}Cf reaction would approximately equal that of the ^{234}U(^{12}C, 5n)^{241}Cf reaction; and (3) the cross-section of the ^{233}U(^{12}C, 3n)^{242}Cf reaction would be slightly less than that for the ^{233}U(^{12}C, 4n)^{241}Cf reaction. A computer calculation based on compound nuclear theory[5] is in progress to detail more accurately these general considerations.

Clearly, as is seen in Fig. 3 and in the spectra of Fig. 1, this type of behaviour is what is indeed observed for the 7·335 MeV peak relative to that of ^{242}Cf. This agreement supports our assignment of the 7·335 MeV α particle to the decay of new isotope ^{241}Cf. We hope to be able to confirm that the 7·335 MeV α peak is indeed a californium isotope by demonstrating that it follows californium chemistry. A cross bombardment of ^{231}Pa with ^{14}N is also planned as confirmatory evidence.

In the bombardment of ^{233}U a new α-decay energy is seen at ~7·58 MeV with a half-life of the order of 1 min. This α energy is not produced in bombardments of ^{238}U, ^{235}U, or ^{234}U, nor does it correspond to any member of the observed decay chains. The energy and half-life are consistent with those expected from α-decay systematics for ^{240}Cf, therefore we tentatively assign this new activity to ^{240}Cf.

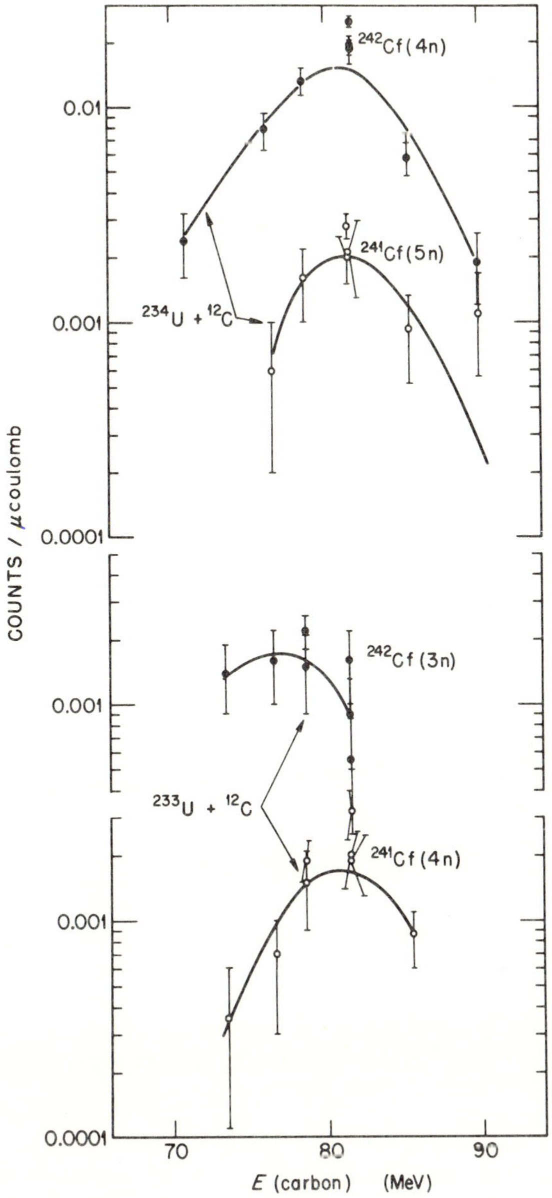

Fig. 3. Excitation functions for the production of ^{242}Cf *and* ^{241}Cf *from* ^{234}U *and* ^{233}U *targets*

Recent studies[6,7] have shown that the divalent oxidation state of ^{256}Md is stable in solution, and that the potential of the Md(II)–Md(III) couple is ~+0·2 volts versus the normal hydrogen electrode (N.H.E.). Additional chemical experiments are being performed to confirm the existence of Md(II) and to obtain a better estimate of the potential of the couple.

EXPERIMENTAL

Beam Characteristics. ^{256}Md was produced by bombarding ^{253}Es with the ORIC α beam. A 40 MeV α beam could be degraded by helium-cooled Al foils to an incident energy on the target of 30 MeV. The maximum beam intensity was 15 eμA. Bombardment times were ~3 hours.

Target. 2–3 mg/cm^2 targets were produced from ^{253}Es containing less than 1% of ^{254}Es and ^{255}Es. The einsteinium was chemically purified by ion exchange chromatography. The dried nitrate was dissolved in 10^{-3} M HNO_3 plating solution and electrodeposited over a 1 cm^2 area onto 1 mil Be foil as described above. The electrical conditions were 5 volts at 25 mA current.

^{256}Md Production. One of the Al absorber foils acted as an entrance beam window to an evacuated reaction chamber/Faraday cup section. The ^{253}Es target was placed in the chamber so that the ^{256}Md atoms produced by the $^{253}Es(\alpha, n)^{256}Md$ reaction and recoiling from the target were collected on a gold leaf placed about 2 cm from the target. After bombardment, the chamber could readily be let up to atmospheric pressure, the gold foil and ^{256}Md atoms taken for chemical studies, and a new gold foil inserted. This system, unlike the gas jet system, does not require a pressure differential across the extremely α-radioactive ^{253}Es (0·1 Ci) target, which increases the probability of rupture. Because of the large amount of α activity the entire bombardment assembly and pumping system was contained in a hermetically-sealed glove-box assembly.

The ^{256}Md atoms were separated from the gold before the reduction experiments by first dissolving the foil in dilute aqua regia (6 M HCl–trace HNO_3) and passing the solution through an anion-exchange resin column. Gold was absorbed on the resin while the ^{256}Md atoms passed through the column.

RESULTS

Initial experiments involved reducing Md(III) with various chemical reductants and then separating the reduced Md(II) from Md(III) and Fm(III) on small extraction chromatographic columns prepared from di(2-ethylhexyl) phosphoric acid supported on Kel-F powder.[6] In dilute acid solution, trivalent ions are retained on the column while univalent and divalent ions are eluted rapidly; thus the column separation provides a good test for the reduction of Md(III) to Md(II). At least partial reduction of ^{256}Md was observed when 0·02 M Cr(II) and V(II) ions (prepared by reaction with Zn-Hg amalgam) were used as reductants; very little or no reduction was obtained with hydrazine, stannous chloride, or hydroxylamine. Chromous ion proved to be the most effective reductant (Fig. 4). Because it decays by electron capture, the simplest method of analysing ^{256}Md is to follow the growth of ^{256}Fm on daughter fission activity. Fission counting of the reduced ^{256}Md fraction showed the growth of ^{256}Fm and very efficient separation of ^{256}Md from ^{256}Fm. Analysis of decay data indicates that about 10^4 atoms of ^{256}Md were made in each bombardment, giving maximum count rates of the order of 50 counts per min. These experiments confirm the existence of a reduced state of ^{256}Md and suggest that the potential of the couple is between +0·3 and 0·0 volts versus N.H.E.

Refinements on the experiment are now being undertaken. The use of chemical reductants alone can only provide rough estimates of the potential of

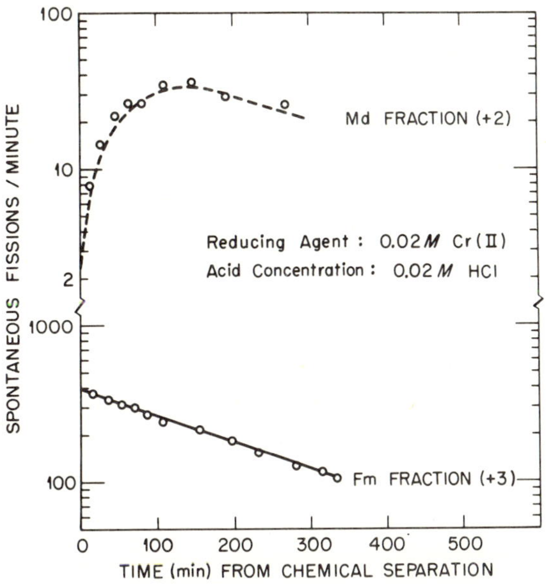

Fig. 4. The spontaneous fission of ^{256}Fm after chemical separation. The top curve confirms that ^{256}Md has been reduced. The line through the top curve is a theoretical fit for an initial ^{256}Md fraction that is completely free of ^{256}Fm, indicating that the separation is very efficient

the couple. Use of a mercury cathode controlled at a known potential (coulometric titrator) is being investigated in an attempt to provide a more accurate adjustment of the potential of the solution; the concentrations of the various oxidation states of the reductants adjust to the applied potential and act as charge-transfer carriers for the ^{256}Md.

3. ELEMENT 104

A 17 mg/cm^2 target of ^{249}Cf, prepared by molecular plating onto 1 mil Be foil from isopropyl alcohol, was bombarded with ^{12}C ions in the gas-jet assembly. The object was to produce element 104 by the $^{249}Cf(^{12}C, 4n)^{257}104$ reaction. This isotope has recently been said to decay by emission of α particles of energies 8·70, 8·78, 8·95, and 9·00 MeV with a half-life of 4·5 s.[8] The first experiments were carried out in order to duplicate the discovery work and to provide a source of 104 atoms for fast chemical (gas chromatographic) experiments.

In these experiments, because of the short half-life, the rabbit was replaced by a shuttle system composed of an Al foil which could be positioned either over the helium gas jet or in front of a Si(Au) α detector by an air piston. As we have only carried out these bombardments for a few hours, we have seen only two α-particle events in the energy range 8·7–9·0 MeV. However, the relative production of ^{250}Fm (7·43 MeV), ^{252}Fm (7·04 MeV), ^{254,255}No (8·10 MeV) and ^{257}104 (8·5–9·0 MeV) is similar to that quoted by the Berkeley group.

REFERENCES

1. Lord, R. S. et al, Heavy Ion Acceleration at ORIC, *Proc. Int. Conf. on Cyclotrons,* Oxford, Sept. 17–19, 1969.
2. MacFarlane, R. and Griffioen, R. D. *Nucl. Instr. Meth.* **24,** 461, (1963).
3. Sikkeland, T. and Ghiorso, A. *Phys. Letters* **24B,** 333, (1967).
4. Field, P. R., Barnes, R. F. Sjoblom, R. K., and Milsted, J. *Phys. Letters* **24B,** 340, (1967).
5. Sikkeland, T. *Proc. of Int. Symposium on Why and How of Nuclides Far off the Stability Line,* 539, Almqvist-Wiksell, Stockholm, (1967).
6. Hulet, E. K. et al, *Science* **158,** 486, (1967).
7. Maly, J. and Cunningham, B. B. *Inorg. Nucl. Chem. Letters* **3,** 445, (1967).

Dependence of the chemical behaviour of molybdenum recoil atoms in metal carbonyls on the energy and the isotope

E. Henrich and G. K. Wolf

Lehrstuhl für Radiochemie, University of Heidelberg, Kernforschungszentrum Karlsruhe, Postfach 3640, Germany

ABSTRACT

Different isotopes recoiling from nuclear reactions sometimes show a difference in their chemical behaviour. In order to study the causes of this difference target elements ranging from molybdenum to indium were irradiated in close contact with $Cr(CO)_6$ with 600 MeV protons. ^{90}Mo and ^{93m}Mo recoil atoms, with energies dependent on the mass difference between target and product, reacted with the $Cr(CO)_6$ forming Mo $(CO)_6$. The yield of this reaction was determined for the individual isotopes and their recoil energy. $Mo(CO)_6$ was also irradiated with 600 MeV protons, fast neutrons and thermal neutrons. The percentage of ^{90}Mo, ^{93m}Mo, and ^{99}Mo having the configuration of the target (retention) was measured for the individual isotopes and the irradiation conditions.

In all cases the retention of ^{93m}Mo was higher than that of ^{90}Mo and ^{99}Mo. The magnitude of this isotope effect was independent of the recoil energy and the irradiation conditions. The existence or non-existence of nuclear levels having sufficient lifetime ($<10^{-12}$s) in the de-excitation cascade of the excited recoiling isotopes is quoted as a cause of their different chemical behaviour.

INTRODUCTION

Atoms recoiling from nuclear reactions in nearly all cases break their chemical bonds. Having lost their excitation energy, they may react again chemically with their environment. The product spectrum resulting from these reactions depends on the chemical environment and the nuclear reaction.

Several studies have shown that different isotopes of the same element sometimes form different product spectra in the same chemical environment; the different isotopes were usually produced by different nuclear reactions.[1-5] Sometimes, when the isotopic composition of the target was suitable, they were generated by thermal neutron capture.[6-10] In some cases an isomeric state exhibits a different chemical behaviour compared with the ground state.[11,12]

This isotope effect must result from conditions which are created by the irradiation and the nuclear reaction, e.g.

(1) Radiation dose received by the target.
(2) Recoil energy of the atoms.
(3) The decay mode of the originally highly excited recoil atoms.

When the isotope effect is studied with isotopes arising from different nuclear reactions, all three factors are of importance. When the different isotopes are produced by the same nuclear reaction, the radiation dose is constant and only the two remaining factors are important. Finally, isomers should have recoil energies which are not very different from the recoil energies of the same isotope in the ground state. Thus, it is probable at least in this case that the decay mode of the recoil atom is the most important factor contributing to an isotope effect, although a contribution of the recoil energy and the radiation dose can not be excluded completely. Nuclear reactions with high-energy particles (spallation) make it possible to study the influence of the recoil energy and of the decay mode of the excited isotopes independently. On one hand the different isotopes are produced under constant radiation conditions by the same nuclear reaction. On the other the dependence of the recoil energy on the mass difference between target and product allows the recoil energy to be varied by using different target elements to produce the same isotopes.

For this reason we irradiated target elements ranging from molybdenum to indium with 600 MeV protons. The targets were in close contact with metal carbonyl catchers. Thus the recoiling molybdenum isotopes ^{90}Mo and ^{93m}Mo could penetrate into the catcher and form molybdenum carbonyl. The percentage of carbonyl formation (retention) was analysed for both isotopes and compared with experimentally-determined recoil ranges (which are proportional to the recoil energies).

In another experiment we irradiated molybdenum carbonyl as a target with 600 MeV protons, fast neutrons and thermal neutrons and determined the retentions of ^{90}Mo, ^{93m}Mo, and ^{99}Mo with respect to the different irradiation irradiation conditions.

The metal carbonyls were chosen as targets and catchers because of their very specific reaction with molybdenum recoil atoms.[15] Preliminary studies which will be described elsewhere[16] showed that in our experiments molybdenum was the only spallation product which reacted with the metal carbonyls.

EXPERIMENTAL

Irradiations. Fig. 1 shows the 'mixed targets', which consisted of several metal foils of thickness 10 mg/cm^2 covered with a 0·5 mm layer of powdered $Cr(CO)_6$ serving as catcher. As target metals, Mo, Rh, Pd, Cd, Ag, and In were chosen. The targets were irradiated for 1 h at room temperature in the external beam of the CERN Synchrocyclotron. The flux of 600 MeV protons was 2×10^{10} p/cm^2 s. The $Mo(CO)_6$ targets consisted of powdered material which was sublimed under vacuum before the irradiation. The following irradiation conditions were chosen:

(1) 200 mg $Mo(CO)_6$ packed in polyethylene foil were irradiated for 1 h and 10 h with 600 MeV protons;
(2) 200 mg $Mo(CO)_6$ wrapped in silver foil (0·2 mm thick) were activated

for 30 min with neutrons produced by the reaction of 50 MeV deuterons from the Karlsruhe cyclotron with a beryllium target. The temperature did not exceed 30°C. The flux was 5×10^{11} n/cm² s;

(3) 20 mg $Mo(CO)_6$ packed in polyethylene foil were activated for 10 min in the thermal column of the Karlsruhe FR2 reactor with a flux of 3×10^{11} n/cm² s. The irradiation temperature was either 60–70°C or dry ice temperature (–79°C).

Chemical separations. The retention was determined by separating the organic from the inorganic fraction either by extraction or by sublimation. For the mixed targets, the $Cr(CO)_6$ layer was separated from the foils and sublimed under vacuum at 90°C. The yield of the sublimation was determined by weight.

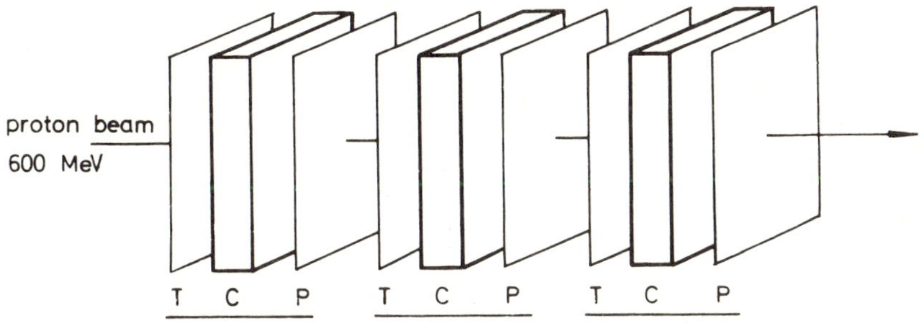

Fig. 1. Arrangement of 'mixed targets' consisting of metal foils as target and $Cr(CO_6)$ *as catcher.* T–*target foil,* Rh, Pd, Ag, Cd, In; C–$Cr(CO_6)$ *powder layer,* 0·5 mm *thick;* P–*PVC foil*

Mo in the residue was separated from the other spallation products as follows: the residue was dissolved in aqua regia and a known amount of Mo(VI) was added. The solution was made 6·5 N with HCl and the Mo extracted with methyl isobutyl ketone and back-extracted with water. After a $Fe(OH)_3$ scavenger precipitation, Mo was precipitated with α-benzoinoxime in 2N HCl, the precipitate extracted with ethyl acetate and back-extracted with 2N ammonia. Acetic acid was added to pH 3 and Mo precipitated with 8-hydroxyquinoline, filtered, dried, weighed for the yield determination and mounted for counting. The sublimate was dissolved in fuming nitric acid and treated in the same way as the residue.

For the $Mo(CO)_6$ irradiated with 600 MeV protons the same procedure was used for the residue. The sublimate could be measured directly because of the absence of interfering ^{48}Cr and ^{48}V activities. $Mo(CO)_6$ targets irradiated with fast and thermal neutrons were sublimed and the sublimate and the residue measured directly, since no interfering activities were present. The targets irradiated with thermal neutrons at dry ice temperature were dissolved in cold $CHCl_3$, the inorganic fraction extracted with 2N nitric acid, and both solutions measured directly.

Measurements. γ-spectra of the Mo fractions in the sublimate and in the residue were recorded with a NaI(Tl) detector. For the fast neutron irradiated targets a planar Ge(Li) detector was used. The photopeaks of ^{93m}Mo, ^{99}Mo, and ^{90}Nb (the daughter of ^{90}Mo) were evaluated. For samples from the thermal neutron irradiations, the total γ-activity was counted with a well type NaI(Tl) detector.

Determination of recoil ranges. The 'thick target, thick catcher' method[17] was used for a crude estimation of the ranges of ^{90}Mo and ^{93m}Mo atoms recoiling during the 600 MeV proton bombardment of different target elements. Mo, Pd, Ag, and In target foils were covered with aluminum catcher foils in forward and backward directions with respect to the proton beam. Each foil was thicker than the range of the recoiling atoms. ^{90}Mo and ^{93m}Mo in the foils were separated and counted according to the procedure described above for the mixed targets.

RESULTS

The results of the experiments are summarised in Tables 1 and 2.

Table 1. PERCENTAGE RETENTION OF MOLYBDENUM RECOIL ATOMS PRODUCED BY 600 MeV PROTONS IN DIFFERENT TARGET ELEMENTS IN CONTACT WITH $Cr(CO)_6$

Target	Mo	Rh	Pd	Ag	Cd	In	*Mean value*
^{93m}Mo-retention %	79·0±2	80·6±2	81·2±2	81·0±2	78·6±2	80·5±2	80·2
^{90}Mo-retention %	70·0±3	70·6±3	–	74·5±3	73·0±3	71·1±3	71·8
Recoil ranges in mg/cm² of the target ^{90}Mo	0·11	–	0·49	0·53	–	0·64	
Recoil ranges in mg/cm² of the target ^{93m}Mo	0·21	–	0·64	0·60	–	0·64	

Table 1 gives the percentage of ^{90}Mo and ^{93m}Mo (the retention) which sublimed together with $Cr(CO)_6$, for the different target elements. In addition the recoil ranges are quoted for all cases where experimental determinations were made. There seems to be no significant difference in the retention values with increasing mass distance between target and product. However, the recoil ranges, which are proportional to the recoil energy, increase by a factor of five from Mo to In. The retention of ^{93m}Mo was about 8% higher than the retention of ^{90}Mo. This difference was not dependent on the target element. From these results we conclude that in our system the retention differs for different isotopes, and that it is independent of the recoil energy in the energy region investigated.

Table 2 shows the results of irradiating $Mo(CO)_6$, which served simultaneously as target and catcher. The bombarding particles varied as well as the irradiation conditions.

Table 2. PERCENTAGE OF THE RETENTION OF MOLYBDENUM RECOIL ATOMS PRODUCED UNDER DIFFERENT IRRADIATION CONDITIONS IN $Mo(CO)_6$

Irradiation conditions	*Protons* 600 MeV 1 h	*Protons* 600 MeV 10 h	*Neutrons* 50 MeV 30 m	*Thermal neutrons* 60°C	*Thermal neutrons* −78°C
^{93m}Mo-retention %	80·6±2	79·7±2	79·6±1		
^{90}Mo-retention %	72·0±2	69·0±2	75·0±2		
^{99}Mo-retention %			70·0±1·5	66±2	40±4

The ^{93m}Mo retention seems to be independent of the type of nuclear reaction and also of the irradiation time. Thus, it is independent of the radiation dose received by the target. The ^{90}Mo retention shows some variation, but since the accuracy of the determination of this value was not very high the variation may be accidental.

A comparison of Tables 1 and 2 shows very little difference between the 'mixed targets' and the $Mo(CO)_6$ targets. As well as ^{90}Mo and ^{93m}Mo, ^{99}Mo is also generated by neutrons. Its retention is a little lower than that of ^{90}Mo and significantly lower than that of ^{93m}Mo. The magnitude of the isotope effect is about the same as for the mixed targets. Our values for ^{90}Mo and ^{99}Mo are in good agreement with the results of Harbottle and Zahn,[18] who bombarded $Mo(CO)_6$ at liquid nitrogen temperature with 2 GeV protons and found 75% retention for ^{90}Mo and 73·7% for ^{99}Mo.

The retention of ^{99}Mo produced with thermal neutrons was a little lower than that produced with fast particles. In this case the irradiation temperature had a pronounced effect on the magnitude of the retention. Thus the $Mo(CO)_6$ irradiated with thermal neutrons showed annealing reactions. Attempts to anneal $Mo(CO)_6$ after irradiation with fast particles failed.

DISCUSSION

We do not wish to speculate in the following section on the formation mechanism of the retention. From our experimental results we propose only to draw conclusions concerning the isotope effect. In all cases investigated in this work ^{93m}Mo showed a higher retention than ^{90}Mo and ^{99}Mo. Of the three factors quoted in the Introduction, the first, viz. different radiation doses received by the targets, can be excluded, because the different isotopes were always produced together in the same target. Thus, ^{93m}Mo originated under exactly the same irradiation conditions as the other Mo isotopes. The second factor, varying energy of the recoil atoms, can also be excluded, for the following reason. Fivefold variations of the recoil energy neither affected the absolute retention values, nor the difference in retention between the individual isotopes. A decrease in recoil energy by several orders of magnitude, as in the case of the thermal neutron irradiations, produced only a small reduction in the retention. It is more probable that this reduction is caused by the different irradiation conditions rather than by a change in recoil energy. Our results, as well as those of other studies, show that $Mo(CO)_6$ irradiated with moderate doses in the thermal column of reactors is sensitive to radiation and thermal annealing. In contrast, the conditions of the fast particle irradiations do not affect the behaviour of the molybdenum recoil atoms in the metal carbonyls. The relatively high doses seem to lead to a saturation effect when variations have no further effect.

There remains the third factor, a difference in the de-excitation mode of the individual excited molybdenum isotopes. This cause, which has been proposed by several authors,[13,14] will now be discussed in more detail. We shall restrict the discussion to solids, since their retentions are mostly high and the product spectra less complex than in the liquid or gas phases. Fig. 2 shows the energy level scheme[21] of ^{60}Co and its isomeric state ^{60m}Co. In producing these isotopes by thermal neutron irradiation of inorganic Co complexes Lazzarini nearly always found[12,22] a higher retention for ^{60m}Co than for ^{60}Co. Keeping in mind that a fraction of ^{60}Co results from the decay of ^{60m}Co during the irradiation, this isotope effect can be interpreted in terms of a 'two-step' reaction. After the nuclear reaction, the ^{60m}Co as well as the ^{60}Co recoil atoms slow down and react in a first step with their chemical environment, leading to a certain percentage retention.

Part of the ^{60}Co atoms, however, results from a secondary nuclear transformation, namely the decay of ^{60m}Co. This highly-converted isomeric transition leads in a second step to destruction of the bonds formed in the first step, followed partly by re-formation. This effect consequently causes a decrease

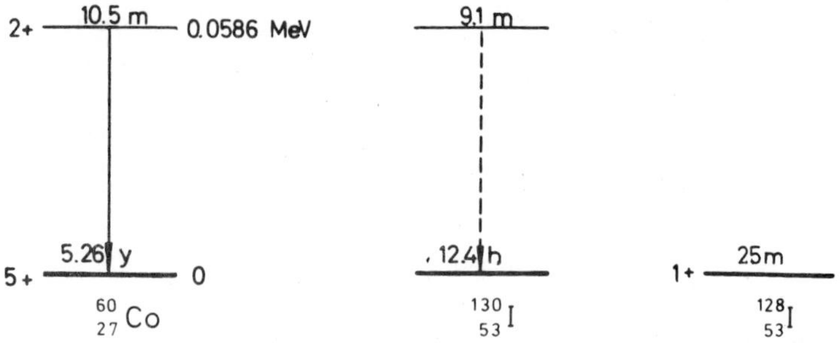

Fig. 2. Energy levels of the isomeric pair ^{60m}Co and ^{60}Co and of ^{128}I and ^{130}I

in retention, i.e. the Co recoil atoms come to rest in a lattice site with a certain defect concentration. This leads to the retention during the first step. In the case of a subsequent isomeric transition, the defect concentration increases in the second step causing mainly a decrease in retention.

A comparison of the behaviour of the isotopes ^{128}I and ^{130}I produced in $NaIO_3$ by thermal neutron irradiation led to similar results. The retention[9] of ^{130}I was ≈40% lower than that of ^{128}I. Corresponding to the interpretation of the ^{60}Co results, the level schemes in Fig. 2 show that part of the ^{130}I is formed by a highly-converted isomeric transition,[23] coming from an isomeric state with 9 min half-life. Thus, a one-step reaction probably causes the retention of ^{128}I and a two-step reaction that of ^{130}I.

^{99}Mo and ^{90}Mo, the level schemes[21] of which are presented in Fig. 3, have no isomers, in contrast to the above examples.

To understand this case, we must pay attention to the fact that recoil atoms, after slowing down in solids, take their position in the lattice[23,24] in $\approx 10^{-12}$ s, whilst the re-formation of their electron shell is completed in $\approx 10^{-8}$ s. This has the consequence that every de-excitation taking place 10^{-12} s after the nuclear reaction or later may have an influence on the defect concentration around the recoil atom.

If it requires more than 10^{-8} s it may again destroy the re-formed electron shell. In principle there is no difference between a partly-converted transition and the de-excitation of an isomeric state. In particular the transitions from rather long-lived low-lying states are often highly converted. Fig. 3 shows that ^{99}Mo has such a low-lying level with a 10^{-5} s life-time, which is sufficient to cause a two-step reaction and the corresponding decrease in retention. We can be sure that the retention is formed in a one-step reaction for ^{93m}Mo, since its spin has the unusually high value of 21/2. Above this level no states with higher spins are known.[25] Thus, converted transitions with life-times long enough for a second step should be very few. This interpretation agrees with the findings of a higher retention for ^{93m}Mo than for the other isotopes. Unfortunately no data concerning the ^{90}Mo levels are available.

The two-step reaction does not yield a higher retention than the one-step reaction for all the systems studied. Comparing ^{80m}Br with ^{82}Br which originates partly from the decay of an isomer of 6 min half-life, a higher retention for ^{82}Br was determined in some compounds.[10,25] But since in other compounds either no difference[26] or a higher retention[8] for ^{80m}Br were found, this case appears

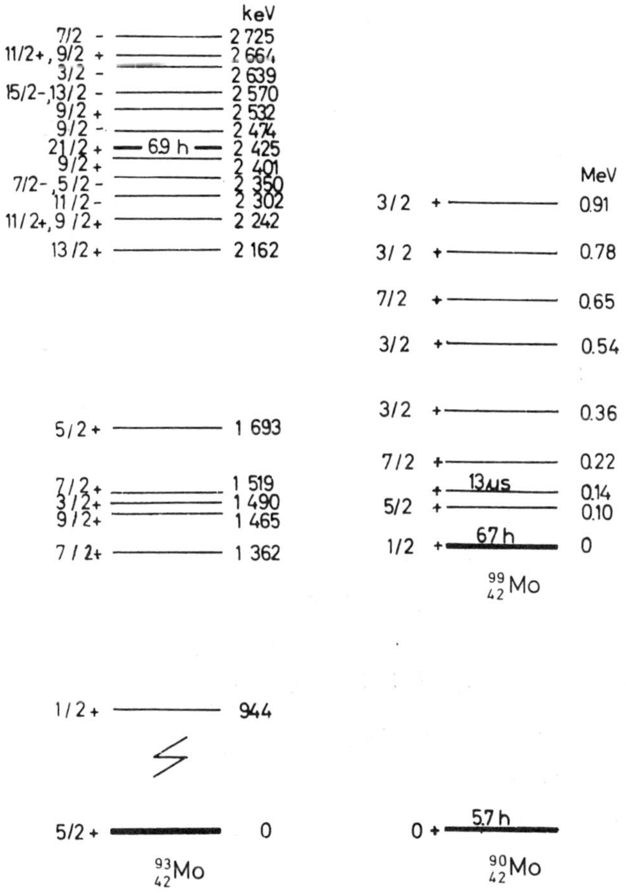

Fig. 3. Energy levels of ^{90}Mo, ^{93m}Mo *and* ^{99}Mo

obscure. In summary, we can say that the existence or non-existence of nuclear levels with sufficient life-time (10^{-12} s), decaying partly by internal conversion in the de-excitation cascade, may cause differences in the chemical behaviour of isotopes recoiling from a nuclear reaction. Predictions about the magnitude and the direction of the isotope effect are difficult in the light of the experimental data available at present.

ACKNOWLEDGEMENTS

We wish to thank CERN and specially Mr. A. Kjelberg for the opportunity to perform part of this work at Geneva. We are indebted to the members of the Nuclear Chemistry Group for their good collaboration. We are grateful to Professor F. Baumgärtner for his interest and support, and to the staff of the cyclotron and the FR2 reactor of the Kernforschungszentrum Karlsruhe for carrying out the irradiations.

REFERENCES

1. Schuler, R. H. and McCauley, C. E., *J. Amer. Chem. Soc.* **79**, 821, (1957).
2. Veljković, S. R. and Harbottle, G., *J. Inorg. Nucl. Chem.* **24**, 1517, (1962).
3. Aten, A. H. W., Lindner-Groen, L., and Lindner, L., Proc. Symp. Chem. Effects following Nucl. Transf., IAEA, Vienna (1964), Vol. II, (125).
4. Wolf, G. K., *Radiochim. Acta* **6**, 39, (1966).
5. Rössler, K., Dissertation, Köln (1967).
6. Libby, W. F., *J. Amer. Chem. Soc.* **62**, 1930, (1940).
7. Jach, J. and Harbottle, G., *Trans. Farad. Soc.* **54**, 520, (1958).
8. Yoshihara, K. and Harbottle, G., *Radiochim. Acta* **1**, 68, (1963).
9. Bellido, A. V., *Radiochim. Acta* **7**, 122, (1967).
10. Campbell, I. G. and Jones, C. H. W., *Radiochim. Acta* **9**, 71, (1968).
11. Apers, D. J., Capron, D. C., and Gilly, L. I., *J. Inorg. Nucl. Chem.* **5**, 23, (1957).
12. E. Lazzarini, *J. Inorg. Nucl. Chem.* **29**, 7, (1967).
13. Wexler, S. and Davies, T. H., *J. Chem. Phys.* **20**, 1688, (1952).
14. Campbell, I. G., *Adv. Inorg. Chem. Radiochem.* **5**, 135, (1963).
15. Baumgärtner, F. and Reichhold, P. Z., *Naturforsch.* **16a**, 945, (1961).
16. Wolf, G. K., Ravn, H., Heuser, H., and Henrich, E., to be published (1970).
17. Sugarman, N., Campos, M., Wielgoz, K., *Phys. Rev.* **101**, 388, (1956).
18. Harbottle, G. and Zahn, U., *Radiochim. Acta* **8**, 114, (1967).
19. Zahn, U., *Radiochim. Acta* **8**, 177, (1967).
20. Zahn, U., Collins, C. H., Collins, K. E., *Radiochim. Acta* **11**, 33, (1969).
21. Lederer, C. M., Hollender, I. M., and Perlman, I., 'Table of Isotopes', John Wiley & Sons, New York, London, Sydney (1968).
22. Lazzarini, E. and Fantola-Lazzarini, A. L., *J. Inorg. Nucl. Chem.* **29**, 895, (1967), *J. Inorg. Nucl. Chem.* **29**, 2161, (1967).
23. Vineyard, G. H., *Discuss. Farad. Soc.* **31**, 7, (1961).
24. Müller, H., *Angew, Chem.* **79**, 128, (1967).
25. Rowland, F. S. and Libby, W. F., *J. Chem. Phys.* **21**, 1495, (1953).
26. Saito, N., Fumitoshi, A., and Sano, H., *Nature* **205**, 388, (1965).

DISCUSSION

H. A. C. McKay (A.E.R.E.): The relative paucity of isotope effects such as the authors describe is surprising. Why are there not many more examples among other elements?

G. K. Wolf: Isotope effects have not been studied very extensively. Most studies in recoil chemistry are done with (n, γ)-produced recoil atoms. In only a limited number of cases does this reaction give more than one nuclide with a half-life sufficient for analysis. Moreover, not all compounds show the same sensitivity to such effects and in some cases the magnitude of the isotope effect may be smaller than the accuracy of the analytical method. We need to study more systems which produce different nuclides by fast particle irradiation, where we are sure to find more effects of this kind.

SESSION 2

Radiation Chemistry

Chairman: G. N. Walton

A comparison of the effects at low and high L.E.T. of ionic scavengers in cyclohexane under irradiation

W. G. Burns and C. R. V. Reed

Chemistry Division, A.E.R.E., Harwell, Berks., England

ABSTRACT

The yields of nitrogen from solutions of nitrous oxide in cyclohexane at varying $[N_2O]$ are compared for γ-radiation and for 2·1 MeV ^4He ions; a similar comparison is made for the yield of HD from solutions of perdeuteroammonia. Nitrous oxide and perdeuteroammonia were chosen as typical scavengers for negative and positive ions respectively; the decrease in $G(H_2)$ and $G(HD)$ measure the respective yields of ions scavenged. The effect of such scavengers in γ-irradiation has previously been interpreted in terms of the distribution in separation of initially-formed positive ions and electrons, and the probability of their reaction with scavenger during their subsequent mutual approach. This interpretation is extended to the effects of dilute scavenger solutions on heavy particle tracks by assuming that the core of numerous low-energy events is dense enough to lead to unscavengeable neutralisation, leaving more widely-spaced high-energy events which are treated similarly to the γ-ray case. The experimental variation with scavenger concentration of G(scavenged ions) at high L.E.T. is greater than that predicted on this basis, and a better correlation was obtained using a treatment of diffusion and reaction which treats the whole track collectively.

INTRODUCTION

The fate of the separated charged entities resulting from the action of ionising radiation in liquids has received much attention recently,[1-8] since this fate is of prime importance in the early events which lead from the absorption of radiation to the formation of stable products.

Experimentally, the effects of charge separation can be studied by measuring radiation-induced electrical conductivity,[9-11], and by determining the effects on product yields of additives intended to react specifically with ions.[2-6,8] Theoretical predictions[1,7,12] are normally concerned with the electron degradation energy spectrum,[12] or the statistics of the production of entities such as 'spurs', 'blobs', and 'short tracks' by secondary electrons.[7,13] The ultimate overall probability of recapture of the electron by positive ions, after thermalisation, governs the G value for free ions as measured in conductivity experiments,[7] and the probability of reaction of ions[8] with the scavenger governs the chemistry of experiments with additives. In the latter case, as well as effects on the ion undergoing reaction, additional considerations are the possibility of differences in the neutralisation reaction, and consequential reactions in the presence of additives, and changed opportunities for reaction with surrounding molecules by the ion whose oppositely-charged partner has been changed by reaction, thus affecting the return time.[14,15]

Experiments with ion scavengers have so far been mainly carried out using lightly-ionising radiation, and it was of interest to use heavy ions from the V.E.C.[16] to extend these to observations with more densely-ionising radiation. The work presented here compares the effects on the γ-radiation chemistry with those on the densely-ionising radiation chemistry of cyclohexane, produced by nitrous oxide and by perdeuteroammonia, added respectively as specific reactants for negative and positive ions. An approximate theoretical approach is given, based on an extension to densely-ionising radiation of treatments of ionic yields by Mozumder and Magee,[7] and of ion scavenging by Hummel,[8] both of which refer to more lightly ionising radiation.

EXPERIMENTAL

Materials. Phillips Research Grade (PRG) cyclohexane was dried with Molecular Sieve 5A, degassed, and distilled under vacuum into the irradiation cells, as previously described.[17] B.O.C. medical grade nitrous oxide was introduced into a vacuum line via a trap at 195°K (solid carbon dioxide and methylated spirits) to remove water, and condensed at 77°K (liquid nitrogen). It was degassed, then distilled from a trap at 195° to one at 77°K, rejecting initial and final fractions. The Bunsen coefficient for solubility in cyclohexane was taken as $\alpha = 3{\cdot}1$.[18] Perdeuteroammonia (Merck, Sharp and Dohme of Canada Ltd.) was degassed and used without further purification. The solubility of ND_3 in cyclohexane was measured using a modification of the Loprest[19] method, which gave $\alpha = 2{\cdot}5$. Solutions of cyclohexene (PRG) and bicyclohexyl (L. Light & Co.) in PRG cyclohexane were used as gas chromatography standards in the liquid product analyses, whilst cylinder hydrogen and atmospheric air were used in the analyses for hydrogen and nitrogen.

Irradiation. Gamma irradiations were carried out in a Ghormley and Hochanadel[20] ^{60}Co γ-source, dose rate $\sim 4{\cdot}0 \times 10^{17}$ eV g^{-1} min^{-1}, using glass cells provided with a suitable breakseal. The gas space was kept to a minimum.

The Harwell Variable Energy Cyclotron[16] was used for irradiations with ^4He ions. The beam collimator was similar to that already described,[17,21] except that there were four stops of 1·5 mm diam., including two 'scraper' stops. The stops were of 6 mm thick graphite, and the insulating material was high-density polythene. The distance from the first stop to the window was 0·85 m. Collimation was checked with the charge collector previously described,[17] which acted as a Faraday cup. The current was optimised on a graphite target, which was then retracted by remote control to start the irradiation.

The cell for V.E.C. irradiations is shown in Fig. 1. The thickness of the nickel window, determined by measuring weight and area before assembly, was 117-119 μg mm^{-2}; this gave ~2·1 MeV in the cell for 7·5 MeV incident ^4He ions,

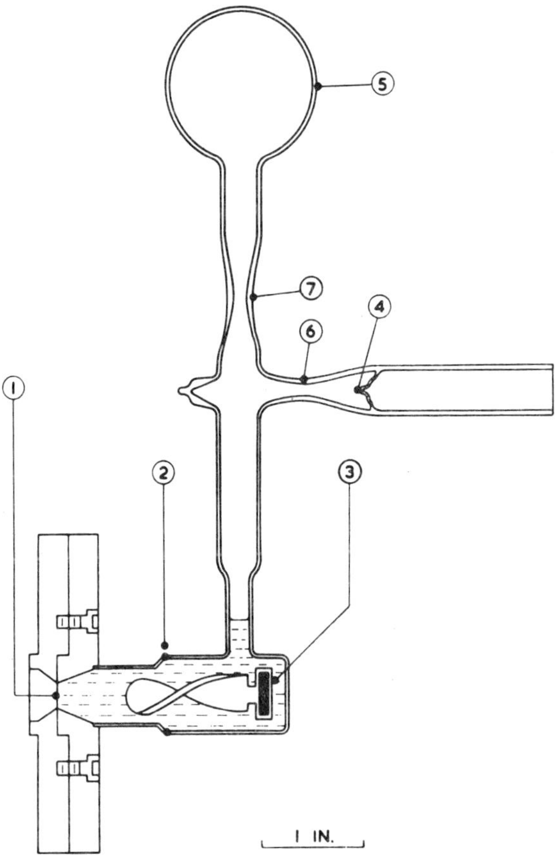

Fig. 1. Irradiation vessel: (1) thin nickel window, (2) glass-to-metal seal, (3) glass-enclosed magnetic stirrer, (4) break-seal, (5) bulb for liquid, (6) and (7) constrictions

and 3·0 MeV in the cell for 8·0 MeV incident ions. The stainless steel flange was connected by a copper-glass 'Housekeeper' seal (2) to the glass body of the cell, which contained a glass-enclosed magnetic stirrer (3). The cell was provided with a break-seal (4) for gas removal, and a bulb (5) with a constriction (7) for sealing off the condensed products.

Dosimetry. Dosimetry for the V.E.C. irradiations was by charge input measurement on a beam of known energy, the latter being calculated from the cyclotron control settings and the published energy-loss data for nickel.[22] The integrated current was measured with calibrated Beam Monitoring Equipment type 3008. The Fricke dosimeter was used for γ-ray dosimetry, taking $G(Fe^{3+}) = 15{\cdot}6$.

Analysis. Analytical procedures were similar to those previously described.[17]

When ND_3 was the additive, the gas fraction volatile at 77°K was analysed for hydrogen using a thermal conductivity detector and a 0·6 m Molecular Sieve 5A column, with argon as carrier gas. When nitrous oxide was the additive both hydrogen and nitrogen were estimated using a Perkin-Elmer model 154B Vapour

Fraktometer and a 1 m M.S. 5A column, with helium as carrier gas, accepting the reduced sensitivity for hydrogen.

Liquid products were measured on an F. & M. model 609 gas chromatograph with a flame ionisation detector, using a 1·2 m column of 5% polyethylene glycol adipate on 60–80 mesh Celite at 95°C for the C_{12} products, and a 7·5 m column of β, β'-oxydipropionitrile on 60–80 mesh Chromosorb P at room temperature for the C_6 fraction. The carrier gas was nitrogen.

RESULTS

(a) *Solutions of nitrous oxide*

(1) *γ-radiation.* Yields of hydrogen and nitrogen from these solutions have already been the subject of many investigations (Fig. 7), so our experiments were concerned with yields of cyclohexene and bicyclohexyl only, with the results shown in Fig. 2. These show a relative increase of 61% in G(cyclohexene) and of

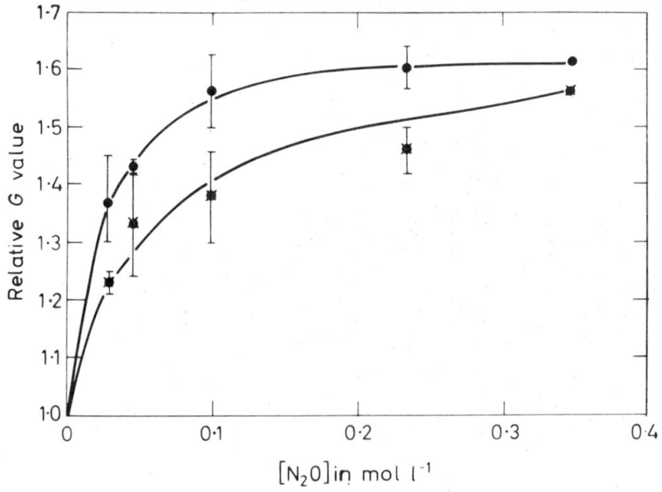

Fig. 2. Relative G *value as a function of* [N_2O]; ●, *cyclohexene*; ✱, *bicyclohexyl.* G *values are compared with those with no additive, (3·05 for cyclohexene, 1·75 for bicyclohexyl). Dose of* ^{60}Co *γ irradiation 4·9 × 10*^{18} *to 1·9 × 10*^{19} *eV g*$^{-1}$

54% in G(bicyclohexyl) at 0·2–0·3M N_2O, which are at variance with the respective increases of 11% and 42% found at the same concentration of nitrous oxide by Sagert and Blair,[3] the only other workers to measure both these products. However, our results confirm the increase of 52% in G(bicyclohexyl) reported by Blackburn and Charlesby;[23] a large value of G(cyclohexene) in the presence of 0·05M N_2O (said to be about twice as large as that in pure cyclohexane) was also reported by Sato, *et al.*[5]

(2) *2·08 MeV* ^{4}He *ions.* These show (Fig. 3) that compared with γ-ray results (Figs. 2 and 7), $G(H_2)$ and $G(N_2)$ are lower and do not approach a limiting

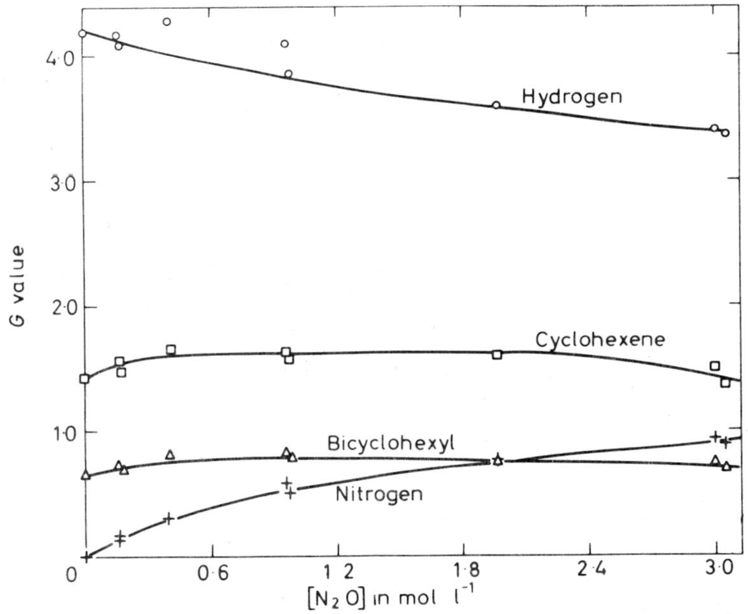

Fig. 3. G *values of hydrogen, nitrogen, cyclohexene, and bicyclohexyl against* [N_2O] *for 2·08 MeV* ^{4}He *ions.* ○, G(H_2); +, G(N_2), □, G(*cyclohexene*); △, G(*bicyclohexyl*). *Dose 3·4–3·8* × 10^{19} *eV* g^{-1}

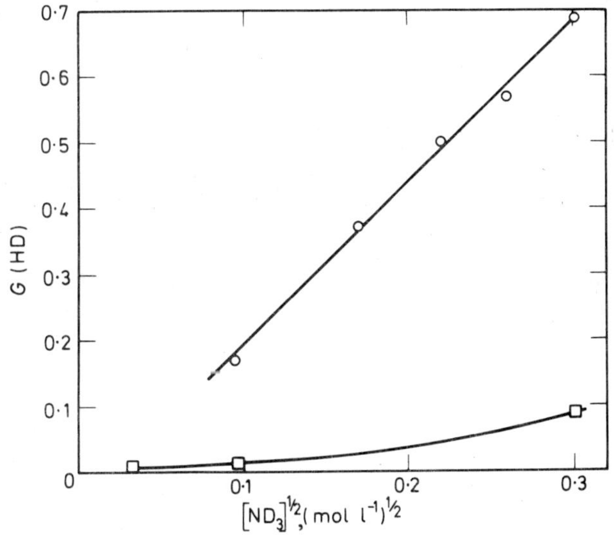

Fig. 4. G(HD) *against* [ND_3]$^{1/2}$. ○, *γ-radiation, dose 5·0* × 10^{19} *eV* g^{-1}; □*3.01 MeV* ^{4}He *ions, dose 5·0* × 10^{19} *eV* g^{-1}

value; the increases in G(cyclohexene) and G(bicyclohexyl) are not so large, and rise to a shallow maximum and then fall with increasing nitrous oxide concentration.

(b) *Solutions of ammonia-d_3*

(1) *γ-radiation.* Yields of HD from solutions of ND_3 are shown in Fig. 4 as a plot of G(HD) against $[ND_3]^{1/2}$, and agree reasonably with those of Williams.[2] G values for hydrogen, cyclohexene and bicyclohexyl were not substantially affected by the additive at the concentrations used, e.g. at $[ND_3]$ = 0·03 to 0·05 M they were 5·23 ± 0·02, 2·77 ± 0·02, 1·80 ± 0·03 respectively, compared with 5·6, 3·05, 1·75[24] found in the absence of additive.

(2) *3·01 MeV* ^{4}He *ions.* Yields of HD, also shown in Fig. 4, are much lower than those for γ-rays. Again, as for γ-rays, G values for hydrogen, cyclohexene, and bicyclohexyl were reduced, but not substantially, having values 4·2 ± 0·1, 1·59 ± 0·1, 0·77 ± 0·02 respectively at 0·93 M ND_3, compared with 4·23, 1·74, 0·76[24] in the absence of additive.

DISCUSSION

Ion scavenging with lightly and densely ionising radiation. The treatment of ion yields, as measured by conductance, given by Mozumder and Magee[7] for a 1 MeV electron, calculates the energy losses due to spurs (entities with energies up to 100 eV), blobs (energies 100–500 eV) and short tracks (energies 500 up to 5000 eV). These entities are convenient geometrically, and as developed for radical reactions[25] by these authors the blobs and short-tracks are self-contained, so that the spurs within them do not contribute to the yield of isolated spurs. However, in using their table of G values for the entities of differing energy produced by a 1 MeV electron (Table 2, ref. 7), allowance has to be made for the secondary, tertiary, etc., formation of extra-spur entities. Their model explains how low-energy electrons may escape from their geminate positive ions to distances further than their calculated range. It is assumed for each blob and short track that, having allowed for extra spur entities, only one ion pair is effective in contributing to the ion-pair yield, and that the others recombine quickly in the force field at the centre of the blob or short track. The contribution to the total ion-pair yield depends on the separation of the ion pair, and that to the scavenged ion yield on this separation and the scavenger concentration.[8]

The force field at the centre of a heavy-particle track will be much greater than in short tracks and blobs, and will be expected to cause recombination of all the spurs produced cylindrically around the track, and of some of the extra-spur entities. The conditions at the extremes of (a) low, and (b) high energy of a secondary electron will be typified by (a) an electron close to the track of positive ions, and surrounded by a negatively-charged cloud, inducing fast recombination, (b) an electron distant from the track and its sheath of near-neighbour electrons, which recombine with positive ions in the track, early in the progress of the distant electron towards the track under the influence of the residual force field. We make the simplifying assumption that the electrons produced by a heavy particle are divided into two independent groups, (a) those with energy below ϵ_0, which recombine quickly, and (b) those with energy above ϵ_0, which give rise to isolated ion pairs with separations appropriate to their energies. The statistics of electron production are such that if ϵ_0 is greater than 100 eV, most electrons will belong to group (a), and the assumption that group (b) electrons behave independently of each other will be more accurate at high ϵ_0 since their average linear density will become lower. In this connection it is worth noting that in accounting for the ion yield and the variation with energy of observed ionisation currents induced by Po α-particles in liquid n-hexane, Richards[26] calculated that only electrons of energy above ~1 keV contributed to ionisation.

The methods used to calculate the yields of scavenged ions are as follows:

(a) *Lightly ionising radiation.* G values for the entities depositing energy, G_i, were taken from Tables 2 and 3, ref. 7; separation distances were taken directly or by graphical interpolation from Table 4, ref. 7, which lists thermalisation separations, r_i, as a function of energy. The probability of reaction with scavenger was calculated by the method of Hummel,[8] from a derivation by Monchick[27] which gives

$$P(q, r_i) = \frac{2\{\exp(-qr_i)\}\{1-\exp(-r_c/r_i)\}}{1+\exp(-2qr_i)+2qr_i^2r_c^{-1}\{1-\exp(-r_c/r_i)\}} \tag{1}$$

where P is the probability of recombination of the ions; $q = (k_s c_s/D)^{1/2}$, where k_s is the rate-constant for reaction of the ion with the scavenger, c_s is the scavenger concentration, D the sum of the diffusion constants for the positive and negative ions; $r_c = e^2/KkT$, where e is the electronic

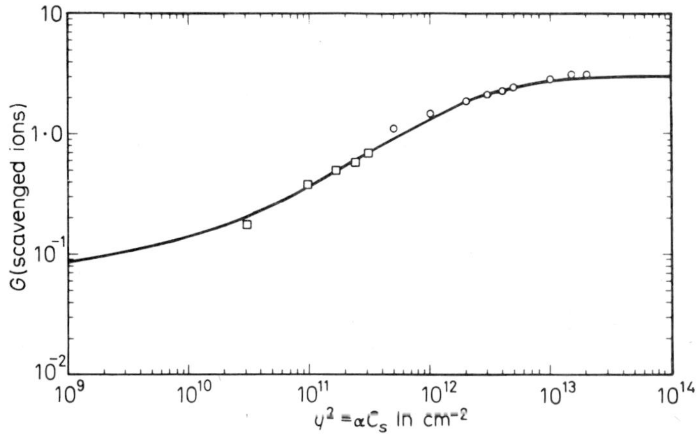

Fig. 5. G(scavenged ions) *against* $q^2 = \alpha c_s$ *for lightly-ionising radiation. Curve from Eqns (1) and (2);* ○, $G(H_2)$ *against* $\alpha\,[N_2O]$ *from Fig. 7,* $\alpha = 0{\cdot}8 \times 10^{-7}$ *cm;* □ $G(HD)$ *against* $\alpha\,[ND_3]$ *from Fig. 4,* $\alpha = 5{\cdot}75 \times 10^{-9}$ *cm*

charge, K the dielectric constant, k is Boltzmann's constant, and T the absolute temperature. For a given q, the G value for scavenged ions is given by

$$G(\text{s.i.}) = \Sigma_i G_i\,(1 - P(q, r_i)) \tag{2}$$

Fig. 5 shows a graph of G(s.i.) against q^2 calculated in this way from (1) and (2).

(b) *Densely ionising radiation.* The maximum energy of secondary electrons, E_m, is taken from the classical formula to be $E_m = 2{\cdot}196E/m$ where E_m is in keV, E is the energy of the primary heavy particle in MeV, and m its mass in atomic mass units. The number of first-generation secondary electrons of energy between ϵ_1 and ϵ_2 per unit energy loss by the primary particle is given by

$$\sigma(E, \epsilon_1, \epsilon_2) = \frac{(\epsilon_2 - \epsilon_1)}{\epsilon_2\,\epsilon_1} 2 \log_e (E_m/I)$$

where I is the mean excitation potential, taken[9] as 0·047 keV. The energy distribution of the first generation of secondary electrons for the whole of the track of the 2·08 or 3·01 MeV ^{4}He ion in question was obtained by

Table 1. *G* VALUES, ENERGIES AND SEPARATIONS FOR SECONDARY ELECTRONS FROM A 3·01 MeV ^{4}He ION

Energy Range keV	*G value, First generation*	*Thermalisation separation* nm	*Average number extra-spur deltas (all subsequent generations)*	*Thermalisation separations for deltas* nm
1·55–1·60	0·000 01	900·00	3·66	13·40
1·50–1·55	0·000 02	900·00	3·55	13·40
1·45–1·50	0·000 03	140·00	3·44	13·30
1·40–1·45	0·000 05	130·00	3·32	13·30
1·35–1·40	0·000 06	120·00	3·21	13·20
1·30–1·35	0·000 08	110·00	3·11	13·15
1·25–1·30	0·000 10	100·00	3·01	13·10
1·20–1·25	0·000 13	90·00	2·92	13·05
1·15–1·20	0·000 15	80·50	2·83	13·00
1·10–1·15	0·000 19	71·20	2·74	12·95
1·05–1·10	0·000 23	63·40	2·65	12·90
1·00–1·05	0·000 27	57·00	2·57	12·83
0·95–1·00	0·000 33	51·30	2·48	12·75
0·90–0·95	0·000 39	46·10	2·39	12·67
0·85–0·90	0·000 47	41·50	2·30	12·60
0·80–0·85	0·000 57	37·50	2·21	12·53
0·75–0·80	0·000 69	34·00	2·12	12·43
0·70–0·75	0·000 84	31·00	2·03	12·35
0·65–0·70	0·001 03	28·50	1·95	12·23
0·60–0·65	0·001 28	26·20	1·86	12·13
0·55–0·60	0·001 60	24·10	1·77	12·03
0·50–0·55	0·002 03	22·20	1·68	11·90
0·45–0·50	0·002 62	20·22	1·59	11·70
0·40–0·45	0·003 46	18·50	1·50	11·50
0·35–0·40	0·004 70	16·97	1·39	11·40
0·30–0·35	0·006 62	15·52	1·29	11·20
0·25–0·30	0·009 82	14·37	1·18	11·10
0·20–0·25	0·015 70	13·20	1·06	11·00
0·15–0·20	0·027 97	12·30	0·0	–
0·10–0·15	0·060 62	11·16	0·0	–

$q^2 = 10^9\ cm^{-2}$, $\epsilon_0 = 100$ eV, G(scavenged ions) = 0·0237 dividing the primary energy into thirty equal steps, and summing appropriately to give the number of secondaries in the energy intervals used by Mozumder and Magee.[7] Table 1 shows this distribution in column 2, where the G values for the electrons have been obtained by dividing their numbers by the primary energy in units of 100 eV. For the first calculations we assume that spurs below ϵ_0 = 100 eV are subject to fast initial recombination, and G values of scavenged ions are obtained for the first generation of secondaries by applying Eqns (1) and (2) to columns 2 and 3 of Table 1. The effect of subsequent generations of secondaries was obtained by using Table 3 of ref. 7, which gives their average number and average energy; from the latter we obtained their thermalisation separations, and calculated the contribution of the extra-spur entities to G (scavenged ions) by applying Eqns (1) and (2) to columns 4 and 5 of Table 1. Table 1 gives a sample calculation for one value of q and of ϵ_0, which was repeated at various values of q and ϵ_0 to give the results shown in Fig. 6.

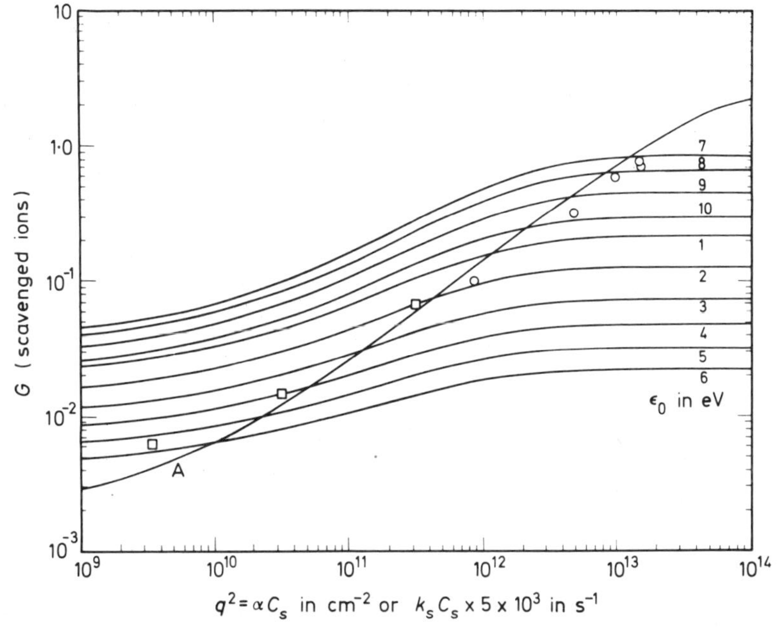

Fig. 6. G(scavenged ions) against $q^2 = \alpha c_s$ for densely-ionising radiation. Curves 1, 2, 3, 4, 5, 6 from Eqns (1) and (2) (ϵ_0 = 100, 200, 300, 400, 500, 600 eV for 3·01 MeV ^4He ions), curves 7, 8, 9, 10 (ϵ_0 = 25, 30, 40, 60 eV for 2·08 MeV ^4He ions). ○, $G(H_2)$ *against* $\alpha[N_2O]$ *from Fig. 3;* □, G(HD) *against* $\alpha[ND_3]$ *from Fig. 4,* α *values as for Fig. 5. Curve* A: *3*S *against* $5 \times 10^3 \times k_s c_s$ *for neutral diffusion and reaction case*

The intended method of application of Figs. 5 and 6 is to obtain an effective value of $\alpha = k_s/D$ by fitting to the curve of Fig. 5 the curve of G(scavenged ions) against $\alpha c_s = q^2$, where c_s and G(scavenged ions) are experimental values for low L.E.T. radiation. Then a plot may be made of G(scavenged ions) for the heavy-particle radiation against $q^2 = \alpha c_s$ and its position on the family of curves of Fig. 6 should show at what energies the secondary electrons contribute to G(scavenged ions). The chemistry of the systems used must be well understood, however, in order to obtain G(scavenged ions), and this is discussed below.

Results with nitrous oxide present. We will use the following reaction scheme as a framework for discussion:

$$C_6H_{12} \rightsquigarrow C_6H_{12}^{+} + e^{-} \quad (1a)$$
$$C_6H_{12} \rightsquigarrow C_6H_{12}^{*} \quad (1b)$$
$$C_6H_{12}^{+} + e^{-} \longrightarrow C_6H_{12}^{*} \quad (2)$$
$$C_6H_{12}^{*} \longrightarrow H\cdot + C_6H_{11}\cdot \quad (3)$$
$$C_6H_{12}^{*} \longrightarrow H_2 + C_6H_{10} \quad (4)$$
$$H\cdot + C_6H_{12} \longrightarrow H_2 + C_6H_{11}\cdot \quad (5)$$
$$2\,C_6H_{11}\cdot \longrightarrow C_6H_{10} + C_6H_{12} \quad (6a)$$
$$2\,C_6H_{11}\cdot \longrightarrow C_6H_{11}{\cdot}C_6H_{11} \quad (6b)$$

$$e^{-} + N_2O \longrightarrow [N_2O^{-}] \longrightarrow N_2 + O^{-} \quad (7)$$

$$O^{-} + C_6H_{12}^{+} \longrightarrow OH\cdot + C_6H_{11}\cdot \quad (8)$$
$$OH\cdot + C_6H_{12} \longrightarrow H_2O + C_6H_{11}\cdot \quad (9)$$

---

or

$$O^{-} + C_6H_{12} \longrightarrow OH^{-} + C_6H_{11}\cdot \quad (10)$$
$$OH^{-} + C_6H_{12}^{+} \longrightarrow H_2O + C_6H_{11}\cdot \quad (11)$$

(1) *Lightly ionising radiation.* The results of other workers on $G(H_2)$ and $G(N_2)$ are shown in Fig. 7. There is better agreement for $G(H_2)$ than for $G(N_2)$, for which the results of Sagert and Blair are the highest,[3] those of Holroyd[28] and of Sherman[29] are the lowest, and those of Sato *et al*,[5] Warman,[30] and the two sets of Scholes *et al*[6,31] form an intermediate group. We agree with the suggestion of Sagert and Blair[3] that a suitable measure of the yield of nitrous oxide molecules which have undergone dissociative reaction is the decrease in $G(H_2)$, [i.e. $\Delta G(H_2)$] plus the sum of the increases $\Delta G(C_6H_{10})$ and $\Delta G(C_{12}H_{22})$ in cyclohexene and bicyclohexyl respectively. This is because suppression of reaction (2) leads to the non-appearance of one H_2 molecule and two cyclohexyl radicals (or a cyclohexene molecule), but instead (8) and (9) or (10) and (11) produce two cyclohexyl radicals. The fact that G(cyclohexene) and G(bicyclohexyl) are higher in the presence of nitrous oxide means that, as Sagert and Blair suggest, dissociative reaction of nitrous oxide occurs in addition to that which prevents (2) from occurring; however, if this is all due to reaction with electrons, (7), then $G(e^{-}) = \Delta G(H_2) + \Delta G(C_6H_{10}) + \Delta G(C_{12}H_{22}) = 2{\cdot}8 + 1{\cdot}8 + 0{\cdot}9 = 5{\cdot}5$, which is too high, based on $W = 24$ eV. A limiting value as high as $\sim 5{\cdot}5$ for dissociated N_2O molecules means that there is no need to postulate additional secondary reactions giving rise to nitrogen, since the limiting $G(N_2)$ is $\sim 5{\cdot}5$, but suggests that some dissociation of nitrous oxide may occur by transfer from non-ionic excitations which do not cause dissociation of cyclohexane

$$C_6H_{12}^{*} + N_2O \longrightarrow C_6H_{12} + N_2 + O \quad (12)$$

The oxygen atom formed in (12) would need to be as efficient as the O^{-} ions

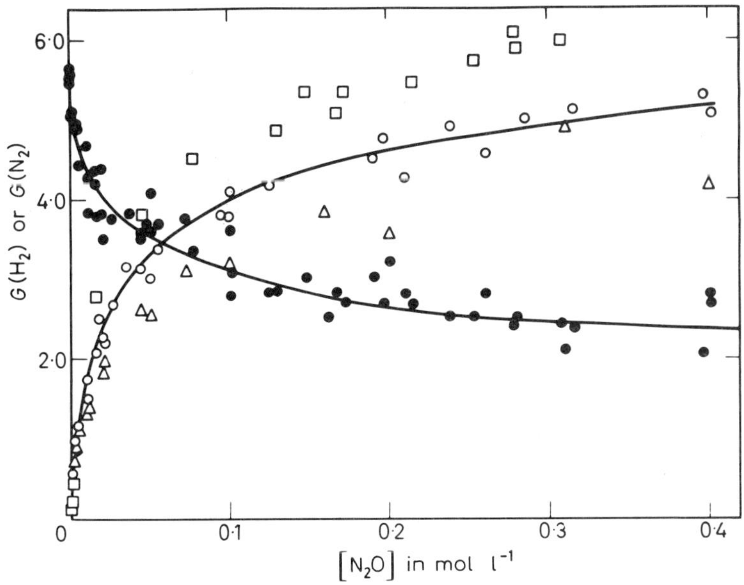

Fig. 7. $G(H_2)$ *and* $G(N_2)$ *against* $[N_2O]$. □, $G(N_2)$, *ref. 3;* ○, $G(N_2)$, *refs. 5, 6, 30, 31;* ▽, $G(N_2)$ *refs. 28, 29.* •, $G(H_2)$, *refs. 3, 5, 6, 28, 29, 30, 31.*

formed in (7) in regenerating cyclohexyl radicals, for instance by successive abstraction of hydrogen atoms, to account for the increase in condensed product yields and the approximate equality of $G(N_2)$ and $\Delta G(H_2) + \Delta G(C_6H_{10}) + \Delta G(C_{12}H_{22})$. The possibility of the transfer of excitation from photolysed liquid cyclohexane to nitrous oxide giving rise to nitrogen has recently been demonstrated by Holroyd,[28] using light of wavelength 147 nm, but this wavelength was sufficient to cause dissociation of cyclohexane, mainly into hydrogen and cyclohexene, and the oxygen atoms formed in (12) did not regenerate cyclohexene.

The increases we have found in cyclohexene and bicyclohexyl are not consistent with an increase only in cyclohexyl radicals, if we accept a disproportionation to combination ratio of ~1·2. A value of 0·9 for $\Delta G(C_{12}H_{22})$ implies an increase in G(cyclohexene) of 1·1 due to cyclohexyl radicals. The additional increase of G (0·7) can be accounted for if the reactions involving the neutralisation of O^- can alternatively produce cyclohexene rather than two cyclohexyl radicals, e.g.

$$O^- + C_6H_{12}^+ \longrightarrow H_2O + C_6H_{10} \qquad (8a)$$

in place of (8) and (9), and if the oxygen atom from (12) can form cyclohexene in a single step. This type of reaction would need to occur for a small fraction (0·7/5·5) of the total reactions of oxygen atoms and O^- ions.

We take $\Delta G(H_2)$ to represent G(scavenged ions) for this additive, bearing in mind that $\Delta G(H_2)$ will give too low a value if some recombination involving $C_6H_{12}^+$ normally occurs without dissociation and the subsequent formation of hydrogen; also that if transfer of non-ionic excitation occurs which normally would give rise to hydrogen, then $\Delta G(H_2)$ will give too high an estimate of G(scavenged ions). Values of $\Delta G(H_2)$ taken from Fig. 7 at $[N_2O]$ = 10, 20, 40, 60, 80, 100, 200, 300, and 400 mM were plotted against αc_s in Fig. 5, and α adjusted to give the reasonable fit shown for $\alpha = 0{\cdot}83 \times 10^{-7}$ cm.

Another possible interpretation is that $G(N_2)$ represents $G(e^-)$ and that the large G values at high $[N_2O]$ are due, at least partly, to scavenging of spurs within blobs and short tracks, a process which the curve of Fig. 5 neglects. Hummel[8] takes this view and obtains $\alpha = 3 \times 10^{-7}$ cm using the results of refs. 5, 29, and 31.

(2) *2·08 MeV* ^{4}He *ions.* The smaller values of $\Delta G(H_2)$ and of $G(N_2)$, compared with γ-rays, are still too large for them to be explicable in terms of reaction (7), even when ϵ_0 is reduced to 60 eV, as can be seen from Fig. 7. If the fall in $G(H_2)$ and the formation of nitrogen were due to reaction (12), then a plot of $[\Delta G(H_2)]^{-1}$ against $[N_2O]^{-1}$ should give a straight line with intercept $1/G(C_6H_{12}{}^*)$ and slope/intercept = $(k_3 + k_4)/k_{12}$. This plot is linear, and gives $G(C_6H_{12}{}^*) = 1{\cdot}6$, and $(k_3 + k_4)/k_{12} = 3$ l mole^{-1}, of the same order of magnitude as the equivalent ratio found by Holroyd[28] for photochemical excitation transfer and excitation transfer during γ-radiation. As with γ-radiation we would need to assume that the oxygen atom from (12) can regenerate the condensed products. In view of the results with added ND_3, it seems more likely that the approximation that only electrons above an energy ϵ_0 are available for reaction does not hold at the high effective concentrations used here. On Fig. 7 we have made plots in which ϵ_0 has been reduced further into the spur region, assuming that the distribution in energy of G values for spurs is the same as for a 1 MeV electron; it can be seen that electrons down to quite low energies are scavenged by nitrous oxide at the concentrations used here. The smaller increases in G(cyclohexene) and G(bicyclohexyl) compared with γ-rays mean that the transfer of electrons or electronic excitations to nitrous oxide in excess of those which normally cause dissociation of cyclohexane molecules is inhibited by the force field of the ^{4}He ion track. The lowering of $\Delta G(C_6H_{10})$ and $\Delta G(C_{12}H_{22})$ at increased $[N_2O]$ is possibly due to secondary reactions of O^- with nitrous oxide, since, compared with the γ-ray case, the smaller yield of O^- ions with ^{4}He ions is formed at larger distances from the spur or track centre, with greater opportunity for secondary reaction than in the γ-ray case. Direct investigation of the gas-phase secondary reactions of nitrous oxide and electrons shows them to be more complex[32,33] than formerly supposed.[30]

Results with perdeuteroammonia. The mechanism of the γ-ray case has been discussed by Williams,[2] who proposes that the relevant reactions are proton transfer, rapid isotopic exchange, and hydrogen abstraction following dissociative neutralisation:

$$XH^+ + ND_3 \longrightarrow ND_3H^+ + X \quad (14)$$

$$ND_3H^+ + ND_3 \rightleftharpoons ND_4^+ + ND_2H \quad (15)$$

$$ND_4^+ + e^- \longrightarrow ND_3 + D\cdot \quad (16)$$

$$D\cdot + C_6H_{12} \longrightarrow HD + C_6H_{11}\cdot \quad (17)$$

Our findings are consistent with this mechanism, which gives G(scavenged ions) = G(HD), and results in the formation of two cyclohexyl radicals, via (14) and

(17), (taking $X \equiv C_6H_{11}$) and one hydrogen molecule (including HD), via (17) for every XH^+ ion undergoing reaction (14). This is the same number of radicals and molecules as would otherwise arise from (2), (3) and (5), the alternative reactions in the absence of additive, thus accounting for the virtually undisturbed values of $G(C_6H_{10})$ and $G(C_{12}H_{22})$. Equivalent yields of hydrogen and $C_6H_{11}\cdot$ radicals are also obtained when $X \equiv C_6H_{12}$, since in this case one cyclohexyl radical is formed in the previous presumed proton-transfer step

$$C_6H_{12}^+ + C_6H_{12} \longrightarrow C_6H_{13}^+ + C_6H_{11}, \qquad (18)$$

giving $C_6H_{13}^+$, which takes part in (14) or is neutralised to give $H\cdot + C_6H_{12}\cdot$ These considerations regarding lack of effect on $G(C_6H_{10})$ and $G(C_{12}H_{22})$ apply at both low and high L.E.T.

Values of G(HD) for γ-radiation which, like those of Williams, give a linear plot (Fig. 4) against $[ND_3]^{1/2}$, when plotted against αc_s on Fig. 5 give a reasonable fit at $\alpha = 5{\cdot}75 \times 10^{-9}$ cm, in fair agreement with the value of $4{\cdot}9 \times 10^{-9}$ cm obtained by Hummel[8] using a similar plot for Williams's[2] results. G(HD) values for 3·01 MeV ^{4}He ions do not give a linear plot for the concentration range used when plotted against $[ND_3]^{1/2}$, and for the lower concentrations are relatively lower, compared with γ-ray values at the same concentrations, than the relative values for G(s.i.) obtained with nitrous oxide in a higher concentration region. The ^{4}He ion results, plotted on Fig. 6 as were those for nitrous oxide, using the α value obtained at low L.E.T., show that the initial assumption of two groups of secondary electrons does not hold, except possibly at very low concentrations. It seems that although for a low-energy electron fast recombination will occur due to the force field of positive ions neighbouring its own geminate partner, and that for a high-energy electron Eqn (1) may be appropriate, there is a continuous gradation between the two situations, in which the effects of neighbouring positive ions become less important at greater distances from the track. The gradation causes the relative change in G(scavenged ions) to be greater for the densely- than for the lightly-ionising case. This contrasts with the predictions of a mechanism with a fixed ϵ_0 (Fig. 6), where the relative variation in G(scavenged ions), between G(free ions) at low scavenger concentration and G(available ions) at high scavenger concentration, is small, and at higher values of ϵ_0 is due principally to effects on the δ-electrons. The fact that there is an approximate continuity of experimental values of G(s.i.) against $q^2 = \alpha c_s$ in Fig. 6 over the two groups of measurements suggests that the higher values of G(s.i.) for nitrous oxide are due to ion scavenging of electrons closer to the track, rather than to a new mechanism such as excitation transfer.

General considerations. A rigorous treatment of the high L.E.T. case would need to take account of the effect on secondary electrons of positive ions neighbouring their geminate partners, both during and after thermalisation, also of electron interactions in the cylindrical geometrical situation of a heavy particle track. In a general treatment of the diffusion and combination reactions of ions in columnar geometry in the gas phase, Wilhelm[34] has shown that the overall collective equation resulting from diffusion under electrostatic forces and combination is

$$(\partial/\partial t)\, n^{\pm} = D\,\nabla^2 n^{\pm} - k n^{\pm 2}$$

where $n^{\pm}$ is the concentration of positive or negative ions, D is the ambipolar ionic diffusion coefficient, k is the recombination rate constant, and ∇^2 is the Laplacian operator. Since this equation is the same as for neutral particles undergoing diffusion and recombination the effect of ionic scavenging would be expected to be given approximately by

$$\frac{\partial c}{\partial t} = D\nabla^2 c - k_S c_S c - kc^2$$

where c is the ionic concentration, k_S and c_S are the rate constant for ion scavenging and the scavenger concentration. This equation has been solved by the approximate method of 'prescribed diffusion' for the initial condition of a Gaussian radial distribution of particles.[35] Curve A of Fig. 6 shows $3S$ from such a solution, where S is the fraction of particles undergoing first-order reaction, plotted against $5 \times 10^3\, k_S c_S$ for the following conditions; r_0, the initial Gaussian half-height radius = 100Å, corresponding to the thermalisation length of a typical electron[7], $k = 2 \times 10^{-9}$ cm^3 molecule^{-1} s^{-1}, $D = 2 \times 10^{-5}$ cm^2 s^{-1}, values for the ionic recombination and diffusion constants found by conductivity measurements.[9] The L.E.T. was taken as 20 eV/Å since the mean value

$$\left[\frac{1}{E_0}\int_0^{E_0} -\left\{\frac{dE}{dx}\right\} dE\right]$$

is equal to 20 eV/Å for both 3·0 and 2·1 MeV ^{4}He ions in cyclohexane, and the G value for ion pairs was taken as 3. The approximate agreement between this curve and the experimental points over a wide concentration range suggests this approach may be at least partially valid. The values of k_S obtainable from the curve are $k_S = \alpha/(5 \times 10^3)$, i.e. $1{\cdot}14 \times 10^{-12}$ cm^3 s^{-1} ($0{\cdot}68 \times 10^9$ 1 mole^{-1} s^{-1}) for (15), and $1{\cdot}6 \times 10^{-11}$ cm^3 s^{-1} ($9{\cdot}6 \times 10^9$ 1 mole^{-1} s^{-1}) for (7), both of which are feasible.

Analogously with the γ-ray results, the value of $G(HD)$ from low concentrations of perdeuteroammonia, ~0·006 2 for 3 MeV ^{4}He ions, is approximately equal to the G value for escaped ions, $5{\cdot}5 \times 10^{-3}$, which can be calculated[36] by the method appropriate for high fields from the results of Richards[26] on n-hexane. These results show that electrons of energy above 600–800 eV escape from the track to the bulk of the liquid. Further work on ion scavenging at low concentrations, in which the primary energy is varied, would be of interest in investigating the conditions for electron escape.

ACKNOWLEDGEMENTS

We gratefully acknowledge the assistance of Dr. J. T. Richards in the early stages of the work on nitrous oxide, and experimental help from Miss S. J. Putman in the work with perdeuteroammonia. We wish to thank Dr. C. B. Amphlett for encouragement and advice.

REFERENCES

1. Freeman, G. R., *J. Chem. Phys.* **46**, 2822, (1967).
2. Williams, T. F., *J. Amer. Chem. Soc.* **86**, 3954, (1964).
3. Sagert, N. H. and Blair, A. S., *Can. J. Chem.* **45**, 1351, (1967).

4. Rajbenbach, L. A., *J. Amer. Chem. Soc.* **88**, 4275, (1966); *J. Chem. Phys.* **47**, 242, (1967).
5. Sato, S., Yugeta, R., Skinsaka, K., and Terao, T., *Bull. Chem. Soc. Japan* **39**, 156, (1966).
6. Munday, C. S., Richards, J. T., and Scholes, G., ed., 'The Chemistry of Ionisation and Excitation', p. 151, Taylor and Francis, London (1967).
7. Mozumder, A. and Magee, J. L., *J. Chem. Phys.* **47**, 939, (1967).
8. Hummel, A., *J. Chem. Phys.* **48**, 3268, (1968).
9. Hummel, A., and Allen, A. O., *J. Chem. Phys.* **44**, 3426, (1966); Hummel, A., Allen, A. O., and Watson, F. H., *J. Chem. Phys.* **44**, 3431, (1966).
10. Freeman, G. R., *J. Chem. Phys.* **38**, 1022, (1963); 39, 988, 1580, (1963).
11. Schmidt, W. F. and Allen, A. O., *J. Phys. Chem.* **72**, 3730, (1968).
12. Santar, J. and Bednář, J., *Int. J. Radiat. Phys. Chem.* **1**, 133, (1969); *Coll. Czech Chem. Comm.* **34**, 1, 311, (1969).
13. Mozumder, A. and Magee, J. L., *J. Chem. Phys.* **45**, 3332, (1966).
14. Mozumder, A., *J. Chem. Phys.* **48**, 1659, (1968).
15. Hummel, A., *J. Chem. Phys.* **49**, 4840, (1968).
16. Lawson, J. D., The Harwell Variable Energy Cyclotron, NIRL/R/85 (1965).
17. Burns, W. G. and Winter, J. W., *Disc. Farad. Soc.* **36**, 124, (1964).
18. Richards, J. T., private communication.
19. Loprest, F. J., *J. Phys. Chem.* **61**, 1128, (1957).
20. Ghormley, J. A. and Hochanadel, C. J., *Rev. Sci. Instrum.* **22**, 473, (1951).
21. Burns, W. G., *Trans. Farad. Soc.* **58**, 961, (1962).
22. Northcliffe, L. C., *Ann. Rev. Nucl. Sci.* **13**, 67, (1963).
23. Blackburn, R. and Charlesby, A., *Nature* **210**, 1036, (1966).
24. Burns, W. G. and Reed, C. R. V., to be published.
25. Mozumder, A. and Magee, J. L., *Radiation Research* **28**, 203, 215, (1966).
26. Richards, E. W. T., *Proc. Phys. Soc.* **A66**, 631, (1953).
27. Monchick, L., *J. Chem. Phys.* **24**, 381, (1956).
28. Holroyd, R. A., *Radiation Chemistry II, Advances in Chemistry Series* **82**, 488, Amer. Chem. Soc., Washington (1968).
29. Sherman, W. V., *J. Chem. Soc.* (A), 599, (1966).
30. Warman, J. M., Asmus, K. D., and Schuler, R. H., *Radiation Chemistry II, Advances in Chemistry Series* **82**, Amer. Chem. Soc., Washington (1968).
31. Scholes, G. and Simic, M., *Nature* **202**, 895, (1964).
32. Moruzzi, J. L. and Dakin, J. T., *J. Chem. Phys.* **49**, 5000, (1968).
33. Parkes, D. A., private communication.
34. Wilhelm, H. E., *J. Chem. Phys.* **47**, 4356, (1967).
35. Ganguly, A. K. and Magee, J. L., *J. Chem. Phys.* **25**, 129, (1956).
36. Hummel, A., Thesis, Free University of Amsterdam (1967).

DISCUSSION

J. Wright (A.E.R.E.): If the same value of r_i is used for conductivity and chemical scavenger calculations, it would be interesting to know whether the random walk treatment is affected by the presence of a potential gradient. Have scavenger studies been made in an electric field?

W. G. Burns: The influence of an electric field in ion scavenging experiments might be an interesting field to study. I know of no experiments with additives in an electric field other than those of Salmon, Dainton, Morrow, and Frankevich, reported in Nature and in High Energy Chemistry. However, in our experiments in which the same distribution of r_i values was used to predict results of conductivity and of ion-scavenging experiments, the electric field is considered to be zero in both cases, since the conductivity experiments were at low field and the results were extrapolated to zero field.

Studies of radiation chemistry and biology with the IPCR 160 cm cyclotron

T. Karasawa, H. Seki, M. Imamura, A. Matsuyama, and M. Matsui

Institute of Physical and Chemical Research (RIKEN), Yamato-Machi, Saitama Pref., Japan

ABSTRACT

A collimated beam (~5 mm diameter) emerges through a thin aluminium foil. Liquid samples are contained in a glass vessel. Part of the beam to be absorbed by the sample is elastically scattered by a thin metallic foil and monitored by a solid state detector.

The 'charge displacement' effect for both protons and He ions increases with the ionisation of the particles and with their range.

Radiation effects in the liquid phase are strongly influenced by stirring and by beam intensity. Stirring was accomplished by a magnetic stirrer or by bubbling air through a capillary.

$G(Fe^{3+})$ values were measured in the Fricke dosimeter for various values of L.E.T. For high intensity and high L.E.T., $G(Fe^{3+})$ is given by

$$G(Fe^{3+}) = \tfrac{3}{2}\, G(H_2O_2) + \tfrac{1}{2}\, G(OH)$$

The incidental effect of beam intensity on radiation killing of bacterial cells due to non-uniform irradiation of the suspension was examined at a constant rate of air bubbling. The following equation is proposed to explain this effect for 26 MeV α-particles;

$$S = \exp \left\{ (-It/q_o)\, [1 - \exp(-kq_o/I)] \right\}$$

where S is the surviving fraction, q_o is D_{37} in nanocoulombs, t is the irradiation time in s, I is the beam current in nA and k is a constant which depends upon the rate of aeration and the size of the effective volume. The values of D_{37} are reported for the various bacterial cells examined.

INTRODUCTION

The IPCR 160 cm variable energy cyclotron accelerates protons, deuterons, $^4He^{++}$ and $^3He^{++}$ ions over a wide range of L.E.T. (linear energy transfer) from

0·25 to 5 eV/A and multicharged C, N, and O ions with L.E.T. varying continuously from 15 to 120 eV/A.

The protons and He ions have been used in studies of radiation chemistry and biology. Special devices are required for such studies which differ from those used in nuclear physics experiments. These have special features which permit accurate determination of the beam current penetrating the sample.

When liquid samples are irradiated, only a small fraction is subjected to bombardment because of the very short range of the particles. The effects of beam current and stirring speed or air bubbling rate were examined in the Fricke dosimeter and for bacterial cells suspended in a buffer medium. The beam current was varied over a range of about three decades.

In the Fricke solution, the maximum intensity of the ion beam beyond which there is no further effect on the oxidation was determined as a function of L.E.T., and a reaction mechanism is proposed for extremely high intensities in the absence of stirring.

In the killing of bacterial cells, the relation between survival fraction and beam intensity for the same dose was examined and expressed by a simple equation, based on the theory of probability for agitation.

IRRADIATION

The experimental arrangement for irradiation with protons and He ions is illustrated schematically in Fig. 1. The end of the evacuated beam pipe was ~10 m from the cyclotron. The beam was collimated with slits to give a spot (~5 mm diam.) on the window of the irradiation cell.

After passing through the collimator, part of the beam was scattered by a thin metallic foil while the main part of the beam emerged into air through

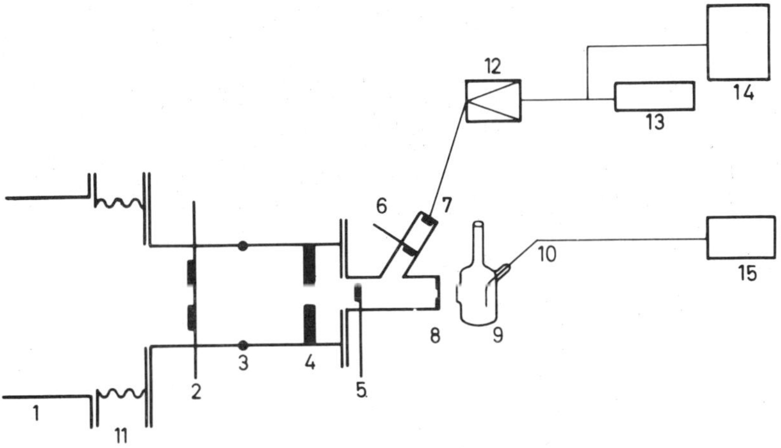

Fig. 1. 1. Beam duct (4 in diam.); 2. First collimator; 3. Equipment for alignment of collimator; 4. Second collimator (5 mm diam.); 5. Thin metallic foil for elastic scattering; 6. Absorber for SSD counter; 7. Solid state detector (SSD); 8. Beam exit window; 9. Irradiation vessel; 10. Platinum wire dipped into solution; 11. Bellows; 12. Amplifier; 13. Scaler; 14. Multi-channel pulse height analyser; 15. Beam current integrator

a thin aluminium window (10μ or 30μ). The irradiation cell was of quartz or Pyrex, 100 mm high and 10 mm O.D., with a 10 mm diam. aperture covered with a thin quartz or mica window cemented by epoxy resin. The thickness of

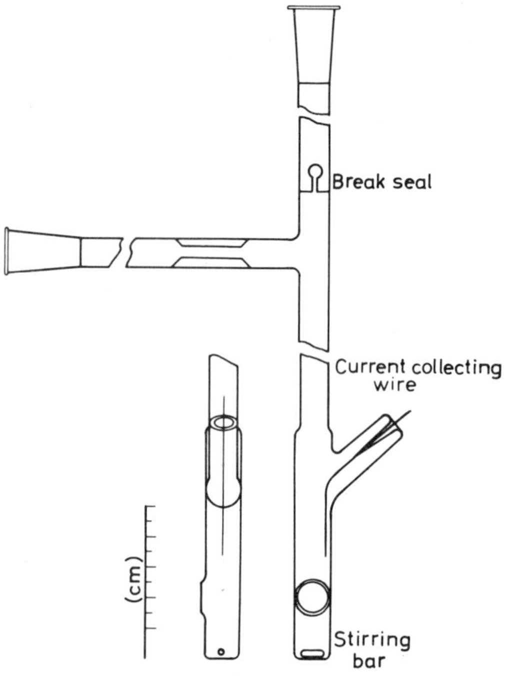

Fig. 2. Irradiation cell

the quartz window was 60–200μ and that of the mica 10μ. The volume of the irradiation solution was 5·0 ml for Fricke's solution and 3·5 ml for bacterial cells suspended in phosphate buffer. The former was stirred by a magnetic stirrer during irradiation, at a speed of 2000 rpm, and the suspension was stirred by bubbling air through a capillary tube at a constant rate of 45 ml/min.

Determination of dose. The beam current was monitored by counting the particles elastically scattered by the thin metallic foil into a solid state detector. This method is the same as that of Kochanny[1] in principle. The arrangement is shown in Fig. 1. Since the foil was very thin, most of the particles passed through it and entered the solution. A small fraction of the beam, proportional to the number passing through the foil, was elastically scattered into the detector. The energy spectrum was displayed on a multi-channel pulse height analyser. The elastic scattering peak was integrated by summing all the counts by a scaler adjusted by the discriminator so that the peaks corresponding to the ground state are counted.

The ratio of the number of particles elastically scattered into the detector to the number of particles passing through the metallic foil and entering the solution was accurately determined by replacing the solution with a Faraday cup. This ratio was determined before and after each run. It remained constant with an error of about ±1%.

The metallic foil was selected so that multiple Coulomb scattering could not affect the beam spot on the cell windows. Reproducible location of the cells was ensured by using an adjustable holder, which was designed to fit over the end of the beam pipe so that the centre of the window coincided with the centre of the emerging beam within 0·5 mm.

The beam energy was determined with a beam analysing magnet when an accurate value was desired, but it was usually determined from the value of the RF voltage and estimated with an error of <2%. The beam energy entering the solution was calculated from the range-energy relation[2], considering the energy losses in the scattering foil, beam exit window, air layer and cell window.

The above methods ensure accurate determination of the absorbed dose with an overall error estimated at <3%.

Comparison of current collected by a wire dipped into solution. The beam current is usually measured by dipping a platinum wire into the solution as a charge collector. The measured current is always somewhat lower than the true value[3].

The ratio (I_s/I_w) of the current measured by the scattering method (I_s) and the dipped wire method (I_w) was examined for several ion beams with different energies. This ratio increases with decreasing ion energy (Fig. 3). Although

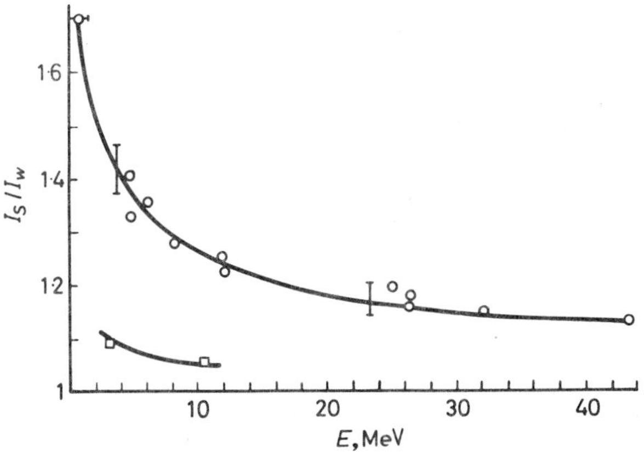

Fig. 3. Ratio I_s/I_w *vs L.E.T.* I_s *is the current measured by the scattering method and* I_w *by the dipped wire.* ○, *He ions* □, *protons*

the reproducibility is only fair, our results agree with those of Allen, Hart and Kochanny within experimental error. When the He ion energy is below 15 MeV the ratio (I_s/I_w) increases rapidly. The ratio increases not only with increasing L.E.T. but also with decreasing particle range.

FRICKE DOSIMETER SOLUTION

Solutions contained 10^{-2} or 10^{-3} M ferrous ammonium sulphate and 10^{-3} M NaCl in aerated aqueous 0·8 N H_2SO_4. All reagents were of c.p. grade and used without further purification. Triply-distilled water was prepared as described by Allen[4]. The ferric ion concentration was determined from the optical density at 304 mμ using $\epsilon = 2175\ M^{-1}\ cm^{-1}$ (23·7°C) and appropriate temperature correction.

The apparent oxidation yield in the air-saturated solution at various beam currents is shown in Fig. 4 for 13 MeV He ions. The yield is constant at

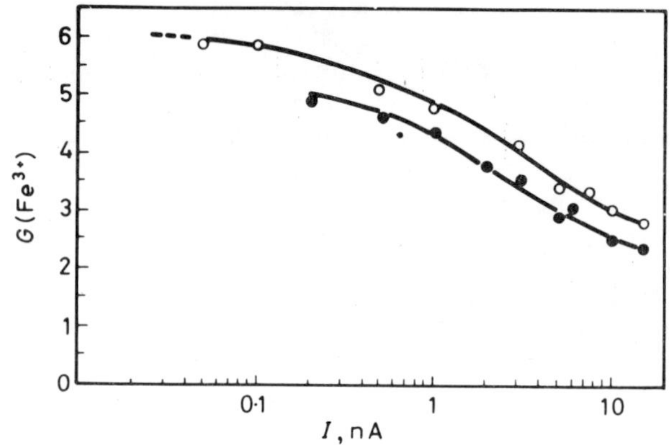

Fig. 4. $G(Fe^{3+})$ *values vs beam current for 13 MeV* He *ions.* ○; *10 mM,* ●; *1mM*

currents below 5×10^{-8} nA and 1×10^{-9} nA in 10 mM and 1 mM solutions respectively.
ly.

Values of $G(Fe^{3+})$ at sufficiently low currents for various values of L.E.T. are shown in Figs. 5 and 6 for 10 mM and 1 mM solutions. The results are compared with those of Schuler and Allen[5] in 10 mM solution for He ions and of Hart *et al*[6] in 1 mM solution for He ions and for protons. There is no significant difference within experimental error. The errors are greater than expected, and sources of error will be investigated further.

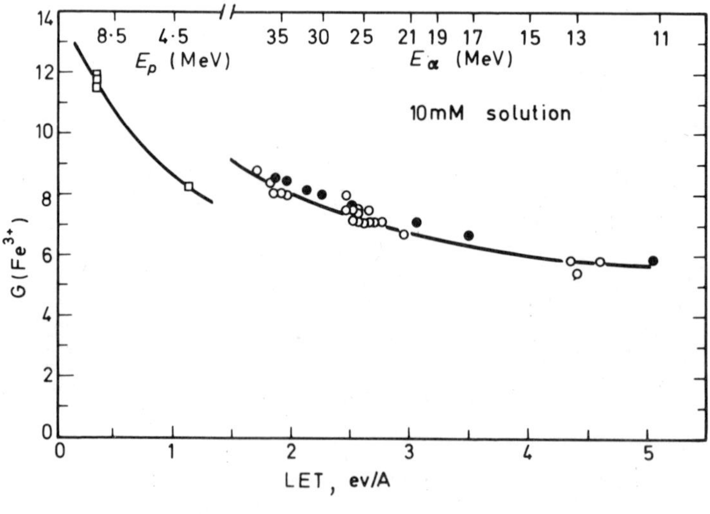

Fig. 5. $G(Fe^{3+})$ *vs L.E.T. in 10 mM solution at low beam current*

The yield in 10 mM solution is higher than that in 1 mM by several per cent for He ions obtained in a simultaneous run, and the difference between yield in 10 mM and 1 mM appears to increase with increasing L.E.T. As with the results of Schuler and Allen,[5] the present data for protons and helium ions do not actually overlap on either of these plots. This difference conflicts with the underlying assumption of the diffusion model, and has been discussed by Faw and Donnert on the basis of the kinetics of secondary electron production.[7]

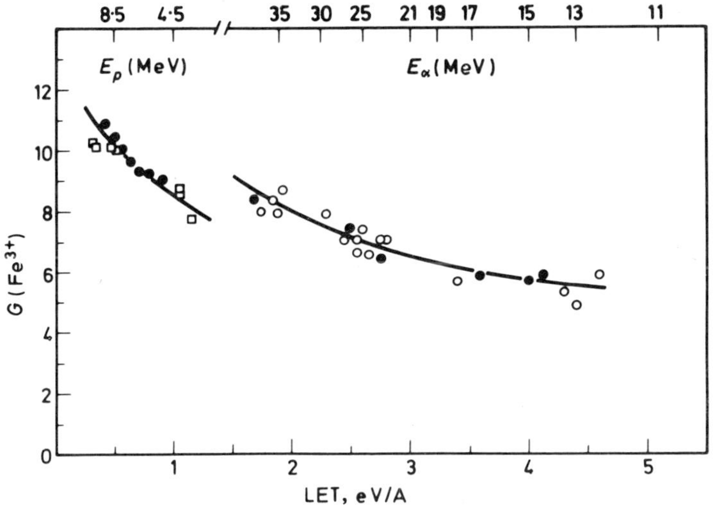

Fig. 6. $G(Fe^{3+})$ *vs L.E.T. in 1 mM solution at low beam current*

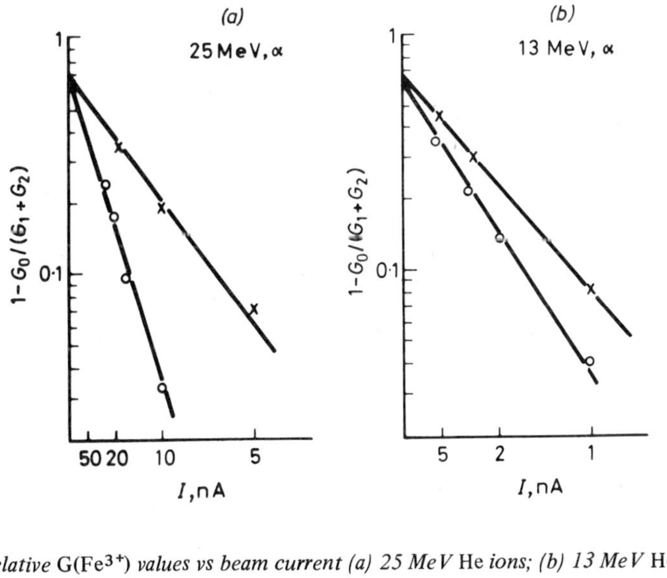

Fig. 7. Relative $G(Fe^{3+})$ *values vs beam current (a) 25 MeV* He *ions; (b) 13 MeV* He *ions.* ×; *1 mM,* ○; *10 mM*

Yields at extremely high currents. The yield is constant at low currents and apparently decreases at higher beam currents (as observed by Allen[5]). $G(Fe^{3+})$ decreases with increasing beam current. The observed value of $G(Fe^{3+})$ may be expressed by the following empirical equation:

$$G_O = G_1 [1 - \exp(-C/I)] + G_2 \tag{1}$$

where C is constant (Fig. 7).

It appears that irradiation without stirring and at relatively high beam current gives the same G values as for stirring at high currents. Values of $G(Fe^{3+})$ obtained without stirring are given in Table 1.

Table 1. $G(Fe^{3+})$ UNDER VARIOUS CONDITIONS (no stirring)

Particle	*Energy (MeV)*	*Concn. of solution (mM)*	*Mean beam current* (nA)					*G* calc.
			0·5	1	2	5	10	
Proton	3·1	1	3·00	–	–	–	–	2·26
		10	3·76	–	–	–	–	
Helium ion	26·0	1	3·71	2·94	–	2·58	–	2·12
		10	4·80	4·62	–	2·76	–	
	25·3	1	1·89	–	1·89	–	–	2·10
		10	2·87	–	2·87	–	–	
	24·3	1	–	–	2·09	–	–	2·07
		10	–	–	2·95	2·77	–	
	16·0	1	–	–	1·75	–	–	1·95
		10	–	–	2·22	–	–	
	13·2	1	1·70	1·57	–	1·62	–	1·85
		10	1·83	1·70	–	1·74	–	

The effect of L.E.T. on the ferrous sulphate oxidation yield may be quantitatively explained by track theory. For low L.E.T. radiation (γ-rays, high-energy X-rays and electrons) the free-radical yields are high and the molecular yields low. At high L.E.T., however, adjacent spurs partly overlap and reactions between primary species take place in the relatively high concentration region of these overlapping spurs. For example, Burch[8] has tried to interpret the L.E.T. effect on $G(Fe^{3+})$ by assuming the following secondary reactions:

$$H + H \longrightarrow H_2 \tag{1}$$
$$OH + OH \longrightarrow H_2O_2 \tag{2}$$
$$H + OH \longrightarrow H_2O \tag{3}$$
$$H_2 + OH \longrightarrow H_2O + H \tag{4}$$

At higher dose rates $G(Fe^{3+})$ is also affected because of reactions of the H, OH and HO_2 radicals with each other.

Under normal conditions, the mechanism proposed by Rigg, Stein and Weiss[9] for subsequent reactions in aerated solutions is as follows:

$$Fe^{2+} + OH \longrightarrow Fe^{3+} + OH^- \quad (5)$$
$$H + O_2 \longrightarrow HO_2 \quad (6)$$
$$HO_2 + H^+ + Fe^{2+} \longrightarrow Fe^{3+} + H_2O_2 \quad (7)$$
$$H_2O_2 + Fe^{2+} \longrightarrow Fe^{3+} + OH + OH^- \quad (8)$$

The overall yield of ferric ion, $G(Fe^{3+})$, is then:

$$G(Fe^{3+}) = G_{OH} + 3G_H + 2G_{H_2O_2} \quad (I)$$

At higher dose rate and high L.E.T., the number of ferrous ions undergoing oxidation via reactions (5) and (7) decreases and $G(Fe^{3+})$ expressed by Eqn. (1) falls. As reaction (6) is rapid ($k_6 \sim 2 \times 10^{10}\ M^{-1}\ s^{-1}$),[8] depletion of oxygen will

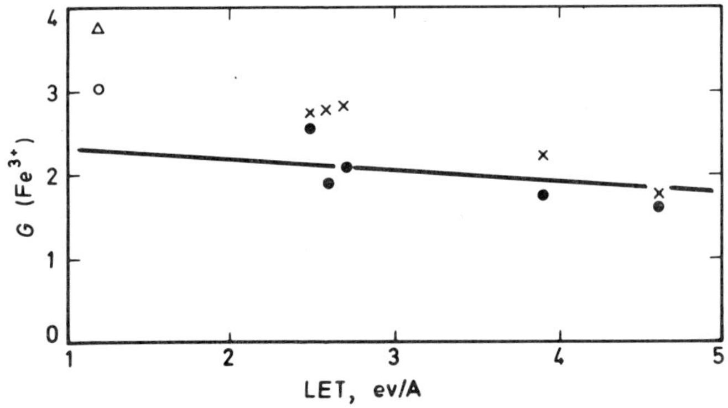

Fig. 8. $G(Fe^{3+})$ *without stirring, at high beam current*

soon take place in the reaction zone at such a high beam current. At high beam currents the fast interspur reactions (1), (2) and (3) will predominate because of the increase in local concentrations of H and OH radicals. The rate constants for these reactions have been reported as:

$$k_1 = (1{\cdot}0 - 1{\cdot}3) \times 10^{10}\ M^{-1}\ s^{-1}$$ (10,11)

$$k_2 = (5 - 9) \times 10^9\ M^{-1}\ s^{-1}$$ (11,12)

$$k_3 = 7 \times 10^9\ M^{-1}\ s^{-1}$$ (10)

The slowest reaction is (8), for which $k_4 = 61{\cdot}9\ M^{-1}\ s^{-1}$. At extremely high currents, ferrous sulphate solution may be largely oxidised by reactions with hydrogen peroxide alone, as in (8). The OH radicals produced via (8), as well as primary OH radicals in the spurs, produce hydrogen peroxide in the dense reaction zone. OH radicals also react as in (2), (3) and (4). Therefore, approximately speaking, three OH radicals produce one molecule of hydrogen peroxide, and in (8) one molecule of hydrogen peroxide oxidises one ferrous ion, producing one OH radical. This leads to the relationship:

$$G(Fe^{3+}) = 1/3\ G(OH) + (1/3)^2\ G(OH) + (1/3)^3\ G(OH) + \ldots = \tfrac{1}{2}G(OH) \quad (9)$$

The total $G(Fe^{3+})$ is thus given by:

$$G(Fe^{3+}) = \tfrac{3}{2}\ G(H_2O_2) + \tfrac{1}{2}G(OH) \quad (II)$$

at extremely high beam currents. Values of $G(Fe^{3+})$ calculated by Eqn. (II) are tabulated in the last column of Table 1 and shown in Fig. 8. With no stirring and at relatively high beam current, $G(Fe^{3+})$ for aerated solutions will be close to that calculated by Eqn. (II); Fig. 8 shows that $G(Fe^{3+})$ obtained without stirring for 1 mM solution shows such a tendency, as mentioned above. Larger *G*-values for 10 mM solutions compared with 1mM solutions may be at least partly explained by some contribution of reaction (5) in more concentrated solutions. The deviation observed for proton irradiations is being studied further.

BACTERIAL CELLS IN LIQUID PHASE

Log-phase cells of *Micrococcus radiodurans* R_1, *Escherichia coli* B/r and *Escherichia coli* $B_{s\text{-}1}$ suspended in 0·067 M phosphate buffer were exposed to He ions. A 3·5 ml aliquot of the cell suspension ($10^7 \sim 10^8$ cells/ml) in a quartz vessel was agitated by bubbling air through a capillary tube at a constant rate of 45 ml/min. He ions were accelerated to 29 MeV and their energy was reduced to 26 MeV at the entrance to the suspension. In preliminary experiments, no effect of cell concentration at the time of irradiation was found on the radiosensitivity of these bacteria (Table 2).

Table 2. CELL CONCENTRATION EFFECT

M.Radiodurans R_1

Flow rate, 45 ml/min; dose, 800 nC

Beam current (nA)	*Viable count*		*Surviving fraction* (%)
	Before irradiation (cells/ml)	*After irradiation* (cells/ml)	
22	$1{\cdot}21 \times 10^7$	$1{\cdot}05 \times 10^6$	8·7
	$1{\cdot}29 \times 10^4$	$1{\cdot}09 \times 10^3$	8·5
12	$2{\cdot}24 \times 10^7$	$5{\cdot}35 \times 10^4$	0·24
	$2{\cdot}09 \times 10^5$	$6{\cdot}77 \times 10^2$	0·35

When bacterial cells suspended in the phosphate buffer were irradiated (referred to as liquid-phase irradiation in this paper), only a relatively small fraction of the bacterial population was subjected to the bombardment, because of the very short range of He ions in an aqueous system. Stirring and beam current should therefore seriously influence the radiation effect. The effect of beam current on radiation killing was examined at a constant rate of air bubbling (45 ml/min), and the results with *M.radiodurans* R_1, *E.coli* B/r and B_{s-1} are shown in Figs. 9(a), (b), and (c). The survival curves of

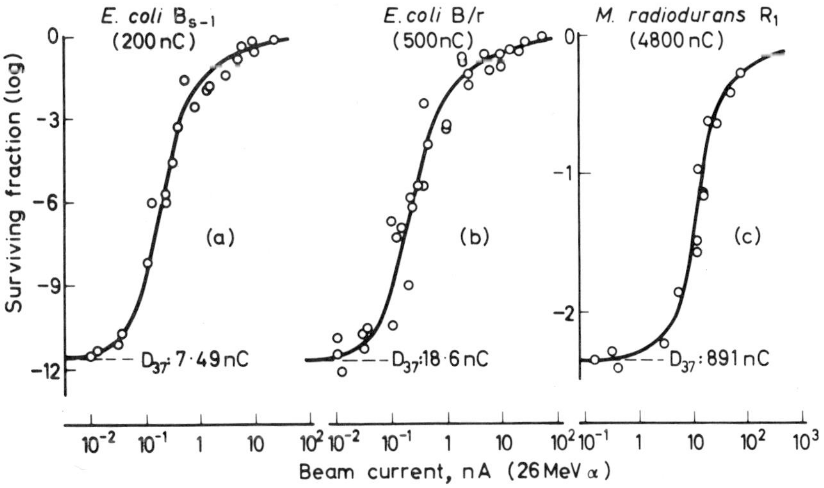

Fig. 9. Surviving fraction vs beam current at constant dose

these three strains for 26 MeV He ions were exponential over the whole range of intensities examined, provided that the particle beam current is sufficiently low.

When the beam current was sufficiently small, the surviving fraction was identical for the same dose, independent of the beam current. However, above a critical beam current, which varied with different bacterial strains, the surviving fraction at a given dose increased with increasing beam current. In

Table 3. AIR BUBBLING EFFECT

M.Radiodurans R_1

Dose 2×10^5 nC; beam current 30 nA; cell concentration $1{\cdot}21 \times 10^7$ cells/ml

Flow rate (ml/min)	*Viable count*	*Surviving fraction* (%)
7·5	$3{\cdot}81 \times 10^6$	31·5
45	$4{\cdot}80 \times 10^5$	3·97
75	$5{\cdot}25 \times 10^4$	0·43

the latter range of beam current, the surviving fraction may be determined not only by the radiosensitivity of the bacterial cells but also by the stirring efficiency. In this range of beam current, the surviving fraction increases as the rate of air bubbling is reduced at the constant beam current (Table 3).

The values of D_{37} (dose required for 1/e = 37% survival) for the liquid-phase irradiation were 329 krads for *M.radiodurans*, 9 ± 2 krads for *E.coli* B/r, and 3·3 ± 1 krads for *E.coli* B_{s-1}, respectively. Estimates of D_{37} may be calculated from the data of Munson *et al*,[13] in which a monolayer of cells on a Millipore filter is bombarded by particles of different L.E.T. (solid-phase irradiation). By integration over the whole range of energy of He ions (0 ~ 26 (incident) MeV). D_{37} in liquid-phase irradiation is found to be 5·5 krads for *E.coli* B/r and 2·0 krads for *E.coli* B_{s-1}. The discrepancy between experimental and calculated values of D_{37} in the liquid-phase irradiation cannot be explained at present, but the ratios of D_{37} for *E.coli* B/r to that for *E.coli* B_{s-1} are in good agreement in both phases. A comparative study of liquid- and solid-phase irradiations using ionising particles of different L.E.T. is in progress.

When the beam current is sufficiently small, the surviving fraction is expressed by the following equation:

$$S = \exp(-It/q_o) \qquad \text{(III)}$$

where S is the surviving fraction, q_o is D_{37} in nanocoulomb, t the irradiation time in s, and I the beam current in nA. When the beam current is large, bubbling cannot replace cells from the irradiated region sufficiently rapidly to maintain uniform irradiation conditions. The effective volume of the irradiated region just behind the window for 26 MeV He ions is estimated to be ~0·001 ml, corresponding to about 1/3500 of the sample volume; thus increasing beam current eventually decreases the killing effect. The beam current I in Eqn. (III) may then be replaced by ηI $(0 < \eta < 1)$. By applying the theory of probability to the agitation produced, the reducing factor η may be expressed as follows:

$$\eta = 1 - \exp(-kq_o/I) \qquad \text{(IV)}$$

where k is a constant having dimensions of s^{-1} and dependent upon the rate of aeration and the size of the effective volume.

Eqns (III) and (IV) may be combined into the following equation:

$$S = \exp\{ (-It/q_o)[- \exp(-kq_o/I)] \} \qquad \text{(V)}$$

this is in agreement with the experimental data on inactivation of bacteria shown in Figs 9(a, b, c).

A comparative study of liquid- and solid-phase irradiations is now in progress. A feature of the device used for solid-phase irradiation is uniform irradiation of the monolayer of bacterial cells on a Millipore filter (type HA, 0·45 μ pore size) attached to the holder, which reciprocates two-dimensionally with constant speed.

The results of sedimentation analysis after He ion irradiation[14] emphasise the capacity of *M.radiodurans* R_1 for the repair of double-strand scissions in DNA, which has been deduced from the analysis after γ-irradiation[15].

ACKNOWLEDGEMENTS

The authors wish to express their cordial thanks to H. Kumagai for encouragement and to K. Matsuda for proposal of the nuclear electronic method of determining beam current.

REFERENCES

1. Kochanny, G. L., Jr., Timnick, A., Hochanadel, C. J., and Goodman, C. D., *Rad. Res.* **19**, 462, (1963).
2. Whaling, W., Handbuch der Physik, Bd. XXXIV, 193, Springer-Verlag, Berlin (1958).
3. Schuler, R. H. and Allen, A. O., *Rev. Sci. Inst.* **26**, 1128, (1955).
4. Allen, A. O., The Radiation Chemistry of Water and Aqueous Solutions, Van Nostrand, Princeton, (1961).
5. Schuler, R. H. and Allen, A. O., *J. Am. Chem. Soc.* **79**, 1565, (1957).
6. Anderson, A. R. and Hart, E. J., *Rad. Res.* **14**, 689, (1961).
7. Faw, R. E. and Donnert, H. J., *Nucl. Inst. Meth.* **58**, 307, (1968).
8. Burch, P. B. J., *Rad. Res.* **11**, 481, (1959).
9. Rigg, T., Stein, G., and Weiss, J., *Proc. Roy. Soc.* **A211**, 375, (1952).
10. Sweet, J. P. and Thomas, J. K., *J. Phys. Chem.* **68**, 1363, (1964).
11. Fricke, H. and Thomas, J. K., *Rad. Res. Suppl.* **4**, 35, (1964).
12. Thomas, J. K., Rabani, J., Matheson, M. S., Hart, E. J., and Gordon, S., *J. Phys. Chem.* **70**, 2409, (1966).
13. Munson, R. J., Neary, G. J., Bridges, B. A., and Preston, R. J., *Int. J. Radiat. Biol.* **13**, 205, (1967).
14. Kitayama, S., Karasawa, T., and Matsuyama, A., *Agr. Biol. Chem.* **33**, 628, (1969).
15. Kitayama, S. and Matsuyama, A., *Biochem. Biophys. Commun.* **33**, 418, (1968).

DISCUSSION

W. G. Burns (A.E.R.E.): Your results for the change of $G(Fe^{+++})$ with current appear to show a dose-rate effect due to inter-reaction of radicals escaping from neighbouring tracks. Have you compared your results with calculations of the currents at which these effects might be expected (e.g. based on Kupperman's article in Vol. 5 of Chemical and Biological Actions of Radiation, edited by Haïssinsky)?

T. Karasawa: At present we have not made any numerical calculations, but only qualitative comparisons.

A comparative study of defect production and changes in surface reactivity induced in oxide powders by high-energy positively-charged particles

R. L. Nelson and M. J. Duck

Chemistry Division, A.E.R.E., Harwell, Berks., England.

ABSTRACT

The adsorption of oxygen on magnesium oxide during bombardment with positive ions is due to several damage processes, both at the surface and within the bulk of the solid. The relative efficiencies of bulk vacancy, surface vacancy and ionization in promoting the adsorption of oxygen have been compared.

The role of ionisation in promoting the adsorption of hydrogen, methane and ethylene on MgO has been studied and mechanisms of charge trapping and charge recombination at the surface have been proposed.

The effect of positive ion bombardment on the surface reactivity of thoria, zirconia and zinc oxide is also due largely to ionisation processes at low total dose but it has not proved possible to identify the role of displacement-induced lattice defects.

INTRODUCTION

The interaction of high-energy radiation with chemical reactions occurring at gas/solid interfaces is a subject of considerable importance both in nuclear reactor technology and in the potential application of radiation in industrial processes. Surface processes such as corrosion and catalysis are complex in themselves, even in the absence of radiation; hence the present work is an attempt to elucidate the mechanisms whereby radiation interacts with surfaces by studying the chemisorption of simple gases at solid surfaces. The solids chosen were ceramic oxides because of their technological interest and the ease of preparation with reproducible surface characteristics.

In the present studies the gas phase was at low pressure ($<133\ \mathrm{Nm}^{-2}$, i.e. 1 torr) so that the ratio of energy absorbed by the solid phase to that absorbed by the gas phase was $\sim 10^5:1$ and the interaction of radiation with the solids is expected to have a dominating effect on the surface reactivity. When charged particles of high energy (>1 MeV) pass through solids their energy is dissipated in the displacement of lattice ions and in ionisation, and both processes may have chemical consequences. Displacement of ions leads to the formation of vacancies and interstitials and the resulting regions of local non-stoichiometry may result in greater reaction with the gas phase, either as a result of the defect sites or from the change in the band structure. In ionisation free charge carriers are released and if these are either formed at the surface or are capable of migrating towards the surface the adsorption of molecules from the gas phase may result.

In the study of ionisation effects γ-sources are convenient, and in recent years Russian workers[1] have correlated a variation in catalytic activity of different solid surfaces with the band structure of the bulk solid. Reactor neutrons have also been used to generate point defects in ceramic oxides and the surface reactivity of these solids on subsequent exposure to different gases has been shown to be due both to surface[2] and bulk[3] point defects. In the present work positive ions from the Harwell Variable Energy Cyclotron (VEC) have been used to irradiate oxides in the presence either of inert or reactive gases to study the relative efficiencies of displacement and ionisation processes in inducing adsorption.

Relatively few studies of heterogeneous systems have involved particle irradiations in accelerators, and these have been confined to the pre-irradiation of solids and their subsequent reactivity in the absence of radiation. Irradiation of Fe_2O_3 with 260 MeV protons[4] leads to an increased rate of solution in acid, while irradiation of NiO[5] reduces the length of the induction period for hydrogen reduction. The catalytic activity of cuprous oxide in the oxidation of carbon monoxide is reduced after irradiation with 3·5 MeV electrons[6] and is correlated with a reduced electrical conductivity. In the above examples point defects in the solid are held solely responsible for the changes in surface reactivity observed.

EXPERIMENTAL

Irradiation cell. This is constructed from silica, apart from the Pt/Mo/Pt current collector, and is shown in Fig. 1 together with the end flanges from the beam collimator. The collimated beam first passes through the nickel foil and then the silica window (0·001 in or 0·002 in thick) of the cell itself before penetrating the powder. The volume between the nickel and silica windows is evacuated independently of the collimator assembly in an attempt to avoid contamination of the main beam line should the thin silica window rupture. The silica cell is positioned reproducibly by means of flanges 1 and 2 and the rear of the cell is subsequently clamped to avoid vibration during cooling of the irradiation zone with compressed air.

In addition to the thin silica window and the current collector the cell has a thin break-seal for subsequent adsorption studies and a narrow (3 mm bore) Spectrosil tube for e.s.r. studies. The oxides are given a final outgassing in the cells with this Spectrosil tubing connected to a high-vacuum system and are sealed off with an over-pressure of pure helium. During oxide preparations which require high-temperature evacuation the exterior section of Pt/Mo probe is preserved in an evacuated section of tubing, and this is only opened to the atmosphere when the sample is ready for irradiation.

In irradiating powders of low apparent density great care must be taken to avoid overheating, otherwise heat-sensitive defects in the powders may be annealed out, and the thin silica window will be subjected to excessive thermal stress. In studies of defect formation in oxides the over-pressure of helium (~20 k Nm^{-1}, i.e. 160 torr) allowed up to 6 watts to be dissipated in 100–200 mg powder without significant annealing of those defects which could be generated by γ-irradiation at 40°C. However, during simultaneous irradiation and

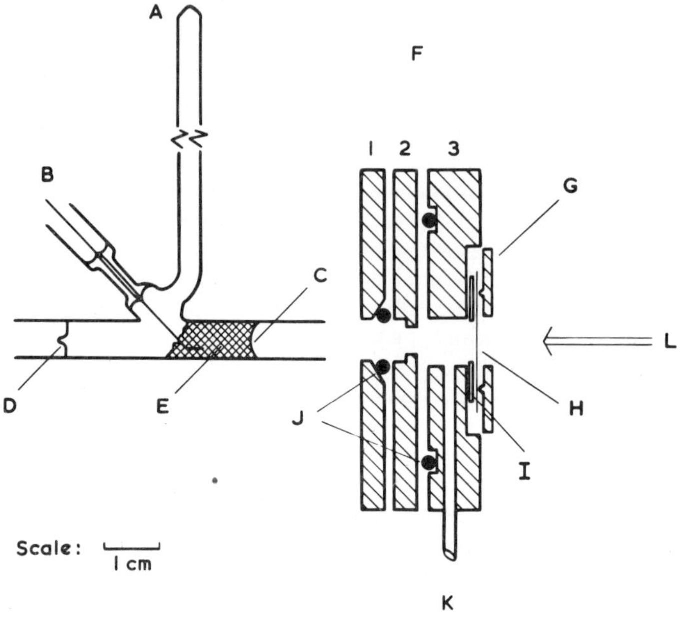

Fig. 1. Silica cell and end-flanges.

A – *Spectrosil tube for e.s.r.*		*B* – Pt/Mo/Pt *probe*	
C – *thin silica window*		*D* – *dome break seal*	
E – *oxide powder*		*F* – *aluminium flanges*	
G – *brass retaining ring*		*H* – *nickel foil (0·000 5 in)*	
I – *indium gasket*		*J* – *'O' rings*	
K – *to vacuum pumps*		*L* – *collimated beam from VEC*	

adsorption studies the helium was necessarily absent, and in the presence of the adsorbate gases at much lower pressures (<133 Nm^{-2}, i.e. 1 torr) overheating occurred at powers greater than 0·1 watt. Overheating was accompanied by luminescence of the powder; this could be observed with closed-circuit television during the irradiation and controlled by reducing the current.

Dosimetry. A solenoid-operated graphite stop was located between the end of the collimator assembly and the aluminium flange with the nickel window. This stop was inserted into the beam at intervals during an irradiation and allowed a measurement of the total current, I_s. Since the oxides studied are of low conductivity, the cell probe (I_c) does not collect all current deposited in the powder, and a fraction leaks through the front of the silica cell to the aluminium flanges (I_f). The cell current is monitored continuously during an irradiation and can be corrected to total dose by means of the ratio I_s/I_c. The correctness of this procedure was checked for each beam used by placing a Faraday cup in the cell position.

For simultaneous irradiation and adsorption studies beam energies were selected such that the range was less than the depth of powder, and published range-energy[7, 8] data were used both for this calculation and to correct the beam energy for losses sustained in passing through the nickel and silica windows.

Adsorption apparatus. The adsorption of simple gases before, during and after irradiation of the oxides was measured in a constant-volume Pyrex system by monitoring the gas pressure with a thermistor detector maintained at 25 ±0·05°C. The adsorption system was connected through a Pyrex-silica graded seal to either the break-seal shown at the rear of the cell in Fig. 1 or a break-seal located in certain cases at the end of the Spectrosil e.s.r. tube. A silica-encapsulated iron breaker was used to open the break-seals when the pressure in the connection to the adsorption apparatus had fallen to ~10^{-4} Nm^{-2} (10^{-6} torr), and the oxide samples were protected from possible contamination by a trap maintained at –78°C.

The methods of surface area determination and e.s.r. examination have been described elsewhere.[2] The values for adsorbed coverage of chemisorbed molecules have been calculated from the monolayer capacity for krypton on the basis that the cross-sections of the former are equal to the cross-section of physically-adsorbed krypton. Although this is not the case the values for surface coverage obtained are useful for direct comparison purposes.

Materials. Helium, oxygen and hydrogen were Grade X (B.O.C. Ltd.); methane and ethylene were standard samples from the National Physical Laboratory. All gases were used without further purification but were always passed through the trap at –78°C before contacting the oxide samples.

Magnesium oxide was prepared either by calcining purified basic carbonate in air at 900°C (specific surface area ~15 m^2g^{-1}) or by decomposing the carbonate in vacuo at 900°C (S.S.A. 50–200 m^2g^{-1}).

Thorium oxide was prepared from Specpure thorium nitrate (Johnson Matthey) by first precipitating either the hydroxide or carbonate, which were subsequently filtered, washed and dried. The hydroxide was decomposed to the oxide in vacuo 700°C (S.S.A. ~0·15 m^2g^{-1}) and the carbonate at 900°C (S.S.A. ~6·1 m^2g^{-1}).

Specpure zinc oxide (S.S.A. ~4 m^2g^{-1}) and zirconium dioxide (S.S.A. ~2 m^2g^{-1}) were used without further treatment. MgO and ZrO_2 powders were evacuated in the irradiation cells at 900°C, ThO_2 at either 700°C or 900°C and ZnO at 500°C, before sealing off either in vacuo or in helium.

RESULTS AND DISCUSSION

A comparison of induced surface reactivity via displacements and ionisation. A detailed study of the efficiencies of different positively-charged particles in generating *F* centres in alkaline earth oxide powders has been published elsewhere.[9] Although the concentration of *F* centres produced in MgO for a constant dose varies with both the mass and energy of the bombarding particle, the efficiency of defect production is constant for a given sample and corresponds to a G_F value in the range 1-5 × 10^{-6} *F* centres per 100 eV absorbed. Similar G_F values are found for CaO and SrO, and the most significant variation in G_F value for MgO powders appears to arise from variation in the average particle size.

The G_F values observed are an order of magnitude lower than those calculated on the basis of a Rutherford scattering model with a displacement threshold, E_d, of 70 eV. The observed G_F values are constant along the track of a positively-charged particle and there is no evidence for a higher efficiency at the end of the track where ionisation losses are unimportant. These observations suggest that stable *F* centres are only formed in collisions involving large (i.e. ≫ 70 eV) energy transfers. G_F values of $1-5 \times 10^{-6}$ for positive particles compare with values of 10^{-3} and 10^{-7} for fast neutrons and electrons respectively, and in these two latter cases the observed G_F values are again at least an order of magnitude lower than calculated values.

In the present study a single sample of MgO (VM56) was first irradiated in helium and the *F* and *S* centres generated were measured by e.s.r. The sample was then exposed to oxygen and irradiated with 10 MeV protons to establish the contribution to surface reactivity from ionisation processes. Fig. 2 presents the dose dependence of *F* and *S* centres at low total doses for 13 MeV proton

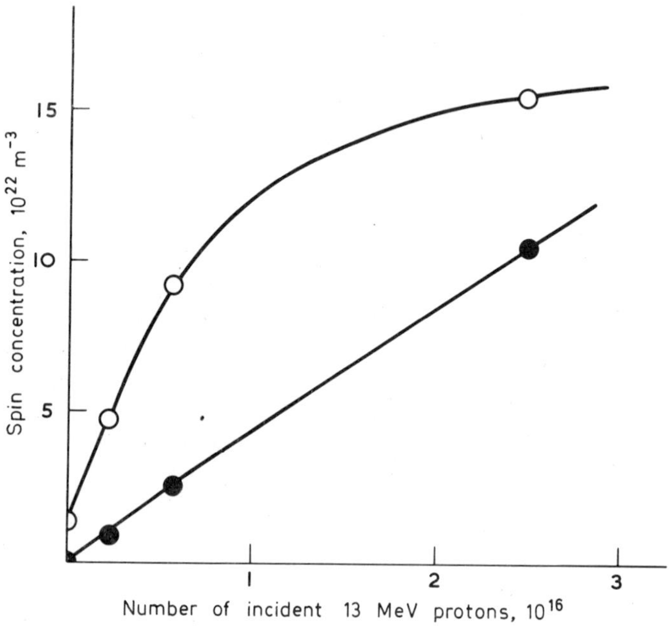

Fig. 2. Dose dependence of S and F centres in MgO
○ – *S centre,* ● – *F centre.*

bombardment. The initial G_F value for *F* centres is $1{\cdot}1 \times 10^{-6}$ and G_S for *S* centres is $4{\cdot}0 \times 10^{-6}$; VM56 is of low specific surface area ($14{\cdot}6\ m^2g^{-1}$) and this accounts for the low G_F value. Although initially higher, the G_S value falls rapidly with increasing dose whereas the G_F value is maintained. This behaviour is similar to the fast neutron dose dependence[10] of *S* and *F* centres in MgO, where the saturation concentrations of both defects are similar at high doses. The *S* centre has been shown[2] to be a surface anion vacancy with a single trapped electron, whereas the *F* centre is a bulk anion vacancy with a trapped electron. Both defects react with oxygen, the *S* centre by transfer of the unpaired electron to an adsorbed oxygen molecule with the formation of an O_2^- ion[11] and the *F* centre by adsorption of oxygen, diffusion of this to the bulk defect and destruction of the anion vacancy.[3] Whereas the reaction of the *S* centres with oxygen is complete, only those F centres close (within ~100 Å) to the crystallite surfaces react, and hence G_F quoted above is an upper limit for G_{FO}, the G value for oxygen adsorption as a result of F centre production.

The finite S centre concentration at zero dose in Fig. 2 results from prior γ-irradiation of the sample, during which electrons were trapped at existing surface anion vacancies. If this pre-irradiation had been conducted on the V.E.C. a very much higher initial G_S would have been observed, but this would have to be assigned to ionisation rather than to displacement-induced surface activity.

After the dose-dependence studies were completed, oxygen was added to the MgO and when adsorption had almost ceased the sample was irradiated with 10 MeV protons. This produced the adsorption shown in Fig. 3, from which a lower limit for an initial G_{-O_2} of 0·8 is obtained. The current used was 0·5 nA and it was not practicable to use a lower current in order to obtain more accurate adsorption kinetics in the early stages. The G_{-O_2} value is calculated on the assumption that all the energy absorbed by the MgO solid is available for the adsorption of oxygen. Studies on the adsorption of oxygen induced by ^{60}Co γ-radiation[12] yield initial G_{-O_2} values in the range 1–10, with the lower values from materials of low S.S.A. (~15 m^2g^{-1}) and the higher values from materials of higher S.S.A.(~200 m^2g^{-1}). This effect of particle size may be due to the greater probability of charge carrier recombination in larger crystallites before the former

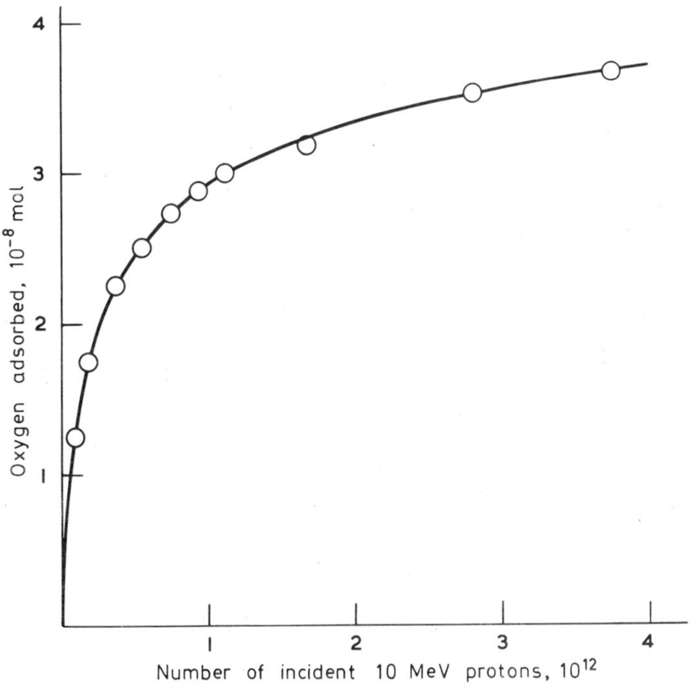

Fig. 3. Adsorption of oxygen on MgO *during proton irradiation*

reach the surface and participate in an adsorption process. On the other hand, free charge carriers generated in small crystallites (~100 Å) have a high probability of reaching the surface, since the G_{-O_2} value of 10 is close to the theoretical maximum value for MgO, which has a band gap of 8·7 eV.

The initial rapid adsorption of oxygen in Fig. 3 is thus probably due to ionisation of MgO lattice ions close to the surface and movement of the free electrons to the surface with the adsorption of oxygen as O_2^-. As the surface acquires negative charge a positive space charge builds up in the sub-surface region, and this dipole layer provides a potential barrier to the subsequent migration of electrons to the surface. The rate of adsorption decreases rapidly and the kinetics obey the equation[10]

$$d[-O_2]/dt = a \exp(-E/RT)\,[-O_2]^2$$

in the early stages. In the later stages, however, the rate of adsorption is faster than theoretical, and this may be due to a diffusion process which takes over at high coverages. The adsorbed coverage of oxygen presented in Fig. 3 is 0·5% but the absolute value for this ionisation-induced adsorption on other samples of MgO will depend on the specific surface area.

The *G* value for *S* centre production also requires to be corrected for specific surface area of the MgO sample, since it is likely that the energy deposited by primary displacement of surface ions is most important in the formation of *S* centres. In VM56 the surface/bulk ratio is 90, so that consideration of *S* centres as resulting from deposition of energy in only the surface layer of ions gives a G_S value of $3{\cdot}6 \times 10^{-4}$. This figure, however, must be an upper limit, since some of the energy deposited in the bulk is also involved (e.g. secondary collisions).

The various processes which lead to adsorption of oxygen during irradiation with positive ions are summarised in Table 1. Although the initial *G* values are

Table 1. MECHANISMS OF OXYGEN ADSORPTION ON MgO DURING IRRADIATION WITH POSITIVE IONS

Mechanism type	*Defects produced*	*Initial G value*	*Comments*
Displacement	*F* centre	$G_{FO} < 1{\cdot}1 \times 10^{-6}$	*G* value almost constant at low dose
Displacement	*S* centre	$4 \times 10^{-6} < G_{SO} < 3{\cdot}6 \times 10^{-4}$	*G* value decreases with total dose
Ionisation	free electrons	$0{\cdot}8 < G_{-O_2} < 10$	*G* value decreases rapidly with total dose.
Enhanced diffusion	?	–	occurs at high adsorbed coverage.

different, the adsorbed coverage at saturation from each of the first three processes is likely to be of the same order (i.e. ~1% monolayer) at the present irradiation temperature (20–100°C), and in all cases this saturation is due to the formation of a potential barrier. The fourth process is capable of reducing such a barrier by diffusion of the adsorbed species into the bulk, and this may be connected with the diffusion responsible for the *F* centre reactivity.

The *G* values for induced reactivity as a result of *F* and *S* centre formation in Table 1 have been deduced from irradiations in the absence of oxygen. Since the *F* and *S* centre vacancies are destroyed by oxygen, the previous reasons for the saturation in the concentration of these defects[10] no longer exist and the processes could continue with constant *G* value were it not for the formation of the potential barrier at the surface.

The radiation-induced adsorption of hydrogen, methane and ethylene. The proton irradiation of MgO in the presence of either hydrogen or methane results in adsorption of both gases with similar kinetics. The initial rates of adsorption correspond to G values in the range 0·1 to 1·0 molecules adsorbed per 100 eV absorbed by the solid. This G value rapidly decreases with total dose, in a similar manner to the adsorption of oxygen, and suggests a mechanism involving excess charge carriers, in this case positive holes, which diffuse to the surface and set up a potential barrier. The adsorption of methane on VM44, a sample of MgO of specific surface area ~25 m^2g^{-1}, during irradiation with 20 MeV protons is

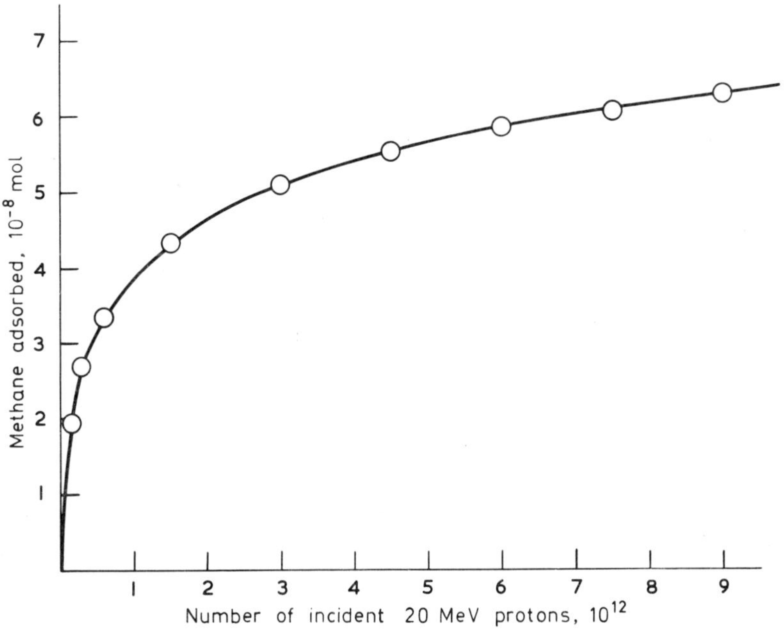

Fig. 4. Adsorption of methane on MgO *during proton irradiation*

shown in Fig. 4. The initial G_{-CH_4} value is 0·6, the total coverage ~0·2% and the adsorption obeys the equation

$$\frac{d[-CH_4]}{dt} = a \exp(-E/RT)\,[-CH_4]^2$$

The positive holes released by ionisation of lattice ions probably react initially with methane at the surface to form CH_4^+, but it has not been possible to detect these radicals by e.s.r. Irradiation of MgO in the presence of hydrogen, however, results in the formation of paramagnetic S_H centres [13] which may be detected by e.s.r. The first stage in this reaction again involves the migration of holes to the surface, with adsorption of hydrogen and subsequent dissociation according to the equation

$$\oplus + H_2 + 2\,O^{2-} \longrightarrow 2\,OH^- + e^-$$

this adsorption of hydrogen prevents the subsequent radiation-induced adsorption of methane and vice versa, clearly indicating the similarity in mechanism and in particular the resulting positively-charged surface and negative space charge. In contrast, the radiation-induced adsorption of oxygen does not inhibit the subsequent adsorption of either hydrogen or methane.

Ethylene is adsorbed by unirradiated MgO, or by MgO which has been irradiated in the presence of O_2, H_2 or CH_4, and the adsorbed coverage is ~0·3%. Additional adsorption occurs on irradiation and the kinetics are unlike those observed with the other gases. Fig. 5 presents adsorption data for the irradiation

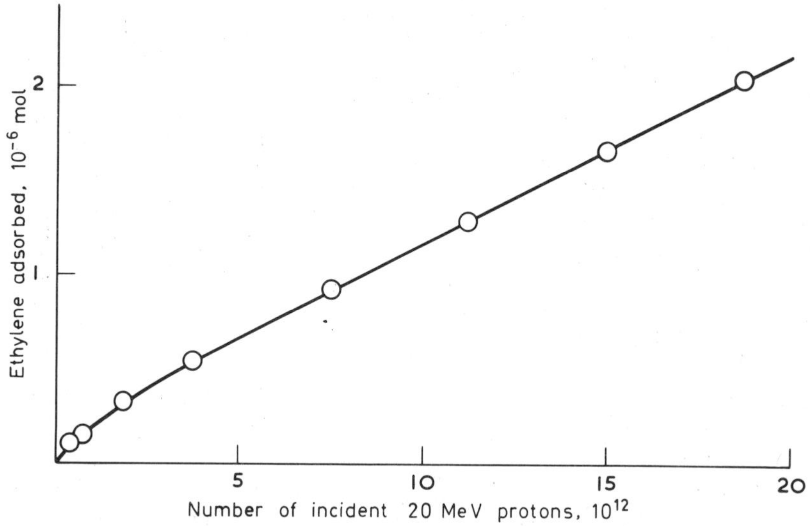

Fig. 5. Adsorption of ethylene on MgO *during proton irradiation*

by 20 MeV protons of the sample of MgO (VM44) which had previously been irradiated in the presence of CH_4. $G_{-C_2H_4}$ is initially ~1·0 molecules of ethylene per 100 eV absorbed by the MgO, but this rapidly drops to a constant value of 0·33. This behaviour suggests that the mechanism of ethylene adsorption does not involve the formation at the surface of a potential barrier to the movement of charge carriers. Instead it appears that both electrons and holes make their way to the surface and recombine to release energy which is utilised in the adsorption of ethylene. This probably occurs via the rupture of the double bond with either adsorption of individual species on many surface sites or formation of adsorbed polymer on relatively few sites. High surface coverages (up to monolayer) have been obtained in the present studies but further work is required to distinguish between the above possibilities. The formation of long (~200 units) polythene chains on alumina surfaces[14] occurs at high ethylene pressures on γ-irradiation. The ethylene pressures were in the range 2 to 2·5 MN m^{-2} (i.e. 20–25 atm.) and $G_{-C_2H_4}$ was ~10^3. $G_{-C_2H_4}$ has been shown[12] to increase with pressure on γ-irradiation of ethylene in the presence of MgO, and it is likely that the increase is principally due to enhancement of the propagation stage rather than to more frequent initiation events.

After radiation-induced adsorption of ethylene on MgO, the surface is much more reactive towards oxygen during irradiation with positively-charged particles than it was in the absence of ethylene. G_{-O_2} is initially between 0·5 and 1·0, and only decreases slowly with total dose until the adsorption shows signs of saturation when the total oxygen coverage is three times that of the previously-adsorbed ethylene coverage. This mechanism of oxygen adsorption probably involves the recombination of electrons and holes at the MgO oxide surface with the energy being utilised in the oxidation of ethylene or polyethylene according to

$$C_2H_4 + 3\ O_2 \longrightarrow 2\ CO_2 + 2\ H_2O$$

Similar G_{-O_2} values have been observed[15] in the γ-induced oxidation of saturated hydrocarbons adsorbed on silica gel and alumina.

In summary, the radiation-induced adsorption of simple gases as a result of ionisation within the MgO lattice may be due to either of two mechanisms. The adsorption of O_2, H_2 and CH_4 on uncontaminated MgO surfaces is due to the preferential trapping of either electrons or holes which have been generated in the sub-surface layers. The initial G value corresponds to the rate of trapping and this decreases rapidly as a charge barrier is formed. The adsorption of ethylene and subsequent adsorption of oxygen are due to arrival at the surface of equal numbers of electrons and holes with the recombination energy leading to polymerisation and oxidation in the absence of any charge barrier. This latter mechanism is of potential importance in catalysis.

The radiation-induced surface reactivity of thoria, zirconia, and zinc oxide. After preparation of zinc oxide with evacuation at 500°C, resonances at g = 2·00 and 1·96 are present in the e.s.r. spectrum. There is considerable conflict in the literature over the assignment of these resonances to specific defects,[16] but the resonance at g = 1·96 is believed to be due, at least in part, to interstitial Zn^{1+} donor sites. Oxygen is adsorbed on ZnO, even before irradiation, to a coverage of ~0·1% with some reduction in the intensity of the donor signal. On irradiation with 20 MeV protons additional adsorption of oxygen occurs with an initial G_{-O_2} of ~0·1; this adsorption rate falls only slowly with dose and only shows signs of saturation at a surface coverage of ~4%. The low initial G_{-O_2} is due to the low specific surface area, but the high coverage obtained before saturation argues that the potential barrier formed by the surface/space charge dipole is lower in semi-conducting zinc oxide than in insulating MgO. The space charge may be more widely dispersed in zinc oxide since the positive holes are capable of being trapped throughout the bulk at donor sites, and this produces a further significant decrease in the resonance at g = 1·96. Enhanced diffusion during irradiation may also account for the higher coverage of adsorbed oxygen.

Considerable adsorption of hydrogen (~5%) occurs on ZnO before irradiation, as is to be expected on an n-type oxide. Radiation produces a further adsorption with a low initial G_{-H_2} of ~0·1 and the rate of this adsorption again drops only slowly with total dose.

If a sample of untreated ZnO which does not possess significant resonances at either g = 2·00 or 1·96 is simultaneously irradiated with protons and evacuated, the spectrum of Fig. 6(a) is developed rapidly and saturates at low dose. The only significant feature is the donor signal at g = 1·96, which probably arises as a result of radiation-induced desorption of oxygen rather than displacement damage in the bulk.

Paramagnetic defects are formed by γ-irradiation of ThO_2, but the concentration and type of defect depend on the temperature of evacuation during preparation [Figs. 6(b) and (c)]. The signals are unaffected by exposure to air and appear to be due to bulk defects, possibly impurities, which are eliminated by high temperature evacuation. Irradiation with protons in the energy range 10–35 MeV does not lead to any further increase in signal intensity so that it would appear unlikely that these signals are connected with the F centre previously claimed[17] in ThO_2. Thermal annealing at 360°C does not completely destroy the present signals, so that relatively deep (>1 eV) traps are involved.

Before irradiation of ThO_2 there is no significant adsorption of oxygen but during irradiation with 27 MeV protons an adsorption occurs with an initial G_{-O_2} of 0·4. G_{-O_2} rapidly decreases with total dose and the adsorption saturates at a coverage of ~1·5%. This adsorption is similar to that on MgO, but the

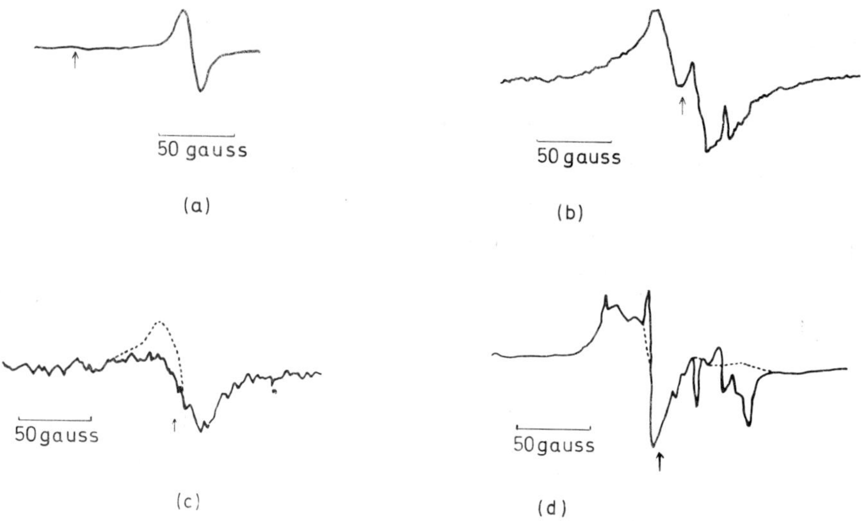

Fig. 6. E.s.r. spectra of irradiated oxides
(a) Spectrum of ZnO *after proton irradiation in vacuo*
(b) Spectrum of ThO_2 *(outgassed at 750°C) after proton irradiation in* He
(c) Spectrum of ThO_2 *(outgassed at 950°C) after proton irradiation in* He *(full line), and also after irradiation in hydrogen (dotted line).*
(d) Spectrum of ZrO_2 *after γ-irradiation in* He *(dotted line) and after proton irradiation in* He *(full line)*
N.B. The field increases from left to right and the arrow denotes the free electron position. All e.s.r. spectra were obtained at 20°C.

adsorption of hydrogen on thoria follows a different pattern with a low initial G_{-H_2} of 0·1 which is followed by a slow continuous process, suggestive of a diffusion mechanism. After irradiation in hydrogen an additional paramagnetic defect is present [Fig. 6(c)], which is likely to be due to a surface centre since it disappears immediately on exposure to air.

Paramagnetic signals are also present in zirconia powder after γ-irradiation. These are complex [Fig. 6(d)] and again probably arise from bulk defects, since they are unaffected by exposure to air. Proton irradiation in vacuo leads to only minor increases in intensity in certain areas of the spectrum, but the e.s.r. signals all anneal out at 20°C and correspond to shallow (<1eV) traps which can readily be repopulated by γ-irradiation.

The adsorption of oxygen induced by proton irradiation is similar to that on MgO and saturates at a coverage of 3%. If ethylene is then adsorbed, the subsequent adsorption of oxygen is greatly enhanced, and coverages of >30% have been obtained after further proton irradiation. Two mechanisms for oxygen adsorption also apply to ZrO_2 as with MgO.

The enhancement in surface reactivity of the oxides ThO_2, ZrO_2, and ZnO by positive ion bombardment shows certain similarities to that in MgO. Ionisation in the bulk of the solid crystallites leads to adsorption due either to electron or hole trapping at the surface or to recombination. A positive role of displacements in promoting adsorption in these oxides has not been proven. In ZnO interstitial and vacancy defects exist before irradiation and are known to be responsible for part, at least, of the adsorption of oxygen and hydrogen. In ThO_2 and ZrO_2 lattice defects, either intrinsic or extrinsic, exist in low concentration but due to the low specific surface area of the material used it has not proved possible to ascertain whether these defects participate in adsorption processes.

Thanks are due to the Harwell V.E.C. staff for positive particle irradiations and to J. W. Hale for the determination of specific surface areas.

REFERENCES

1. Zhabrova, G. M., Vladimirova, V. I., Kadenatsi, B. M., Kazanskii, V. B., and Pariiskii, G. B., *J. Catalysis* **6**, 411, (1966).
2. Nelson, R. L., Tench, A. J., and Harmsworth, B. J., *Trans. Faraday Soc.*, **63**, 1427, (1967).
3. Nelson, R. L. and Tench, A. J., *Trans. Faraday Soc.* **63**, 3039, (1967).
4. Simnad, M. T. and Smoluchowski, R., *J. Chem. Physics* **23**, 1961, (1955).
5. Simnad, M. T., Smoluchowski, R., and Spilners, A., *J. Appl. Physics* **29**, 1630, (1958).
6. Stevens, N. J., Mason, E. A., and Reid, R. C., *J. Physic. Chem.* **66**, 2613, (1962).
7. Barkas, W. H. and Berger, M. J., 'Tables of energy losses and ranges of heavy charged particles', N.A.S.A. SP-3013, (1964).
8. Northcliffe, L. C., in 'Studies in penetration of charged particles in matter', NAS-NRC Publication 1133 (1964) p.173.
9. Nelson, R. L. and Duck, M. J., I.A.E.A. Symposium on Radiation Damage in Reactor Materials, Vienna, Paper SM-120/B-4 (1969).
10. Nelson, R. L., Hale, J. W., Harmsworth, B. J., and Tench, A. J., *Trans. Faraday Soc.* **64**, 2521, (1968).

11 (a). Tench, A. J. and Nelson, R. L., *J. Chem. Physics* **44**, 1714, (1966).
(b). Tench, A. J. and Holroyd, P., *Chem. Comm.* 471, (1968).

12. Nelson, R. L., Hale, J. W., and Harmsworth, B. J., unpublished results.
13. Tench, A. J. and Nelson, R. L., *J. Colloid and Interface Sci.* **26**, 364, (1968).
14. Kolbanovsky, Yu. A., Polack, L. S., Popov, V. T., Shaburkina, V. I., and Shakhrai, V. A., 20th Int. Congress of Pure and Applied Chemistry, I.U.P.A.C., Moscow, Paper B43, (1965).
15. Kohn, H. W., *J. Physic. Chem.* **68**, 3129, (1964).

16 (a). Kokes, R. J., Proc. 3rd Int. Congress on Catalysis, Amsterdam, Paper 125, (1964).
(b). Hahn, D., Nink, R., and Severin, R., *Physik kondens. Materie* **5**, 371, (1966).

17. Neeley, V. I. Gruber, J. B., and Gray, W. J., *Physic. Rev.* **158**, 809, (1967).

DISCUSSION

W. Myers (Ohio State University Hospitals): Did you assess the adsorption that may have resulted from the intense surface ionisation due to positron emission by 110-min ^{18}F produced in the $^{18}O(p, n)^{18}F$ (possibly $^{18}O(d, 2n)^{18}F$?) reaction? Might the surface reactivity also have resulted from 2·05-min ^{15}O formed in the $^{16}O(p, d)^{15}O$ [or $^{16}O(d, t)^{15}O$?] reaction?

R. L. Nelson: No specific experiments were made to assess the importance of positron emission in promoting ionisation processes at the surface. The above transmutations have much lower cross-sections than the normal ionisation processes or even than lattice displacement processes, and a *G* value much greater than the theoretical maximum would be necessary if the suggested positron mechanism were responsible for oxygen adsorption. The similarity of the adsorption process at low proton doses to that on γ-irradiation alone argues that normal bulk ionisation can account for most of the reaction with oxygen.

G. N. Walton (A.E.R.E.): Can you infer anything about the chemical potential of the oxygen vacancies or trapped electrons? Would the potential at the surface be sufficient to decompose water in, for example, steam? How does the potential of an oxygen vacancy in MgO differ from that in ZrO_2?

R. L. Nelson: In irradiated MgO the electron at a bulk oxygen ion vacancy (*F* centre) is in a trap at least 4·9 eV deep, whereas the analogous surface trap (*S* centre) is only 2·5 eV deep. Water adsorption during irradiation has not been studied, but if charge carriers were to recombine at adsorbed water sites the energy released (8·7 eV) would certainly be capable of decomposing water. The products might recombine unless selective desorption occurred. Since ZrO_2 (band gap 5·0 eV) is less ionic than MgO, the corresponding traps would be expected to be more shallow. There is no evidence for an *S* centre in ZrO_2.

W. G. Burns (A.E.R.E.): it may be relevant to Mr. Walton's question that H_2 is produced on dissolving alkali halide crystals containing *F* centres in water.

An apparatus for the proton irradiation of supercritical water in a flow system

W. R. Marsh

Chemistry Division, A.E.R.E., Harwell, Berks., England

ABSTRACT

An apparatus is described for the irradiation of water at pressures up to 400 bar (6000 psi) and temperatures up to 600°C by high-energy protons from the Harwell Variable Energy Cyclotron.

The irradiation vessel is a pressure vessel fabricated from Nimonic 80A alloy, one end of which is thinner than the normal wall thickness and acts as a window for the particle beam. Water is compressed and delivered into the externally-heated irradiation vessel and flows down a pre-heater before entering the irradiation zone. After irradiation the water is cooled before passing to the analysis system.

Irradiation is carried out with ~45 MeV protons, which penetrate the end window and emerge with a mean energy of ~7 MeV. Dosimetry is by charge-input measurement on a beam of measured energy. The beam current is sampled with a beam chopper which interrupts the beam for about 1 s in a 5 s cycle, and the beam energy is measured by introducing into the collimated beam a disc 0·12 in thick of Nimonic 80A (to simulate the vessel window) behind which is a solid state detector connected to appropriate electronics for monitoring the emergent beam energy.

The sampling system allows a measured volume to be delivered into a conventional gas-handling vacuum apparatus with minimal disturbance to flow. Dissolved product gases are measured by gas chromatography. At a steady current, steady concentrations of products, dependent on flow-rate and production *G* value, were obtained.

Under supercritical conditions the low dielectric constant (<10) might be expected to lead to high concentrations of reactive intermediate species near the track centre, resulting in higher molecular yields. The net yields of hydrogen measured with pure water ($G = \sim 5$) in preliminary experiments were higher than at room temperature, but could be suppressed to negligible values by small concentrations of oxygen and ammonia ($<5 \times 10^{-2}$M), suggesting that fast back reactions occur at the high temperatures.

INTRODUCTION

The radiation chemistry of water has been studied extensively at room temperature[1,2] but only to a limited degree at higher temperatures. This paper describes an apparatus designed to study the radiolysis of water under supercritical conditions. The critical constants of water are $T_c = 374°C, P_c = 221{\cdot}2$ bar, $\rho_c = 0{\cdot}32$ g ml^{-1}, and isotherms in the neighbourhood of the critical point are shown in Fig. 1.

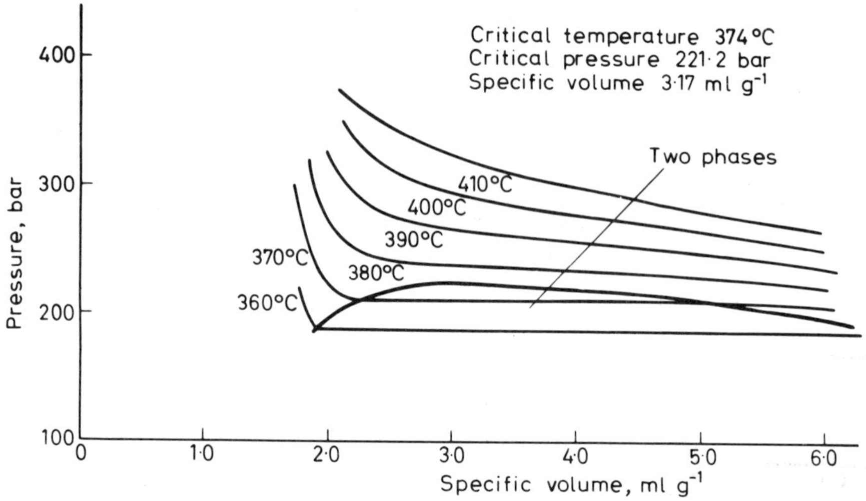

Fig. 1. Isotherms for water in the critical region

Experiments on supercritical water are of technological interest, and should also help in the interpretation of aqueous radiation chemistry by an extension of the conditions of the medium.

Radiolysis of water leads to the formation of the primary species OH, e_{aq}, H_3O^+ and H in the reactions

$$H_2O \rightsquigarrow H_2O^+ + e^- \tag{1}$$

$$H_2O^+ + H_2O \longrightarrow H_3O^+ + OH \tag{2}$$

followed by solvation to $H_3O^+_{aq}$. The spatial distribution of $H_3O^+_{aq}$ and OH is essentially the same, whereas the electrons ejected in the ionization process, which are mainly of low energy, become thermalised and solvated to e^-_{aq}, after travelling a few nm, and have a wider spatial distribution.

In addition to these major processes others occur such as dissociation of excited molecules formed by electron impact

$$H_2O \rightsquigarrow H_2O^* \tag{3}$$

$$H_2O^* \longrightarrow H + OH \tag{4}$$

These primary reactive species which are distributed inhomogeneously in small zones of high concentration then react with each other to give the molecular products H_2 and H_2O_2

$$OH + OH \longrightarrow H_2O_2 \tag{5}$$

$$e^-_{aq} + e^-_{aq} \longrightarrow H_2 + 2\,OH^-_{aq} \tag{6}$$

$$H + H \longrightarrow H_2 \tag{7}$$

$$e^-_{aq} + H \longrightarrow H_2 + OH^-_{aq} \tag{8}$$

In addition the reaction $e^-_{aq} + H_3O^+_{aq} \longrightarrow H + H_2O$ (9)

provides another source of H atoms.

At the same time back-reactions also take place

$$H + OH \longrightarrow H_2O \tag{10}$$

$$OH^-_{aq} + H_3O^+_{aq} \longrightarrow 2H_2O \tag{11}$$

$$e^-_{aq} + OH \longrightarrow OH^-_{aq} \tag{12}$$

$$\left.\begin{aligned} &e^-_{aq} + H_2O_2 \longrightarrow OH^-_{aq} + OH && (13)\\ &H + H_2O_2 \longrightarrow H_2O + OH && (14)\\ &OH + H_2 \longrightarrow H_2O + H && (15)\end{aligned}\right\}\ \text{chain back-reaction}$$

and it is the competition between all these reactions, as well as diffusion away from the zones of high concentration followed by reaction with solutes present, which determine the final composition of the system. Hence any change in conditions, e.g. temperature and dielectric constant, which could alter the spatial distributions, rate constants and diffusion coefficients could lead to changes in the final product composition. In water the dielectric constant varies from 81 at 25°C to 9·7 at 370°C and 3·9 at 600°C; consequently under supercritical conditions the solvated electron, e^-_{aq}, would be under a higher electrostatic field force from positive ions and would not be spread out over as wide a radius as at room temperature, and this might result in a greater concentration of H_2O^* or H_3O^* being formed in the reactions

$$H_2O^+ + e^- \longrightarrow H_2O^* \longrightarrow H + OH \tag{16}$$

or

$$H_3O^+ + e^- \longrightarrow H_3O^* \longrightarrow H + H + OH \tag{17}$$

This higher concentration of H_2O^* near the track centre could lead to greater overall decomposition by 17 and in the reaction

$$H_2O^* + H_2O^* \longrightarrow H_2 + H_2O_2 \tag{18}$$

Compared with room temperature, the overall effects may therefore be

(1) $G(e^-_{aq})$ reduced,[3]

(2) greater overall decomposition,

(3) less difference between $G(H + e^-_{aq})$ and $G(OH)$ and therefore similarly between $G(H_2)$ and $G(H_2O_2)$.

In addition, at a lower density the radical yield will tend to be higher, since the initial track or spur radii depend on the range of low-energy secondary electrons, which will be greater, and therefore the molecular yields will be lower. Another important effect is that with no scavenger present the measured molecular product yields may be smaller due to increases in the rates of the chain back-reactions; the rate constant of reaction 14 is comparatively low at room temperature and is therefore likely to increase with temperature.[2]

In addition to providing this basic information, the work is of technological interest in view of the possibility of using supercritical water as a reactor coolant.

In order to carry out heavy particle irradiations on a medium under these severe conditions it is necessary to employ a pressure vessel as the irradiation vessel, part of which must be thin enough to serve as a window to allow the passage of a particle beam. A particle beam is used for accurate dosimetry and the Variable Energy Cyclotron is well-suited for this purpose, since it can provide a beam of sufficient energy (protons of energy greater than ~42 MeV are necessary to penetrate the end window); in addition the energy may be varied in order to cover the range of neutron energies present in a reactor.

EXPERIMENTAL

The experiments are carried out in a once-through flow system. Water is compressed to high pressure and delivered to the irradiation vessel which is located in a furnace mounted on the end of a beam line. The water is pre-heated before entering the irradiation zone, and after irradiation it passes through a cooler and then to the analysis system.

The apparatus is shown in Fig. 2; all components are made from stainless steel with the exception of the glass reservoir for the compressor, and the pressure vessel, which is of Nimonic 80A alloy. Low-pressure water is fed to the compressor—a Metripump metering pump, or a Madan Air Hydro-Pump—and compressed to high pressure. The pump outlet leads directly to a manifold to which are attached (1) a pressure-relief valve, which is adjustable and set to control the pressure in the system, excess pressure being blown off through the valve and water at low pressure returned to the reservoir; (2) a low-pressure switch, which is a safety device, pre-set to operate at a pressure lower than the control pressure in the system and to shut down the pump in the event of any failure in the high pressure system; (3) a pressure transducer connected to a millivolt recorder to provide a record of the pressure, and (4) a 0–690 bar (0–10 000 psi) Bourdon gauge. High-pressure water is pumped through stainless steel capillary tubing (about 15 m) to the irradiation vessel inside the furnace, after which it passes through a cooler before returning via capillary tubing to the analysis system. Here the pressure is reduced to about 10 bar (150 psi) by means of a Hymatic gas-operated reducing valve before entering the analysis system, which is contained between this valve and the pressure-relief valve set at 7 bar (100 psi) Water flows through the sampling loop via an 8-port Loenco valve

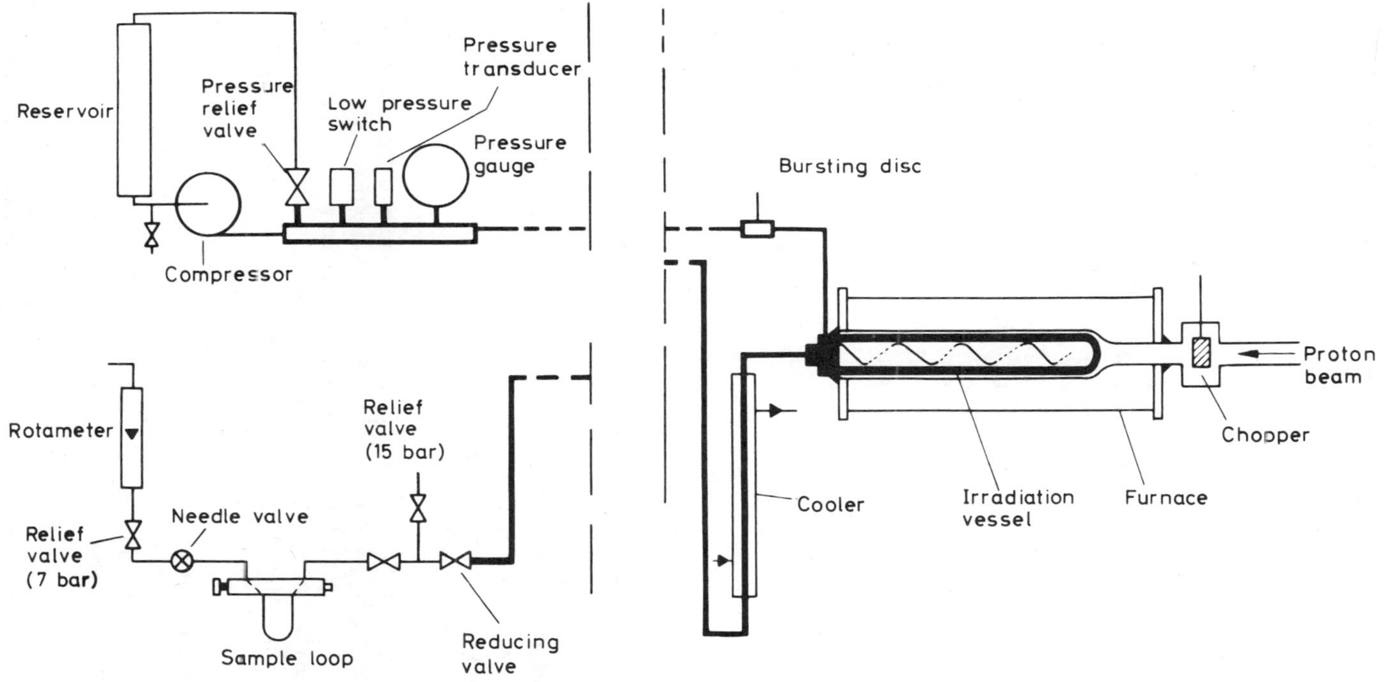

Fig. 2. Flow diagram

through a relief valve and then through a rotameter to waste. The flow-rate through the system is controlled by the 'Circle-Seal' micro-calibrated needle valve.

A bursting disc (420 bar, 6000 psi), is built into the high-pressure system and another pressure relief valve set to about 15 bar (220 psi) protects the analysis system.

The Irradiation Vessel. This is fabricated from a special ingot of Nimonic 80A alloy and is subjected to careful heat treatment after the final machining in order to improve its high-pressure performance. Water enters the vessel (Fig. 3) then passes down an outer helical pre-heater section before arriving at the irradiation zone. The return flow is via a central hole through the pre-heater section which also contains a high-grade chromel–alumel thermocouple, clad in stainless steel. The end of the vessel, which is also the window, is much thinner, 3·05 mm (0·120 in), compared with the wall thickness of 9·53 mm (0·375 in) and is accurately machined in order to give a uniform thickness of metal parallel to the axis over the centre 6 mm. diam., so that all particles in the collimated proton beam incident on the window penetrate the same thickness of material. Because of the special nature of the vessel and the uncertainty involved in calculations of the vessel dimensions, particularly the end window, a similar vessel was hydraulically tested to destruction and withstood a pressure of 5724 bar (83 000 psi) before failing. This pressure agreed with the calculated value and so justified the assumptions made during calculations. Another feature of the design is the high-pressure seal, which in addition to sealing the vessel also directs the flow of inlet water down the helical pre-heater; many variations of design and material were tried before finally adopting a gold-plated nickel gasket. The vessel is housed inside a vacuum furnace located on the end of a beam line.

Beam Line Assembly. The furnace is constructed with three separate heating zones, each with its own power supply and controller. The two outer zones are controlled by Ether Transitrol Temperature Controllers and the centre zone by a Leeds-Northrop Proportional Controller, which is activated by the thermocouple in the flowing water inside the vessel. A further controller gives protection against overheating by switching off the power to the furnace in the event of a temperature excursion above a pre-set value.

Protons from the collimator pass into the housing (Fig. 4) which contains a beam stopper, an absorber slide, and a semi-conductor detector assembly. The beam stopper, which is a carbon disc sufficiently thick to completely stop the highest energy protons, is also used for beam current monitoring during the preliminary setting-up procedure. The absorber slide has two nickel discs mounted on it which may be used to degrade the energy of the beam. The semi-conductor detector is mounted behind a disc of Nimonic 80A alloy machined from the same ingot used for the pressure vessel and of the same thickness, 3·05 mm (0·120 in), as the end window of the vessel. This assembly is used to determine the energy spectrum of the proton beam after passage through the alloy, which is a necessary preliminary to each experiment.

Following this section the beam enters the 'chopper' housing immediately in front of the pressure vessel. Since the vessel cannot be insulated electrically from its surroundings the usual procedure of beam current monitoring by charge-input measurement is not possible, and so the beam current on an insulated target is regularly monitored. An aluminium beam stopper (in the form of a Faraday cup) is attached to a compressed air valve operating on a 5 s cycle. The stopper is driven into the beam for approximately 1 s then driven out and

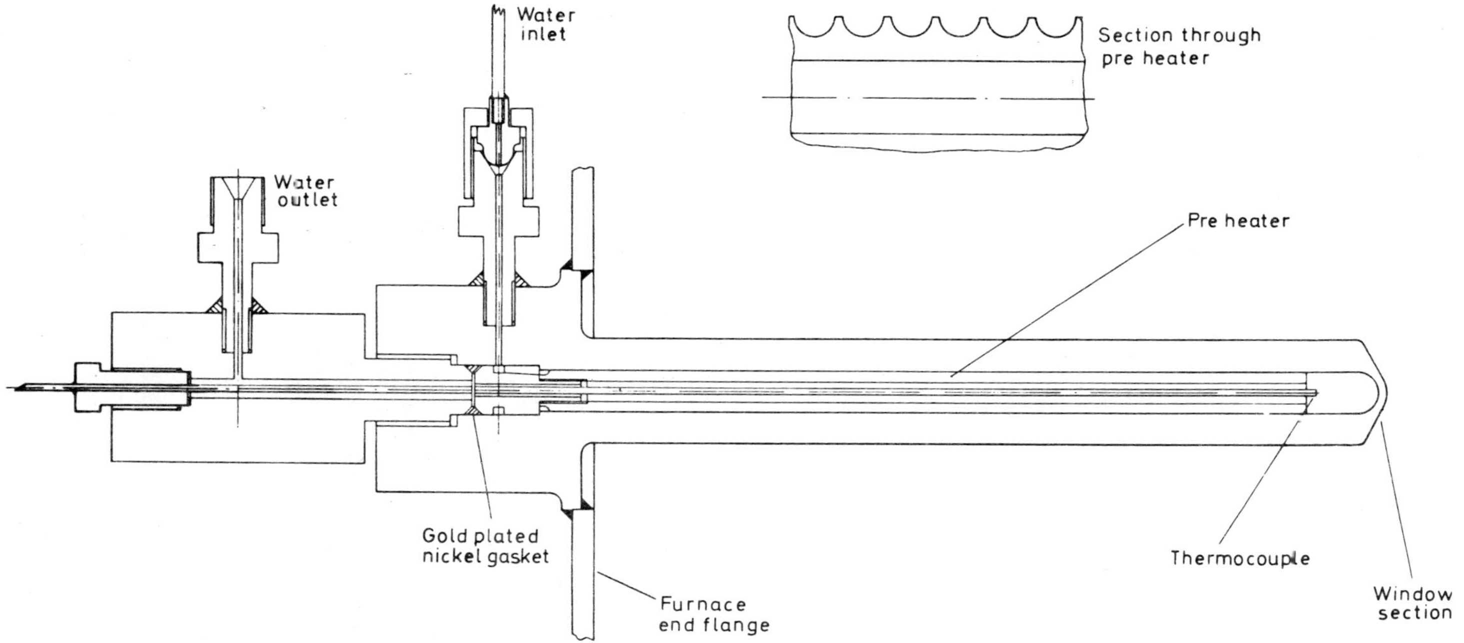

Fig. 3. Irradiation vessel

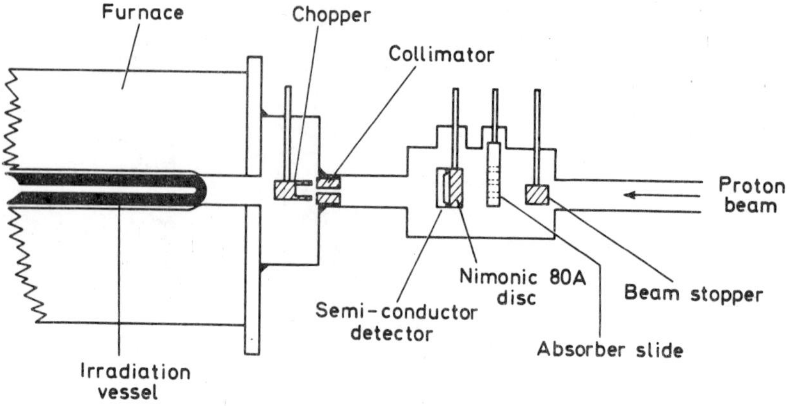

Fig. 4. Beam line assembly

held out for the remainder of the cycle. A separate experiment was carried out using an insulated target of the same cross-sectional area as the cell window in place of the irradiation vessel and a calibration factor determined by monitoring this target and the 'chopper' simultaneously.

Gas Analysis. This is carried out by periodically discharging the contents of the sample loop into an evacuated trap cooled to –78°C. Here the water is frozen out and the non-condensible gases pumped away by a Toepler Pump. After collection, the gases are transferred to the sample loop of a gas chromatograph and analysed. A Perkin-Elmer F11 Gas Chromatograph with hot wire detector is used with various combinations of carrier gas and separation column depending on the components to be analysed. For the analysis of hydrogen, argon or nitrogen is used as carrier gas, with the separating column a 2 m length of 6 mm bore stainless steel tubing packed with 30–60 mesh 5A Molecular Sieve. The same column is used for the separation of oxygen and nitrogen, but with helium as carrier gas. For the analysis of carbon dioxide a 2 m stainless steel column of 6 mm bore, packed with 80–120 mesh Porapak 'Q', is used with helium as carrier gas. Calibration of the chromatographs is carried out by injecting known volumes of the pure gases.

Alternative methods for measuring dissolved gases are being examined, including a semi-continuous method, whereby the dissolved gases are stripped continuously into an inert gas which is subsequently analysed by gas chromatography.

RESULTS AND DISCUSSION

Dosimetry. The method involves measurement of the charge deposited by a beam of known energy. The charge deposited is calculated from the charge measured on the 'chopper' by applying the calibration factor. The energy of the transmitted beam is obtained using the semi-conductor detector mounted behind the Nimonic 80A disc which simulates the window of the irradiation vessel, and which is previously compared with another detector mounted inside the vessel.

With constant energy input at constant flow rate a steady concentration of products should result.

Fig. 5 shows a typical spectrum of particles entering the liquid after passing

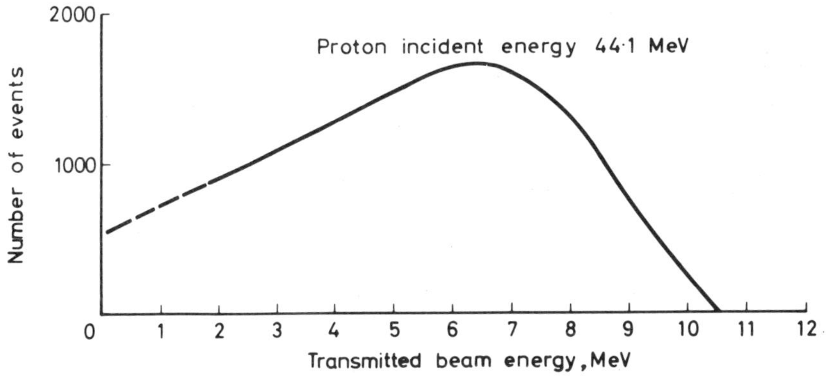

Fig. 5. Energy spectrum of emergent beam

through the vessel window, the area under the curve giving the total number of particles.

The yield of a product formed by irradiating with a non-uniform beam, i.e. a beam having an energy spread, can be calculated as follows, if the G value is known as a function of energy.

Let the number of particles entering the liquid be n, where $n = n_1 + n_2 + --- + n_n$; n_1, n_2, etc. are the number of particles of differing energies $E_1, E_2, --- E_n$, in MeV, each of which gives rise to its own product yield (G value), $g_1, g_2, --- g_n$.

If M is the total number of product molecules formed

$$M = (n_1 g_1 E_1 + --- + n_n g_n E_n) \times 10^4$$

now let

$$f_1 = \frac{n_1}{n}, f_2 = \frac{n_2}{n}, \text{etc.}$$

then

$$M = n(f_1 g_1 E_1 + f_2 g_2 E_2 + --- + f_n g_n E_n) \times 10^4$$

Since values of f are obtainable from Fig. 4, a dosimetric system in which g is known as a function of energy can be used as a check on the charge measurement, since

$$n = \frac{M}{(\Sigma f_i g_i E_i)} \times 10^4 \qquad \text{(A)}$$

If we now define an effective mean energy $\bar{E}$, where $\bar{E} = \Sigma f_i E_i$, we will have its corresponding G value $\bar{G}$ given by

$$M = n(\Sigma f_i g_i E_i) \times 10^4 = n \bar{G} \bar{E} \times 10^4$$

i.e.

$$\bar{G} = \frac{(\Sigma f_i g_i E_i)}{\bar{E}}$$

and therefore

$$n = \frac{M}{\bar{G}\bar{E}} \times \frac{1}{10^4} \qquad \text{(B)}$$

With flow conditions steady at F ml min^{-1}, a constant current A should give rise to a constant concentration of product molecules M ml^{-1}. If we consider Eqn (B) to give the number of particles, n, delivered to 1 ml, then since the number of particles = current (a) × time (s) × 6·28 × 10^{18} we have

$$n = \frac{A}{F} \times 60 \times 6{\cdot}28 \times 10^{18} \qquad \text{(C)}$$

and combining Eqns (B) and (C) we have

$$\bar{G} = \frac{M}{E} \times \frac{1}{10^4} \times \frac{F}{A} \times \frac{1}{60} \times \frac{1}{6{\cdot}28 \times 10^{18}}$$

so giving effective G values of the products. A chemical dosimeter has been used to test the method of dosimetry. The yield of carbon dioxide from an oxygenated solution of 10^{-2}M formic acid in 10^{-3}N sulphuric acid has previously been determined as a function of proton energy.[4] Experiments were therefore carried out using this dosimeter with 44 MeV protons incident on the cell, and the concentration of carbon dioxide was measured under constant energy input and flowrate. The data of Anderson and Hart[4] was fitted into Eqn (A), and found

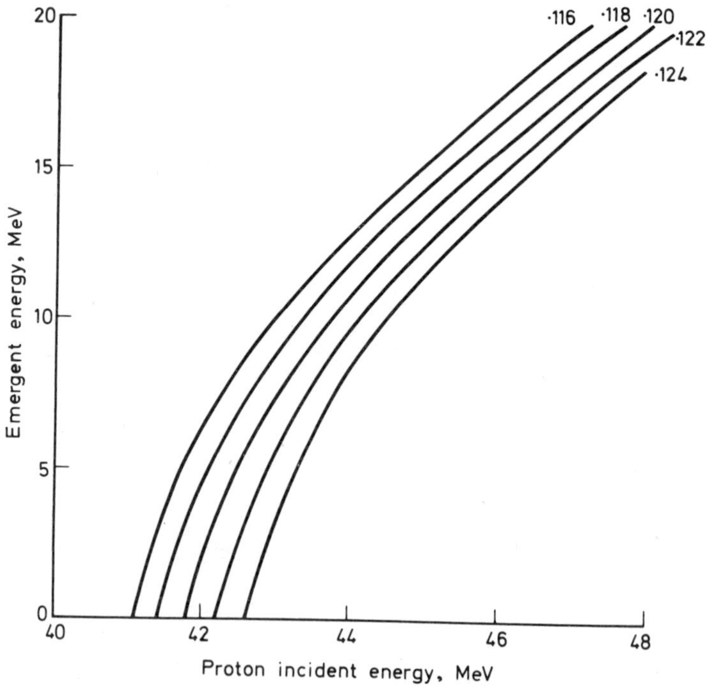

Fig. 6. Emergent Energy versus Incident Proton Energy

to be a factor of 4 higher than results based on charge input measurements using Eqn (C).

The possibility that the measured current is too low is discounted, since an experiment carried out with the final collimator disc polarised a few hundred volts positive (to remove secondary electrons) failed to give any increase in current. Similarly, the neutron and X-ray fluxes generated in the window material by the high-energy proton beam are too small to result in a significant increase in energy input to the solution.

A correction is necessary for the different emergent energy distributions due to the absorber disc and to the pressure vessel. Measurements so far made at lower energies, necessitated by the small depletion depth (500 μm) of available detectors which were small enough to fit inside the vessel, indicate that the mean energy inside the vessel may be up to 20% greater than that obtainable from Fig. 5. A thicker detector for the internal measurements is being obtained to extend these measurements, and the thickness of the absorber will be reduced to give more exact correspondence between the two distributions. Fig. 6 shows the relationship between incident and emergent energies for various thicknesses of Nimonic 80A alloy calculated from range-energy data,[5] and it is apparent that small differences in thickness of the alloy can result in large variations in the emergent energy, especially in the lower energy range. The graph also shows incidentally that it is necessary to measure the energy spectrum for each experiment since the beam energy can only be specified to ±½%.

The high measured value of $G(CO_2)$ with formic acid solutions does not appear to be due to large physical effects, or to large errors in physical measurements, and so we need to consider the chemistry of the system, which may be represented as follows:[4]

$$H_2O \rightsquigarrow H + OH \qquad (1)$$

$$H_2O \rightsquigarrow \tfrac{1}{2}H_2 + H_2O_2 \qquad (2)$$

$$OH + HCOOH \longrightarrow HCOO + H_2O \qquad (3)$$

$$H + HCOOH \longrightarrow HCOO + H_2 \qquad (4)$$

$$HCOO + O_2 \longrightarrow CO_2 + HO_2 \qquad (5)$$

$$H + O_2 \longrightarrow HO_2 \qquad (6)$$

$$HO_2 + HO_2 \longrightarrow H_2O_2 + O_2 \qquad (7)$$

In oxygenated solution $G(OH) = G(CO_2)$. However, in the presence of, for example, ferrous ions the following reactions can occur[1]:

$$Fe^{2+} + HO_2 + H^+ \longrightarrow Fe^{3+} + H_2O_2 \qquad (8)$$

$$H_2O_2 + Fe^{2+} \longrightarrow Fe^{3+} + OH^- + OH \qquad (9)$$

Equations (8), (9), (3) and (5) constitute a chain process which could lead to oxidation of large quantities of formic acid with subsequent production of carbon dioxide. In the experiments described, the presence of a low concentration of Fe^{2+} or Cr^{2+} or other oxidisable cation might exist, with the possibility of a higher yield of carbon dioxide. In order to test this, a standard formic acid solution was prepared, and irradiated in a 1000 Ci ^{60}Co γ-ray source using the syringe method,[4] Some of the solution was also passed through the high-pressure system under conditions similar to those used for the cyclotron irradiations, before being irradiated in the cobalt source. An increased yield by a factor of about 2 was obtained in the latter case, and it seems that this dosimeter is not suitable for use in metal systems.

Other chemical dosimeters are therefore being considered, e.g. ethylene, nitrous oxide, cyclohexane, and benzene. The hydrogen yield from ethylene and its variation with dose and L.E.T. has been studied,[6] although only at low pressure (15–45 cm Hg), whereas nitrous oxide has been studied at high pressure[7,8] but using γ-radiation. The hydrogen yields from benzene[9] or cyclohexane[10] and their variation with L.E.T. are known; these systems have the added attraction of a high density ($\sim$0·9 g ml^{-1}), although problems could arise due to inability to stir the liquid during irradiation.

In addition to the dosimetry, a few experiments have been carried out on other water systems, mainly with the aim of gaining experience under super-critical conditions. The experiments were carried out using helium-saturated water at 450°C and 275 bar (4000 psi) at a flow rate of 9 ml min^{-1}, with 46 MeV protons incident on the vessel. The measured yield of hydrogen was $G(H_2) \sim 5$. A similar experiment, using a 5 x 10^{-3} M solution of ammonium hydroxide saturated with oxygen, gave a much lower yield of hydrogen which showed little increase above the blank.

ACKNOWLEDGEMENTS

Are due to Mr. K. A. Major and Mr. R. L. Stevenson for the design of the pressure vessel, and to Dr. A. R. Anderson and Mr. J. V. Best for the design of the flow circuit. I also wish to thank Mr. R. A. Barton for practical assistance, and Dr. C. B. Amphlett and Dr. W. G. Burns for encouragement and advice.

REFERENCES

1. Allen, A. O., 'The radiation chemistry of water and aqueous solutions', D. Van Nostrand Company Inc., New York, (1961).
2. Kupperman, A., 'Radiation Research', Ed. Silini, p. 212, North Holland Publishing Company, Amsterdam, Holland, (1967).
3. Freeman, G. R. and Fayadh, J. M., *J. Phys. Chem.* **43**, 86, (1965).
4. Anderson, A. R. and Hart, E. J., *Radiation Research* **14**, 689, (1961).
5. Williamson, C. F., Boujot, J., and Picard, J., 'Tables of Range and Stopping Power of Chemical Elements for Charged Particles of Energy 0·5 to 500 MeV', Report C.E.A. 3042, (1966).
6. Anderson, A. R. and Best, J. V. F., *Nature* **216**, 576, (1967).
7. Sears, J. T., *J. Phys. Chem.* 73, 1143, (1969).
8. Burns, W. G. and Marsh, W. R., *Trans. Farad. Soc.* **65**, 1827, (1969).
9. Burns, W. G. and Marsh, W. R., *Trans. Farad. Soc.* **64**, 2375, (1968).
10. Burns, W. G., Marsh, W. R., and Reed, C. R. V., *Nature* **218**, 867, (1968).

DISCUSSION

G. N. Walton (A.E.R.E.): Have you had any difficulties in controlling the supercritical condition in your experiments? At the critical point the density of water changes very rapidly with temperature, and the heat exchange at a heater surface also changes rapidly. There seems to be a short distance between the heater and the irradiation volume, and control difficulties would be reflected in instability and drifting temperature; have you experienced this?

W. R. Marsh: The water temperature is continuously monitored by the thermocouple in the flowing stream; when the control temperature is attained the temperature of the system remains steady provided that the flow rate remains constant. Operating the sample valve changes the temperature because the water flow is temporarily stopped, but the system soon recovers; sufficient time is allowed for the system to stabilise before sampling. The type of pump used causes a small cyclic variation in pressure, but we have not had sufficient experience of the system to know whether this is important.

SESSION 3

Isotope Production

Chairman: Dr. H. A. C. McKay

Initial expcrience with a 30-in isochronous cyclotron for medical use

R. S. Tilbury, J. P. Mamacos, and J. S. Laughlin

Division of Biophysics, Sloan-Kettering Institute for Cancer Research, 425 E. 68th Street, New York, N.Y. 10021

ABSTRACT

Several methods have been developed for the production of radionuclides using a 30 in diam. isochronous cyclotron at the Sloan-Kettering Institute for Cancer Research. The radionuclides are converted into radiochemicals or radiopharmaceuticals for use in research and diagnosis of cancer. Alternative reactions have been investigated for the production of ^{18}F, ^{123}I, ^{121}I, ^{52}Fe, ^{11}C and other radionuclides. Methods of production and chemical processing of radioactive gases, O_2, N_2, CO, and CO_2 labelled with ^{15}O, ^{13}N, and ^{11}C, have been developed.

Target chambers for the irradiation of gases, solids, and liquids have been designed and constructed. The liquid target chamber used for ^{18}F production contains 10 ml of water, is constructed mainly of titanium, and provides a catalyst chamber for the recombination of hydrogen and oxygen radiolytic gases. The solid target chamber used for the production of ^{52}Fe, ^{48}Cr and the irradiation of any solid material in vacuo is water cooled to withstand $> 50\,\mu A$ beam current. The gas target chambers are designed for the irradiation of gases or gaseous mixtures, or for the irradiation of solids (e.g. charcoal, boric oxide, tellurium metal) in a gas stream. An internal target probe has been designed and constructed with efficient water cooling for the irradiation of solids inside the vacuum chamber of the cyclotron, to take advantage of the higher beam currents available and of the variation of particle energy with radius.

Yields to date with the radioactive gases are:

$^{15}O_2$ 3 mCi/l, $C^{15}O_2$ 1 mCi/l, ^{11}CO 0·1 mCi/l, $^{11}CO_2$ 0·1 mCi/l, and $^{13}N_2$ 1 mCi/l.

Nitrogen-13 gas has been used clinically in conjunction with our scintillation camera system. The yield of ^{52}Fe as ferrous citrate has been 1 μCi/μAh and we have produced 200 μCi batches. Larger amounts will be possible when the internal target probe is in operation. ^{18}F is produced twice weekly in 100 mCi amounts using 40 min irradiations at 25 μA; the maximum yield is 600 mCi in a two-hour irradiation. Radionuclide impurities detected have been ^{48}V and ^{7}Be

in the p.p.m. range. No tritium (<1 p.p.m.) was detected. Methods have been developed which ensure pyrogen-free and sterile solutions for direct administration to patients.

INTRODUCTION

A small versatile cyclotron[1] installed in the sub-basement of the Sloan-Kettering Institute became operational in December, 1967. Since that time methods have been developed for the production of some eight radioisotopes and about twelve radiochemical or radiopharmaceutical preparations. The cyclotron itself is a 30 in isochronous machine, model CS-15, built by the Cyclotron Corporation. It will accelerate deuterons to 8 MeV, protons or α-particles to 15 MeV, and ^{3}He particles to 23 MeV. Internal beam currents in excess of 100 μA and external beam currents in excess of 50 μA are routinely available. The layout of the cyclotron facility is shown in Fig. 1. High density concrete shielding

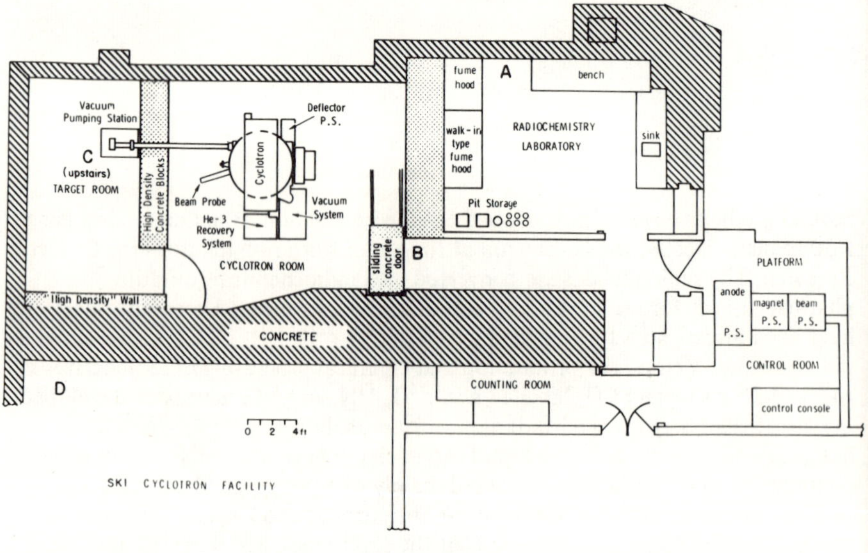

Fig. 1. Plan of the S.K.I. cyclotron laboratory. There is also 4½ ft of high-density concrete in the ceiling. The laboratory is located two floors below street level and has a rock floor

(thickness 4–5 ft) is placed on all sides of the cyclotron and in the ceiling. Targets may be irradiated either in the external position at the end of a 12 ft beam pipe located in the target toom, or internally in the position labelled 'beam probe', in order to take advantage of the higher beam currents available internally and also of the variation of particle energy with radius.

EXPERIMENTAL AND RESULTS

Target chambers have been designed and constructed for the irradiation of liquids, solids and gases, and of solids in the presence of gases, all as external target systems, and an internal target system has been built. The liquid target chamber (Fig. 2) is used mainly for the production of ^{18}F by the irradiation of water. It consists of a titanium target plate with a 3/16 in recess, 2 in diam.,

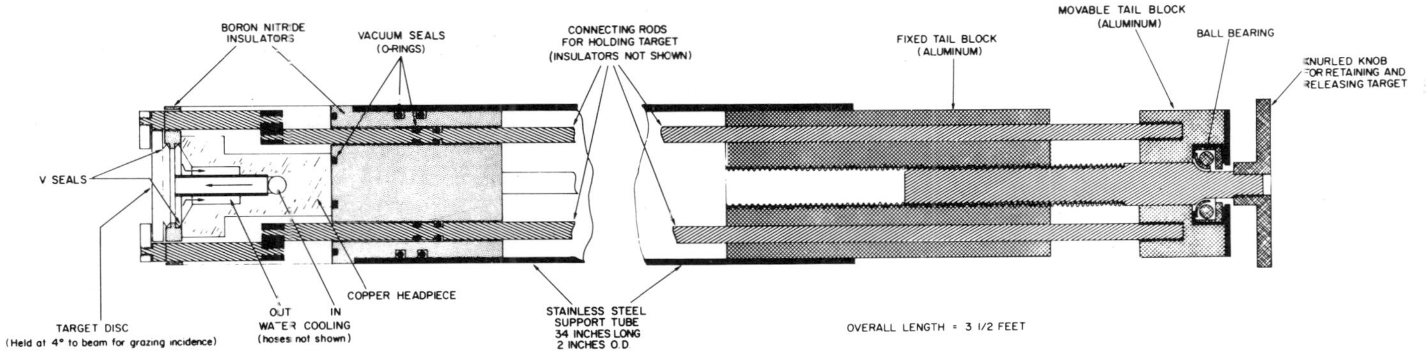

Fig. 4. Cross-section of the internal target probe for the S.K.I. cyclotron. The target is released by rotating the knurled knob. The internal beam strikes the target at grazing incidence in a line perpendicular to the paper

filter with tellurium metal powder sintered onto the beam side of it. Helium carrier gas is passed through the filter and out to the processing system. Both the body of the target chamber and the back face are cooled with water.

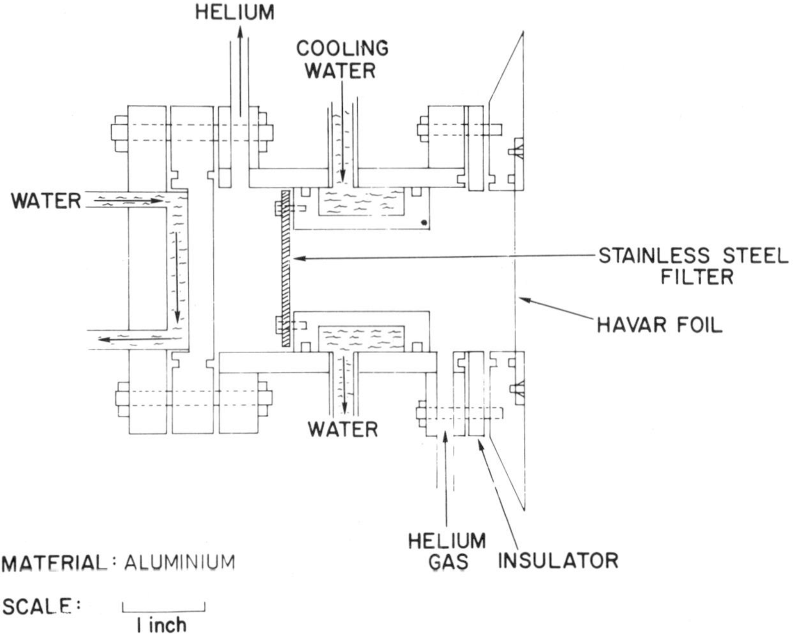

Fig. 5. Cross-section of the target chamber for ^{123}I *and* ^{13}N *production. The target (e.g.* ^{122}Te) *is placed on the stainless steel filter and irradiated from the right*

Havar foil is used to retain the helium gas. For ^{13}N production the stainless steel filter is replaced with a porous graphite disc. Behind the disc are placed ¼ in diam. carbon balls. The deuteron beam strikes the carbon to form ^{13}N, which is swept out of the target chamber by a suitable carrier gas.

^{52}Fe is produced by irradiating natural chromium electroplated onto copper discs. Either the internal or the external target can be used for this irradiation, with 23 MeV ^{3}He particles. The thick target yield is about 1 μCi/μA. About two hours after the irradiation the chromium target is etched with concentrated hydrochloric acid. Etching is accelerated by applying +6 V to the chromium target and passing a current of ~4 A through the solution to a platinum electrode. The etch solution is then centrifuged and the solution passed through an ion-exchange column of Biorad AG 1 X8, 200–400 mesh. The column is 1 cm diam. × ~20 cm long. The column is then washed with 12N HC1 until all the green chromium colour has gone, followed by successive elutions with 25 ml of 9M, 6M, and 3M HC1. Ferric chloride is then eluted from the column with 25 ml 1N HC1. The ferric chloride solution is then reduced with ascorbic acid solution and made isotonic by adding 3% sodium citrate solution. The final 10 ml of ferrous citrate solution is sterile filtered into a multidose vial and is ready for injection.

^{48}Cr has been prepared by irradiating titanium with ^{3}He particles; the yield was 10 μCi/μAh. ^{51}Cr is also formed as an impurity, to the extent of ~10% of the ^{48}Cr at the end of irradiation.

Radioactive gases are produced in the system shown schematically in Fig. 6. Starting with one of four primary gases we use one of three different cyclotron

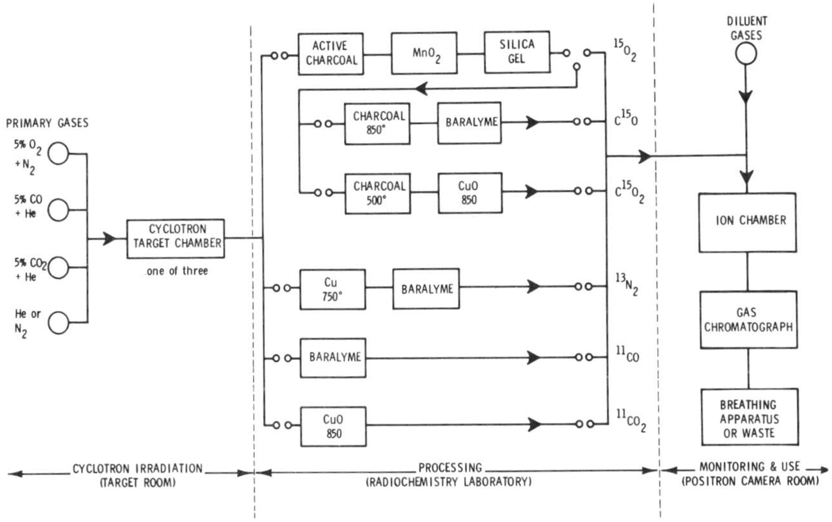

Fig. 6. Schematic diagram of the system for the production of the radioactive gases O_2, CO, CO_2, *or* N_2 *labelled with* ^{15}O, ^{11}C *or* ^{13}N

target chambers, depending on which of the radionuclides ^{15}O, ^{13}N or ^{11}C is desired. The gas is passed through the system for chemical purification and conversion to the required gaseous product and then to an ion chamber to measure its activity. The gases are analysed by gas chromatography. Yields of the radioactive gases have been 3mCi/l for oxygen, 1mCi/l for CO and CO_2 labelled with ^{15}O, 0·1 mCi/l for ^{11}C using deuterons on boric oxide and 1·3 mCi/l for ^{11}C using protons on boric oxide. ^{11}C can be converted to CO or CO_2 as desired. ^{13}N is produced at ~1mCi/l by deuterons on carbon.

CONCLUSIONS

We have developed methods for the production of about eight radionuclides using a small isochronous cyclotron in a medical institution. About 13 radiochemical or radiopharmaceutical products have been produced from these radionuclides. ^{18}F is produced once or twice a week and is used clinically for bone scanning. Ferrous citrate labelled with ^{52}Fe has been used clinically for scanning the erythropoietic system, and the radioactive gases have been used for lung function studies in conjunction with our positron camera. The practicality of using a small cyclotron in a large medical institution has been demonstrated.

ACKNOWLEDGEMENT

This work is supported in part by the National Cancer Institute grant No. CA-08748 and by the U.S. Atomic Energy Commission contract AT(30-1)910.

REFERENCES

1. Laughlin, J. S., Mamacos, J. P., and Tilbury, R. S., 'Isochronous Cyclotron Installation for Radionuclide Production', *Radiology* (in press).
2. Tilbury, R. S., Dahl, J. R., Mamacos, J. P., and Laughlin, J. S., 'Fluorine-18 Production for Medical Use by Helium-3 Bombardment of Water', submitted for publication to the *Int. J. App. Rad. and Isotopes,* 1969.

DISCUSSION

H. D. Maccabee (M.R.C., Harwell): What is the function of the 'Marman flange' shown in your diagram?

R. S. Tilbury: This couples the target chamber to the end of the beam tube under vacuum. Target chambers can be changed rapidly by using identical flanges. Flanges are coupled by a quick-release V-shaped clamp, with a captive O-ring providing the vacuum seal.

K. V. Ettinger (Birmingham University): What catalyst do you use to recombine hydrogen and oxygen?

R. S. Tilbury: Finely-divided palladium metal supported on alumina pellets (~⅛ in × ⅛ in diam.), from Engelhart Industries (Deoxo Model D.) We have found appreciable quantities of ^{18}F to be adsorbed onto the pellets, and plan to use instead a platinum wire coil heated to red heat.

J. C. Clark (M.R.C., Hammersmith): What are the target length and beam current for $^{15}O_2$ production?

R. S. Tilbury: About 8 in and ~20μA.

M. A. Chaudhri (M.R.C., Hammersmith): I would like to comment on the units in which you have expressed various yields. The quantity mCi/μAh is so sensitive that one cannot use it to compare yields from the Hammersmith and Sloan-Kettering machines unless the beam intensities and other conditions are identical.

R. S. Tilbury: I agree, but it has been used in the past. I hope that a recommendation for a suitable unit will arise from the round-table discussion at this meeting.

^{123}I production for pharmaceutical use

V. J. Sodd, J. W. Blue†, and K. L. Scholz**

ABSTRACT

^{123}I has been suggested as the radioisotope of iodine most suitable for low-exposure *in vivo* measurements.[1] For the past two years the Nuclear Medicine Laboratory of the Bureau of Radiological Health, DHEW, has investigated various nuclear reactions and cyclotron techniques for producing ^{123}I for radiopharmaceutical use.

Both indirect and direct methods of ^{123}I production were evaluated. In the indirect method ^{123}I was produced from the decay of 2·1 h ^{123}Xe, which was produced in a gas-flow target by several nuclear reactions (Blue and Sodd,[2] following paper).

The second method, which is discussed in this paper, includes two reactions which produce ^{123}I directly, viz. ^{121}Sb(^{4}He, 2n) ^{123}I and ^{122}Te(d, n) ^{123}I.

The former reaction was studied at the Lewis Research Center (L.R.C.) 60 in cyclotron by bombarding 98·4% isotopically-enriched ^{121}Sb powder targets with 25·2–36·5 MeV α particles. The ^{123}I yield from ^{121}Sb targets (19–70 mg/cm^2) ranged from ~0·2 to 0·9 mCi/μAh. The ^{124}I, ^{125}I, and ^{126}I contaminants could be limited to 0·5, 0·02, and 0·002% respectively at the end of bombardment by controlling the thickness of the target (~15 mg/0·7 cm^2) and the energy of the α-particle beam.

The second reaction was studied at the L.R.C. machine by bombarding 95·4% isotopically-enriched ^{122}Te powder targets with 6·4–9·0 MeV deuterons. The ^{123}I yield from targets of thicknesses between 18 and 45 mg/cm^2 ranged from 0·002–0·13 mCi/μAh. At a bombarding energy of 6·8 MeV, the maximum levels of ^{124}I, ^{125}I, ^{126}I, ^{130}I, and ^{131}I contaminants were 0·09, 0·02, 0·04, 0·9, and 0·09% respectively.

Also presented are Ge(Li) γ-ray spectra of ^{123}I, and chemical procedures for isolating radiopharmaceutical ^{123}I from the target material and for the recovery of ^{121}Sb.

*Nuclear Medicine Laboratory, Cincinnati General Hospital, D.M.R.E., B.P.H., U.S. Department of Health, Education and Welfare, Cincinnati, Ohio, U.S.A.

†Lewis Research Center, National Aeronautics and Space Administration, Cleveland, Ohio, U.S.A.

INTRODUCTION

Background. This work describes cyclotron techniques for the production of ^{123}I using the $^{121}Sb(^{4}He, 2n)$ ^{123}I and the $^{122}Te(d, n)$ ^{123}I reactions. The production of this isotope has been prompted by its recognition as the most ideal radioisotope of iodine for low-exposure *in vivo* measurements.[1] It has been calculated,[3] for example, that the ^{123}I radiation dose to a patient from a human thyroid uptake test is only a few per cent of that received with the more commonly used ^{125}I and ^{131}I. This reduction results from the fact that ^{123}I emits far less charged-particle radiation than does ^{131}I, and its physical half-life is only 0·9 and 6·9% of that of ^{125}I and ^{131}I, respectively. In addition, Wagner and Emmons,[4] who have introduced formulae to determine the ideal radiopharmaceutical as one with a physical half-life no longer than necessary to ensure maximal activities in the organ studied at the optimal observation time, have found ^{123}I to be 'more ideal' than ^{125}I and ^{131}I. Furthermore, the 159 keV α-ray of ^{123}I approaches an optimum energy that is low enough to be efficiently counted with thin crystals and allow use of high resolution collimators, but high enough to obtain necessary tissue penetration. All these desirable characteristics make this nuclide in demand for nuclear medical applications.

Preliminary clinical results[5,6] of the use of ^{123}I for *in vivo* thyroid uptake measurements suggest that while this radionuclide is advantageous for low-exposure measurements, the ^{124}I contamination (0·5 - 1·0%) prevents full use of its capabilities as a high-resolution scanning agent.

Existing methods. To date, most of the ^{123}I used in nuclear medical applications has been produced by two nuclear reactions. Beaver, Hupf, Eldridge, and Holley[7-9] have produced it at the ORNL 86 in cyclotron by the 15·5 MeV-proton bombardment of isotopically-enriched ^{123}Te according to the reaction $^{123}Te(p, n)$ ^{123}I. Goolden, Glass and Silvester[10] report its production at the Medical Research Council Cyclotron by the ^{4}He bombardment of natural antimony according to the reaction $^{121}Sb(^{4}He, 2n)$ ^{123}I. Both of these methods result in yields of ^{123}I that are adequate for nuclear medical applications but give contamination with other iodine isotopes having half-lives greater than that of ^{123}I.

The purity of ^{123}I produced by the $^{123}Te(p, n)$ ^{123}I method is limited by ^{124}Te, ^{125}Te, ^{126}Te, and ^{130}Te contaminants in the ^{123}Te target. On proton bombardment these target impurities produce ^{124}I, ^{125}I, ^{126}I, and ^{130}I. The highest purity at which ^{123}Te is available without extraordinary and economically impractical enrichment procedures is ~79%. Using targets of this purity, the contamination of ^{123}I with ^{124}I, ^{125}I, and ^{126}I has been reported[9] to be ~0·9, 0·08, and 0·3%. The ^{130}I contamination is not reported.

The purity of the ^{123}I produced by the $^{121}Sb(^{4}He, 2n)$ ^{123}I reaction on natural antimony is limited by two factors. First, the 42·8% ^{123}Sb in natural antimony produces ^{125}I via the $^{123}Sb(^{4}He, 2n)$ ^{125}I reaction. Second, the energy of the α particle and the target thickness must be controlled so that the $(^{4}He, 2n)$ reactions predominate, because $(^{4}He, n)$ reactions on ^{121}Sb and ^{123}Sb produce ^{124}I and ^{126}I respectively, and $(^{4}He, 3n)$ reactions on ^{123}Sb produce ^{124}I.

Silvester, Sugden, and Watson (see ref. 10) were able to control the energy of the ^{4}He beam on natural antimony targets so that the ^{124}I and ^{126}I content was kept down to 0·8 and 0·16% respectively. They also obtained ^{125}I contamination from ^{4}He reactions on the ^{123}Sb in the natural antimony at levels of 1·3%.

Myers and Anger[1] have attempted to reduce the ^{125}I and ^{126}I contamination in ^{123}I by using targets of ^{121}Sb powder enriched to 99·4%. They report only slight contamination of their ^{123}I with other radio-iodines, but do not give detailed results.

Present work. For the past two years, the Nuclear Medicine Laboratory of the U.S. Department of Health, Education, and Welfare has investigated various nuclear reactions and cyclotron techniques for producing ^{123}I for radiopharmaceutical use. Both indirect and direct methods of ^{123}I production were evaluated. In the indirect method pure ^{123}I ($^{124}I < 0{\cdot}005\%$) results from the positron decay of 2·1 h ^{123}Xe, which is produced in a gas-flow target via either the $^{122}Te(^{4}He, 3n)\ ^{123}Xe$, $^{122}Te(^{3}He, 2n)\ ^{123}Xe$, or $^{123}Te(^{3}He, 3n)\ ^{123}Xe$ reaction. The results of these reactions and a description of the gas-flow target are reported by Blue and Sodd.[2]

The direct method of producing ^{123}I described in this paper includes the nuclear reactions $^{121}Sb(^{4}He, 2n)\ ^{123}I$ and $^{122}Te(d, n)\ ^{123}I$. Both reactions produce adequate yields of ^{123}I of sufficient purity to be used clinically for *in vivo* thyroid measurements.

The detailed results of several bombardments of 98·4% isotopically-enriched ^{121}Sb targets with 25·2-36·5 MeV ^{4}He particles are presented. The use of highly-enriched ^{121}Sb targets greatly reduces ^{125}I and ^{126}I contamination. The only contamination problem which remains is the production of ^{124}I via the $^{121}Sb(^{4}He, n)$ reaction. Consideration is given to the proper energy of the ^{4}He beam and to the thickness of the target at which the ^{124}I contamination can be kept to a minimum.

We also present the results of ^{123}I production by bombarding 95·4% isotopically-enriched ^{122}Te with 6·4–9·0 MeV deuterons.

The distillation of ^{123}I from the dissolved target powder and its preparation as an orally-administered radiopharmaceutical are described, as is also the chemical procedure for the recovery of enriched ^{121}Sb.

EXPERIMENTAL

Irradiations. The bombardments were done on the 60 in cyclotron at the Lewis Research Center (L.R.C.) NASA, which accelerates α particles to 42 MeV and deuterons to 20 MeV. Fig. 1 shows the diagram of the target holder used at the L.R.C. Cyclotron. The metallic target powder was packed into a 3/8 in cavity milled into the aluminum target plate and covered with an absorber foil. The area of the cavity was slightly larger than that of the aperture in the carbon collimator, so that only that portion of the holder containing target material was irradiated. The volume of the cavity was just large enough to contain the compacted powder, so that when the plate was covered and in a vertical position the powder distribution remained constant. The absorber foil prevented escape of target material into the cyclotron vacuum system and also attenuated the beam energy to the desired value. The aluminium target holder was bolted to a water-cooled 6 in copper plate, which was bolted onto the end of a cyclotron beam tube. The irradiations were performed in vacuum.

The area and position of the incident beam was determined before each bombardment by televiewing its fluorescence on a zinc sulphide window positioned on the beam tube in place of the target assembly. The beam was adjusted in area to a 0·5 in square with roughing slits so that it completely

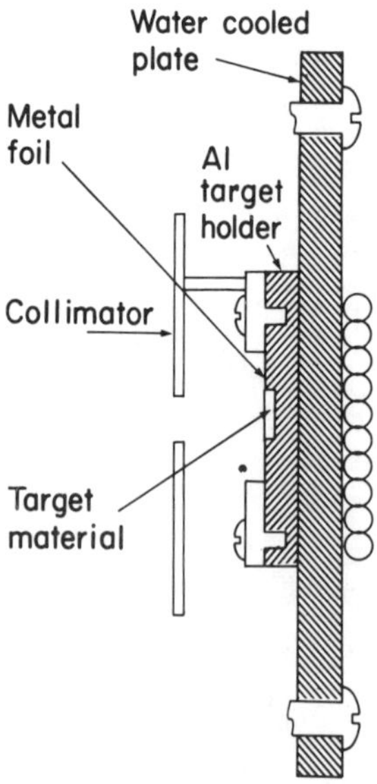

Fig. 1. L.R.C. cyclotron target holder

covered the area of the target collimator. However, after some bombardments, it was obvious from the uneven discoloration of the covering foil and of the condition of the target powder that the target was not hit uniformly.

Counting considerations. The identification and assay of γ-emitting radionuclides employed a 1600-channel pulse-height spectrometer with a 4·2 cm^3 Ge(Li) detector. The resolution (FWHM) of the spectrometer for the 661·6 keV γ-ray of ^{137}Cs was 3·2 keV. The energy was calibrated with ^{57}Co (121·98 keV), ^{137}Cs (661·64 keV) and ^{60}Co (1173·2 and 1332·5 keV). Radioiodine identification was accomplished by both energy and half-life analysis. The counting efficiency of the spectrometer was determined for 1 ml liquid samples in plastic vials centred directly on the detector. Table 1 lists the energies and abundances of γ-rays used in standardisation and assay calculations.

The counting rate of a given radioiodine was determined by computer integration under its respective γ-ray photopeak. The computer integration

Table 1. ENERGIES (E_γ) AND ABUNDANCES* (Ab_γ) OF γ-RAYS OF STANDARD RADIONUCLIDES AND OF RADIOIODINES

Standard	E_γ keV	Ab_γ %	*Radioiodine*	E_γ keV	Ab_γ %
^{125}I	35	7	^{125}I	35	7
^{141}Ce	145	48	^{123}I	159	83
^{139}Ce	165	80	^{131}I	364	82
^{131}I	364	82	^{126}I	388	34
^{134}Cs	605	98	^{130}I	418	35
^{137}Cs–^{137m}Ba	662	85	^{124}I	602	67

* Table of Isotopes, 6th Ed., Lederer, C. M., Hollander, J. M., and Perlman, I., Wiley, New York (1967).

program subtracted the Compton continuum (i.e. a 'valley to valley' background) from each photopeak. This same procedure was used to determine the efficiency of the spectrometer for γ-ray emitting standards.

Separation of ^{123}I. The distillation apparatus used to separate ^{123}I is shown in Fig. 2. The target powder (10–50 mg) was scraped from the target holder and transferred with the aid of ~0·5 ml iodide carrier solution (200 μg I/ml) to a 300 ml Erlenmeyer flask. The standard Erlenmeyer flask was connected through ground glass joints to a 100 ml separating funnel, a 12 in water-cooled condenser and a 300 ml bubbling receiving vessel. The powder was dissolved in 12 ml hot conc. H_2SO_4, and the solution cooled and diluted with 50 ml distilled water. Iodine then was distilled into the receiving vessel containing 25 ml 0·5 N NaOH. The yield of this distillation was determined by γ counting the relative amounts of radioiodine in distillate and residue. The average distillation yield from

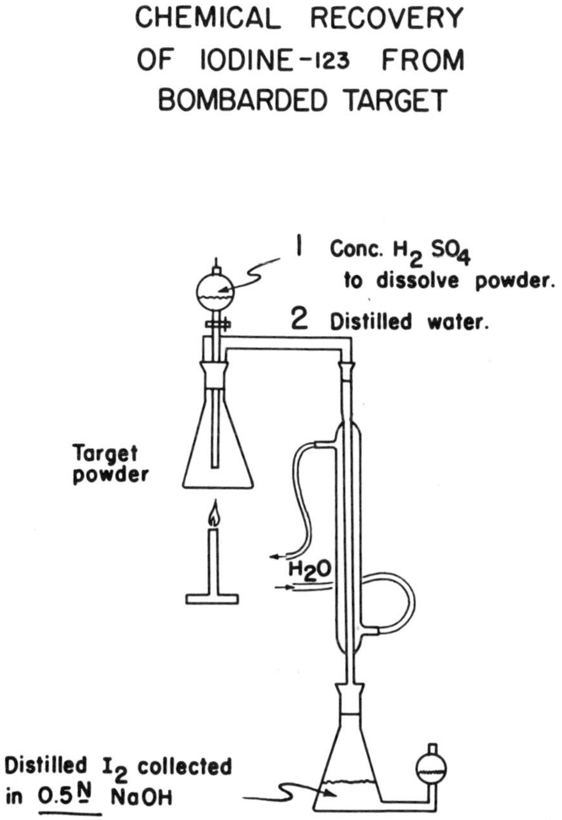

Fig. 2. ^{123}I *distillation apparatus*

fifteen ^{121}Sb targets was 91·2 ± 3·1%. High, but erratic, iodine yields ($\gtrsim$ 75%) were obtained when the same procedure was used without carrier.

Recovery of target material. When the availability and cost of the target materials were considered important factors, the enriched ^{121}Sb and ^{122}Te were recovered from composite residues. ^{122}Te was recovered according to the procedure of Humpf, Eldridge and Beaver[9] while ^{121}Sb was recovered by the following procedure.

The acidic residue from the iodine distillation, containing about 100 mg of ^{121}Sb, was diluted with distilled water, neutralised with NH_4OH, and treated with 3 ml of hydrazine hydrate ($N_2H_4 \cdot H_2O$). The solution was adjusted in acidity to pH 1 with H_2SO_4 and saturated at room temperature with gaseous H_2S. The Sb_2S_3 precipitate was collected by centrifugation and washed with 1 N H_2SO_4. The sulphide was then dissolved in 10 ml of concentrated HC1 and diluted with 10 ml of distilled water, and filtered through a glass-fibre filter. The clear solution was treated with chromous chloride solution to reduce antimony to the metal, which was then collected by centrifugation, washed with ethyl alcohol and dried at 100°C. The chemical yield determined with natural antimony and ^{125}Sb tracer was 94·8 ± 1·5%.

RESULTS

$^{121}Sb(^4He, 2n)^{123}I$: Some typical Ge(Li) spectra of an iodine fraction distilled from an ^{121}Sb target are shown in Fig. 3. Figs. 3(a–d) were obtained 22 h, and 6, 18 and 50 days, respectively after bombarding 98·4% enriched ^{121}Sb with 31 MeV α particles. Figs. 3(b–d) were plotted at factors of 10^1, 10^2, and 10^3

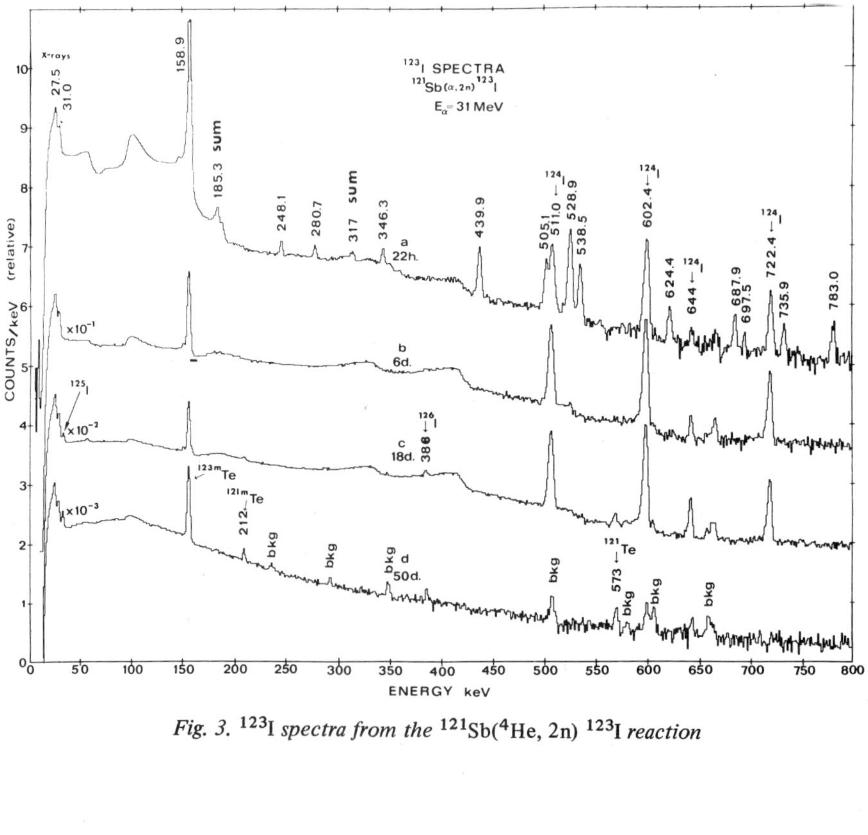

Fig. 3. ^{123}I *spectra from the* $^{121}Sb(^4He, 2n)^{123}I$ *reaction*

respectively, lower in relation to Fig. 3a to aid in visual comparison. No major activity was found above 800 keV.

Identified in Fig. 3(a) are the 27·5 and 31·0 keV X-rays and the 158·9, 248·1, 280·7, 346·3, 439·9, 505·1, 528·9, 538·5, 624·4, 687·9, 697·5, 735·9, and 783·0 keV γ-ray photopeaks of ^{123}I. The photopeaks at 185·3 and 317 keV are random sums of abundant photons. The 185·3 keV peak is a sum of the 158·9 keV γ-ray photon of ^{123}I and the X-ray photon of tellurium; the 317 keV peak is a sum of two 158·9 keV photons. With lower activity sources these sum peaks are not seen. Also identified in Fig. 3a are the 511·0 keV annihilation peak and the 602·4, 644 and 722·4 keV γ-ray photopeaks of ^{124}I. The ^{124}I contamination from the ^{121}Sb(^{4}He, n) ^{124}I reaction was 0·70% at the end of this particular bombardment.

In Fig. 3(b) only 0·06% of the 13·3 h ^{123}I and 37% of the 4·2 d ^{124}I remained in the sample, as evidenced by the persistence of the 158·9 and the 511·0, 602·4, 644, and 722·4 keV photopeaks respectively.

The presence of 60·2 d ^{125}I and 12·8 d ^{126}I was confirmed in Fig. 3(c) by the identification of photopeaks at 35·4 and 388 keV respectively. The ^{125}I and ^{126}I contaminations from the ^{123}Sb(^{4}He, 2n) ^{125}I and ^{123}Sb(^{4}He, n) ^{126}I reactions on the 1·6% ^{123}Sb impurity in the target were $1{\cdot}5 \times 10^{-2}$ and $2{\cdot}6 \times 10^{-3}$%, respectively.

In Fig. 3(d) the presence of 118 d ^{123m}Te, 154 d ^{121m}Te and 17 d ^{121}Te were identified by photopeaks at 158·9 and 212 and 573 keV respectively. ^{123m}Te was produced from the ^{121}Sb(^{4}He, d) ^{123m}Te reaction, and was found mainly in the target residue, although some was carried over into the distillate. ^{121m}Te and ^{121}Te resulted either directly from ^{121}Sb(^{4}He, p3n) ^{121}Te and ^{121m}Te reactions, or indirectly from the decay of ^{121}I produced via ^{121}Sb(^{4}He, 4n) ^{121}I.

The results of bombarding thin (~15 mg/0·7 cm^2) and thick (~50 mg/0·7 cm^2) targets of 98·4% enriched ^{121}Sb with α particles of energy 25·2–36·5 MeV are listed in Tables 2 and 3, respectively. These show the bombardment energy, weight of target per 0·7 cm^2, ^{123}I yield (mCi/μA-h) and the percentages of ^{124}I, ^{125}I and ^{126}I contaminants. All yields are corrected to the end of the bombardment. The energies listed are 'average' energies, i.e. the average of the incident and exit energy of the beam traversing the target. The uncertainty is the difference between the average and the extreme values. The superficial density of the target (mg/cm^2) can be calculated by dividing its weight by the target area, 0·7 cm^2. All bombardments were made with beam currents of 4 ± 2 μA for 3–5 h. The absolute ^{123}I yields from bombardments 23, 30, and 33 are not listed because the current integrator was not operating during those runs. However, the percentages of ^{124}I, ^{125}I, and ^{126}I contaminants are listed because these relative values are independent of the total beam charge on the target.

Table 2. RADIOIODINE YIELDS FROM THE ^{121}Sb(^{4}He, 2n) ^{123}I REACTION ON THIN (~15 mg/0·7 cm^2) TARGETS OF ^{121}Sb

No.	*Energy* MeV	*Targets* wt., mg/0·7 cm^2	^{123}I mCi/μAh	^{124}I %	^{125}I 10^{-2}%	^{126}I 10^{-3}%
4	28·4 ± 1·2	15·0	0·14	0·54	1·4	2·3
9	33·3 ± 1·1	15·8	0·23	0·62	1·5	1·9
5	33·4 ± 1·0	15·0	0·18	0·53	1·3	2·9
7	33·5 ± 0·9	13·4	0·16	0·54	1·4	2·3
10	36·5 ± 1·0	15·0	0·08	1·35	1·4	2·2

Table 3. RADIOIODINE YIELDS FROM THE $^{121}Sb(^{4}He, 2n)\,^{123}I$ REACTION ON THICK (~50 mg/0·7 cm^2) TARGETS OF ^{121}Sb

No.	*Energy* MeV	*Targets* wt., mg/0·7 cm^2	^{123}I mCi/μAh	^{124}I %	^{125}I 10^{-2}%	^{126}I 10^{-2}%
2	25·2 ± 4·0	50·0	0·69	0·89	1·4	2·3
18	31·0 ± 3·4	49·6	0·48	1·04	1·6	3·0
20	31·0 ± 3·4	49·8	0·47	0·70	1·5	2·5
23	31·0 ± 3·4	50·0	*	0·68	1·4	–
30	31·0 ± 3·4	50·8	*	0·70	–	–
32	31·0 ± 3·4	49·9	0·80	0·65	1·2	2·0
33	31·0 ± 3·4	50·6	*	0·69	1·3	2·1
50	31·0 ± 3·4	50·3	0·58	0·74	1·4	2·2
3	31·1 ± 3·3	46·5	0·93	0·84	1·5	3·0
41	31·5 ± 3·4	50·0	0·55	0·76	1·3	2·7
42	31·5 ± 3·4	50·2	0·38	0·87	1·8	2·9
46	31·5 ± 3·4	50·8	0·50	0·86	1·6	2·1
48	31·5 ± 3·4	50·8	0·75	1·04	1·4	3·2

*Current integrator in error

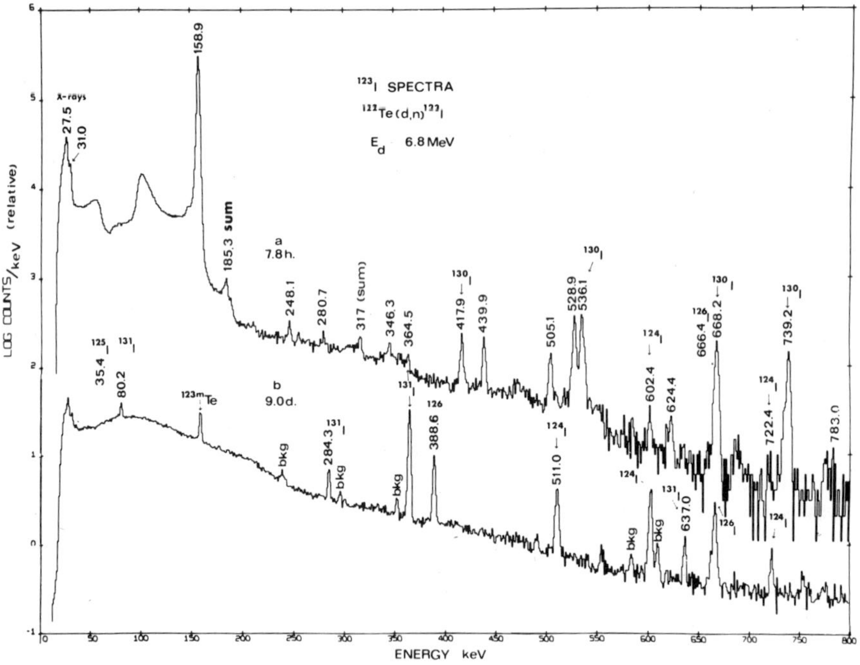

Fig. 4. ^{123}I spectra from the ^{122}Te (d, n) ^{123}I *reaction*

^{122}Te(d, n) ^{123}I: Two typical Ge(Li) spectra of a radioiodine fraction distilled from a 95·4% enriched ^{122}Te target are shown in Fig. 4. Figs. 4(a) and (b) were obtained 7·8 h and 9 days respectively after the bombardment. The deuteron energy of this bombardment was 6·8 ± 0·5 MeV. No major activity was found above 800 keV. Identified in Fig. 4(a) are the 27·5 and 31·0 keV X-rays and the 158·9, 248·1, 280·7, 346·3, 439·9, 505·1, 528·9, 624·4 and 783·0 keV γ-ray photopeaks of ^{123}I. The 185·3 and 317 keV photopeaks are sums of abundant photons as described above for Fig. 3(a). Also identified are the γ-ray photopeaks of ^{124}I at 602·4 and 722·4 keV, and of ^{130}I at 417·9, 536·1, 668·2, and 739·2 keV. The ^{124}I and ^{130}I contaminations, which resulted principally from ^{123}Te(d, n) ^{124}I and ^{130}Te(d, 2n) ^{130}I reactions on the tellurium impurities in the target, were $8{\cdot}3 \times 10^{-2}$ and 0·92% respectively at the end of this bombardment.

Identified in Fig. 4(b) are the γ-ray photopeaks of ^{125}I at 35·4 keV, ^{126}I at 388·6 and 666·4 keV, and ^{131}I at 80·2, 284·3, 364·5, and 637·0 keV. The ^{125}I, ^{126}I, and ^{131}I contaminants, which result mainly from ^{124}Te(d, n) ^{125}I, ^{125}Te(d, n) I^{126}I, and ^{130}Te(d, n) ^{131}I reactions, were $2{\cdot}1 \times 10^{-2}$, $3{\cdot}9 \times 10^{-2}$, and 0·07% respectively at the end of this bombardment. Also identified in Fig. 4(b) is the 511 keV annihilation photopeak caused primarily by ^{124}I. The photopeak at 158·9 keV was caused by residual ^{123}I and 118 d ^{123m}Te, the latter persisting long after the 13·3 h ^{123}I has disappeared. ^{123m}Te was produced by decay of ^{123}I or was carried over during the distillation despite the presence of hold-back carrier.

The radioiodine yields from 6·4–9·0 MeV-deuteron bombardments of 95·4% enriched ^{122}Te are listed in Table 4. Also listed for each bombardment are the 'average' energy of the deuteron beam, the weight of the target per 0·7 cm^2, and

Table 4. RADIOIODINE YIELDS FROM THE ^{122}Te(d, n) ^{123}I REACTION ON 95·4% ^{122}Te

No.	*Energy* MeV	*Targets* wt., mg/0·7 cm^2	^{123}I 10^{-2} mCi/μAh	^{124}I 10^{-2}%	^{125}I 10^{-2}%	^{126}I 10^{-2}%	^{130}I %	^{131}I %
39	6·4 ± 0·3	12·5	0·19	7·2	1·6	4·1	0·97	0·15
34	6·8 ± 0·5	17·3	1·5	8·3	2·1	3·9	0·92	0·07
38	6·8 ± 0·5	16·3	10·1	8·5	1·3	4·4	0·86	0·09
29	6·9 ± 0·4	15·8	6·2	7·2	2·1	3·4	0·82	0·08
28*	7·3 ± 1·0	31·5	13·2	19·2	1·2	6·6	0·96	0·07
31	9·0 ± 0·3	12·6	8·1	24·0	1·3	9·6	2·5	0·09

*80·4% ^{122}Te was bombarded; yields corrected to 95·4% ^{122}Te

the percentages of ^{124}I, ^{125}I, ^{126}I, ^{130}I, and ^{131}I contaminants. The bombardments were made with beam currents ranging from 4 to 7 μA for 1–6 h. Bombardment 28 was actually done on 80·4% enriched ^{122}Te. The results in Table 4 for this run were corrected to those which would be obtained from bombardment of 95·4% enriched ^{122}Te. The corrections were made assuming that ^{123}I, ^{124}I, ^{125}I, ^{126}I, ^{130}I, and ^{131}I were produced from ^{122}Te(d, n) ^{123}I, ^{123}Te(d, n) ^{124}I, ^{124}Te(d, n) ^{125}I, ^{125}Te(d, n) ^{126}I, ^{130}Te(d, n) ^{131}I, and ^{130}Te(d, 2n) ^{130}I reactions respectively.

Errors. For the most part, the errors in the experiment resulted from counting statistics and from uncertainties in spectrometer efficiency and irradiation parameters. Counting errors at the 95% confidence level for ^{123}I and ^{124}I were $< \pm 1\%$; for ^{125}I, ^{126}I, ^{130}I, and ^{131}I were normally $< \pm 5\%$. In a few instances, when the counting rates of small quantities of contaminants were encountered, the errors were as high as ±25%. Errors in spectrometer efficiency were estimated to be < ±3%.

The greatest errors in the experiment were due to uncertainties in the irradiation parameters. These uncertainties result from non-uniformities in the incident beam density, unevenness of the target thickness, misalignment of the incident beam over the entire area of the target, and incomplete transfer of target powder from the target holder to the distillation apparatus for radioiodine separation. The magnitude of these errors can be approximated from a statistical analysis of results from similar bombardments. The arithmetic mean and its standard deviation were calculated from the results of twelve α-particle bombardments of ^{121}Sb shown in Table 3. The average ^{123}I yield was 0·60 ± 0·18 mCi/μAh; the average ^{124}I, ^{125}I, and ^{126}I contaminants were 0·79 ± 0·13%, (1·5 ± 0·2) X 10^{-2}% and (2·6 ± 0·4) X 10^{-3}% respectively.

DISCUSSION

^{121}Sb(^{4}He, 2n)^{123}I: The ^{123}I yields in Table 2 from thin ^{121}Sb targets appear to be similar for bombardments with 28·4–33·5 MeV α particles. The ^{123}I yield at this energy normalised to target weights of 15 mg was 0·14–0·22 mCi/μAh. This energy range is between ~25 MeV, where the cross-section for the ^{121}Sb(^{4}He, 2n) reaction should be a maximum, and ~34 MeV, where that for the ^{121}Sb(^{4}He, 3n) reaction should be a maximum. Energies higher than 25 MeV were used to lessen the probability of the ^{121}Sb(^{4}He, n) ^{124}I reaction. This was advantageous since the ^{121}Sb(^{4}He, 3n) reaction produced 3·5 m ^{122}I. The ^{123}I yield from the bombardment at higher energy, 36·5 MeV, is lower.

The lowest ^{124}I contamination was 0·54–0·62% at bombarding energies from 28·4 to 33·5 MeV on ~15 mg targets. ^{124}I contamination increased from ~0·6 to 1·35% when the bombarding energy was increased from ~33·4 to 36·5 MeV. This relative increase at higher energy results from a decrease in the probability of the ^{121}Sb(^{4}He, 2n) ^{123}I reaction and an increase in the probability of the ^{123}Sb(^{4}He, 3n) ^{124}I reaction.

The ^{123}I yields from ~50 mg/0·7 cm^2 ^{121}Sb targets listed in Table 3 were 0·38–0·93 mCi/μAh at an α-particle bombarding energy of ~31 ± 3 MeV. The average ^{123}I yield at this energy was 0·60 ± 0·18 mCi/μAh. The average yield (0·012 mCi/μAh-mg) for 50 mg targets is the same as that for 15 mg targets.

The ^{124}I yields listed in Table 3 were 0·65–1·06%, with an average yield of 0·79 ± 0·13%. This average ^{124}I yield was approximately 30% higher than that with the 15 mg targets. The increase in the ^{124}I yield resulted from the increased probability of the ^{121}Sb(^{4}He, n) ^{124}I reaction occurring as the α-particle energy was attenuated in traversing the thicker targets. The approximately linear relationship of ^{123}I yield vs the weight of ^{121}Sb targets from 15 to 50 mg, however, indicated that penetration through the 50 mg/0·7 cm^2 target thickness did not attenuate the incident α-particle energy (~34 MeV) below that which caused the desired ^{121}Sb(^{4}He, 2n) ^{123}I reaction.

Because of the linear increase of ^{123}I with target thickness, the ^{123}I yield per μAh could probably have been increased further by using ^{121}Sb targets greater than 50 mg. Unfortunately, the ^{124}I contamination was increased in the thicker targets. Because minimum ^{124}I contamination is advantageous if ^{123}I is to be used for low-dose radiopharmaceutical use,[5,6] bombardment of targets heavier than ~50 mg was not attempted.

The average ^{125}I contamination listed in Table 3 from 50 mg ^{121}Sb targets irradiated with ~31 MeV α particles was (1·5 ± 0·2) × 10^{-2}%; the average ^{126}I contamination was (2·6 ± 0·4) × 10^{-3}%. These percentages were not significantly different from those obtained from ~15 mg targets at bombardment energies from 28·4 to 36·5 MeV.

$^{122}Te(d, n)\ ^{123}I$: The ^{123}I yields listed in Table 4 from 6·8–7·3 MeV deuteron bombardment of 15·8–31·5 mg ^{122}Te targets, i.e. from bombardments 28, 29, 34, and 38, varied considerably, from (1·5–13·2) × 10^{-2} mCi/μAh. The large variation in yield from the $^{122}Te(d, n)\ ^{123}I$ reaction is probably related to the fact that the calculated coulomb barrier[11] for the reaction is ~8 MeV. Assuming that the cross-section of the $^{122}Te(d, n)\ ^{123}I$ reaction is a smooth function of the excitation energy,[12] it was calculated that the cross-section should be approximately a maximum at the theoretical threshold for the $^{122}Te(d, 2n)\ ^{122}I$ reaction, i.e. ~7 MeV. However, because the coulomb barrier opposes the approach of the deuteron to the ^{122}Te nucleus at this 'preferred' energy, the cross-section for the reaction will probably be a very steep function of the energy. This supposition is borne out by the 0·19 × 10^{-2} mCi/μAh ^{123}I yield at 6·4 MeV. This yield is 10–100 times less than those obtained at <1 MeV higher bombardment energies. The large variations in ^{123}I yields under seemingly identical conditions are probably due to the unavoidable variations in the thicknesses of the powder targets, which in some instances reduced the excitation energy significantly below the coulomb barrier.

The ^{123}I yield at 9·0 MeV deuteron bombarding energy, ~1 MeV higher than the calculated coulomb barrier, was approximately the same as those at 6·8–7·3 MeV. The yields of ^{124}I, ^{126}I, and ^{130}I, however, have increased 2–3 fold at this higher excitation energy because of the increased cross-sections of the $^{124}Te(d, 2n)\ ^{124}I$, $^{126}Te(d, 2n)\ ^{126}I$, and $^{130}Te(d, 2n)\ ^{130}I$ reactions respectively. The results show that the highest and purest yields of ^{123}I were obtained with ~7 MeV deuteron bombardment of ~15 mg 95·4% ^{122}Te targets.

Comparison of direct methods. A comparison of the ^{123}I yields and radioiodine contaminants from the $^{122}Sb(^4He, 2n)\ ^{123}I$ reaction on ~50 mg ^{121}Sb targets at ~31 MeV and the $^{122}Te(d, n)\ ^{123}I$ reaction on ~16 mg ^{122}Te targets at ~6·9 MeV is made in Table 5 with those reported[9] for the $^{123}Te(p, n)\ ^{123}I$ reaction on ~250 mg ^{123}Te targets at ~16 MeV. The ^{123}I yield from the $^{121}Sb(^4He, 2n)\ ^{123}I$ reaction averages about 30% higher than the yields reported for the $^{123}Te(p, n)\ ^{123}I$ reaction; the ^{124}I contaminations are essentially the same for both methods. The $^{121}Sb(^4He, 2n)\ ^{123}I$ method yields 6 times less ^{125}I than

Table 5. RADIOIODINE YIELDS FROM VARIOUS METHODS

Reaction	^{123}I mCi/μAh	^{124}I %	^{125}I %	^{126}I %	^{130}I %	^{131}I %
$^{121}Sb(^4He, 2n)\ ^{123}I$	0·4–0·9	0·7–1·0	0·01–0·02	0·002–0·003	ND	ND
$^{122}Te(d, n)\ ^{123}I$	0·06–0·10	0·07–0·09	0·01–0·02	0·03–0·04	0·8–0·9	0·07–0·09
$^{123}Te(p, n)\ ^{123}I$*	0·4–0·6	0·8–1·0	0·07–0·08	0·2–0·3	–	–

ND=not detectable
*Hupf, Eldridge, and Beaver

the ^{123}Te(p, n) ^{123}I method, and more important, 100 times less ^{126}I, which is a high-energy β- and γ-ray emitter.

The ^{123}I yield from the ^{122}Te(d, n) ^{123}I reaction is ~12% of that from the ^{121}Sb(^{4}He, 2n) ^{123}I and the ^{123}Te(p, n) ^{123}I reactions. However, in its favour, the ^{122}Te(d, n) ^{123}I method produces about 10 times less ^{124}I contamination than either of the other two methods, and also low concentrations of ^{125}I, ^{126}I and ^{131}I. The greatest contaminant from the ^{122}Te(d, n) ^{123}I method is 0·8–0·9% ^{130}I. Even though the half-life of ^{130}I (12·3 h) is slightly less than that of ^{123}I, its high energy and abundant γ-rays (417·9 keV, 35%; 536·1 keV, 99%; 668·2 keV, 100%; and 739·2 keV, 87% as seen in Fig. 4) may interfere with high resolution *in vivo* scanning.

CONCLUSION

Both the ^{121}Sb(^{4}He, 2n) ^{123}I and the ^{122}Te(d, n) ^{123}I reactions produce ~99% pure ^{123}I in sufficient quantities for radiopharmaceutical use. The ^{123}I yield of the ^{121}Sb(^{4}He, 2n) ^{123}I reaction at 33 MeV is about 7–9 times greater than the ^{122}Te(d, n) ^{123}I reaction at 7 MeV but its ^{124}I contamination is about 10 times greater.

In comparing three methods which produce ^{123}I directly, the ^{121}Sb(^{4}He, 2n) ^{123}I method has a 30% higher yield and less radioiodine contamination, notably ^{126}I, than the ^{123}Te(p, n) ^{123}I method.

The ^{122}Te(d, n) ^{123}I method has about 12% of the ^{123}I yield, but only 20% of the longer lived radioiodine contaminants, as compared to the other two methods, and may be preferred where purity is desired.

ACKNOWLEDGMENT

The authors wish to thank Wayne Smith and other members of the L.R.C. Cyclotron staff for their co-operation, interest, and helpful suggestions during the course of these experiments.

REFERENCES

1. Myers, W. G. and Anger, H. O., *J. Nucl. Med.*, 3, 183, (1962).
2. Blue, J. W. and Sodd, V. J., following paper.
3. Rhodes, B. A., Buddemeyer, E. U., Stern, H. S., and Wagner, H. N., *J. Nucl. Med.* 7, 385, (1966).
4. Wagner, H. N. and Emmons, H., 'Radioactive Pharmaceuticals', Proc. of Symposium at ORINS, held 1–4 Nov., 1965, Eds., Andrews, G. A., Kniseley, R. M., and Wagner, H. N., April, 1966. CONF-651111.
5. Wellman, H. N., Mack, J. F., Friedman, B. I., and Saenger, E. L., *J. Nucl. Med.* 9, 359, (1968).
6. Wellman, H. N., Mack, J. F., Saenger, E. L., and Sodd, V. J., *J. Nucl. Med.* 10, 381, (1969).
7. Beaver, J. E. and Holley, W. L., *Trans. Am. Nucl. Soc.* 8, 335, (1965).
8. Eldridge, J. S., White, J. C., and Lyon, W. S., ORNL-3889, Anal. Chem. Div. Ann. Progr. Rep. for period ending 15 Nov., 1965, p. 48, Jan. (1966).

9. Hupf, H. B., Eldridge, J. S., and Beaver, J. E., *Intern, J. Appl. Radn. and Isotopes* **19**, 345, (1968).
10. Goolden, A. W. G., Glass, H. I., and Silvester, D. J., *Br. J. Radiol.* **41**, 20, (1968).
11. Blatt, J. M. and Weisskopf, V. F., 'Theoretical Nuclear Physics', 16, John Wiley and Sons, Inc., New York (1952).
12. Ibid., p. 490.

DISCUSSION

M. A. Chaudhri (MRC, Hammersmith): Did I understand you correctly that you use thin targets instead of thick ones, which naturally give much higher yields? If so, can you please give your reasons for doing so?

V. J. Sodd (NASA): Thick targets (~20 mg cm^{-2}) were used for the $^{122}Te(\alpha, n)^{123}I$ reaction at ~7 MeV. For the $^{121}Sb(\alpha, 2n)$ ^{123}I reaction the ^{123}I yield increased linearly with increasing target thickness from 20 to 70 mg cm^{-2}. However, with the thicker targets ^{127}I was also produced by $^{121}Sb(\alpha, n)$ reactions with He ions which had been degraded in energy while traversing the target. For this reason targets thicker than 70 mg cm^{-2} were not used, even though higher ^{123}I yields might have been obtained.

^{123}I production from the β^+ decay of ^{123}Xe

J. W. Blue* and V. J. Sodd†

ABSTRACT

Two arrangements are presented for the production of ^{123}Xe which appear to be well suited to the production of large amounts of ^{123}I. The tellurium target need not be chemically dissolved and therefore can be re-used for subsequent bombardments. Both arrangements use the reactions $^{122}Te(\alpha, 3n)$, $^{122}Te(^3He, 2n)$, or $^{123}Te(^3He, 3n)$ to produce ^{123}Xe, which is then physically separated from the tellurium target by diffusion from the microcrystal, and subsequently transported in a helium gas stream to low-temperature traps. After a time the ^{123}I produced from the β^+ decay of ^{123}Xe is removed from the trap. Direct labelling of organic molecules is possible, and macroaggregated serum albumin was labelled with ^{123}I merely by contacting a solution with the contents of the trap.

INTRODUCTION

Background. This work is a study of ^{123}I production from the decay of ^{123}Xe. The interest in this method is due to the very high purity of the ^{123}I produced and the possibility of adapting the method to produce significant amounts of ^{123}I. It may therefore become important if ^{123}I is extensively substituted for ^{131}I in medical diagnosis.

^{123}I has several advantages over ^{131}I. Because of the shorter half-life and the decay by electron capture, the ratio of the $^{123}I/^{131}I$ radiation exposure to the patient for equal initial amounts of activity is 1/40. The lower energy of the γ-ray of ^{123}I (159 keV) compared to the principal γ-ray of ^{131}I (364 keV) permits the use of less bulky collimators, which are more effective.

The advantage of lower radiation exposure with the use of ^{123}I is only fully realised if contamination by other radioiodines is small. For example, a 1% initial contamination of ^{123}I with ^{124}I doubles the patient radiation exposure. Another paper in this conference[1] reviews the various methods of ^{123}I production.

* Lewis Research Center, National Aeronautics and Space Administration, Cleveland, Ohio, U.S.A.

† Nuclear Medicine Laboratory, Cincinnati General Hospital, DMRE, BRH, U.S. Department of Health, Education and Welfare, Cincinnati, Ohio, U.S.A.

It is concluded that all charged-particle reactions produce contamination from other radioiodines ranging from 0·5 to 1%. For the production of ^{123}Xe by ^{122}Te(α, 3n), ^{122}Te(^{3}He, 2n), or ^{123}Te(^{3}He, 3n) reactions, the production of radioiodine contaminants is also of this order, but ^{123}Xe is readily separated from the contaminants, and subsequent decay of ^{123}Xe yields high-purity ^{123}I. The only xenon isotope that poses any problem as a radiocontaminant is ^{125}Xe. The long lifetime of ^{125}Xe (17 h) relative to ^{123}Xe (2·1 h) keeps the yield of ^{125}I low. In addition, ^{125}I has such low-energy radiations that it does not interfere with those of ^{123}I. The important factor accounting for the low levels of contamination of ^{123}I produced from the decay of ^{123}Xe is that the even xenon isotopes 124 and 126 are stable and thus ^{124}I and ^{126}I, which might otherwise be serious contaminants, are eliminated.

Two methods of ^{123}Xe *production.* An arrangement for producing ^{123}I via ^{123}Xe has been described.[2] This arrangement is designated the generator method, because ^{123}Xe is removed from the bombardment region continuously during the bombardment.

A second arrangement for ^{123}Xe production involves a bombardment phase followed by a xenon removal phase, and will be designated the cyclic method. Both methods were studied using the reaction ^{122}Te(α, 3n) at the full energy, 42 MeV, of the Lewis Research Center cyclotron. These methods are also applicable to ^{3}He bombardment of ^{122}Te or ^{123}Te.

The present paper describes these arrangements and our recent results.

EXPERIMENTAL

Generator method. The generator arrangement consists of a continuous helium flow system, in a closed loop. A layer of enriched tellurium powder target, 20–40 mg/cm^2, is held against a porous metal plate by a screen (Fig. 1). The incident α-particle beam penetrates a thin metal foil window used to separate

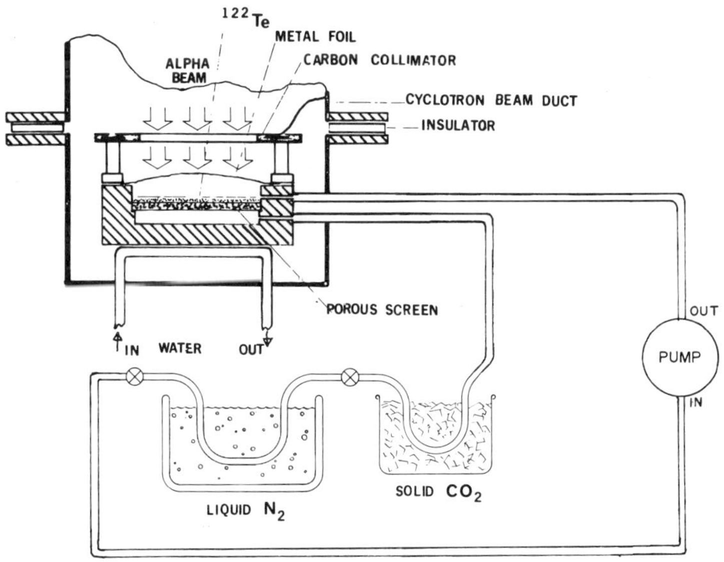

Fig. 1. Generator method in the closed loop configuration

the cyclotron vacuum system from the target chamber. A low flow of helium gas through the tellurium powder and porous plate carries ^{123}Xe produced by the nuclear reaction into the dry ice trap and then into the liquid nitrogen trap.

The purpose of the closed loop system was to conserve helium when the system is operated with the gas flow at a rate such that some cooling of the powder target was obtained. A heat transfer study showed that at a flow rate of 5 standard l/min the bulk gas temperature at the exit of a trap 50 cm long × 3 mm diam. was nearly the temperature of the wall of the trap. Therefore it appeared to be possible to cool the powder target with the gas flow and also to cool the gas to a sufficiently low temperature for the radioxenon to stick to the wall. However, experiment showed that at flow rates sufficient for cooling, the radioiodines did not stick to the wall of the dry ice trap, and the amount of ^{124}I in the liquid nitrogen was excessive. At 5 l/min the gas residence time in the dry ice trap was apparently too short to allow complete radioiodine diffusion to the wall. When the flow rate was reduced to 0·02 l/min the dry ice trap is 100% efficient, but target cooling is not obtained. At such low flow rates the conservation of helium is not important, so the single-pass system shown in Fig. 2 was used. This arrangement eliminates the complication of a

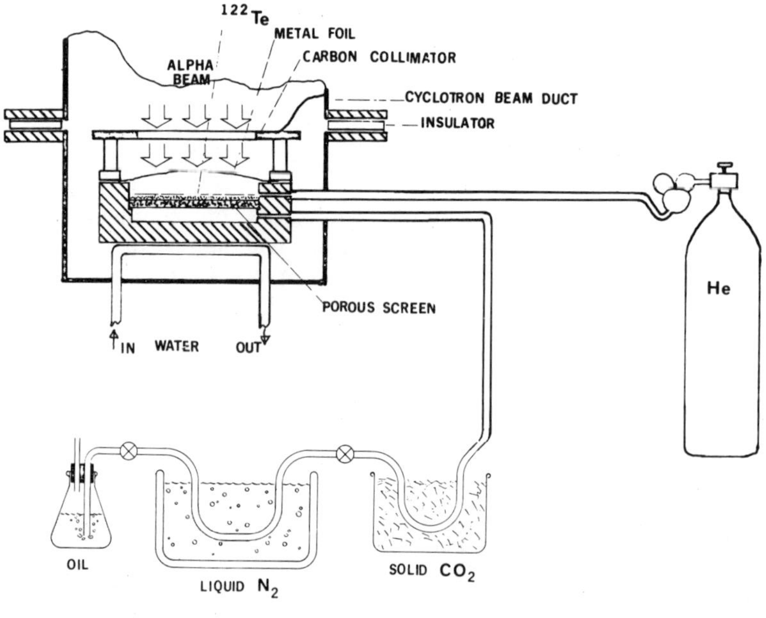

Fig. 2. Generator method in the single pass configuration

pump and the need to evacuate the closed system, since the single-pass system can be flushed out.

Optimum bombardment period for generator. Assuming a constant beam current on a fixed amount of tellurium target, then there is a constant production rate, R, for the number of ^{123}Xe atoms produced per unit time. Of these atoms, some fraction, f, will be released from the target to reach the liquid nitrogen trap, where they adhere to the wall. The number of ^{123}Xe atoms in the trap as a function of the bombardment time, T, is

$$N^{Xe} = \frac{fR}{\lambda_1}(1 - e^{-\lambda_1 T}) \qquad (1)$$

In addition, there are ^{123}I atoms in the trap produced by ^{123}Xe decay during the bombardment. The ratio, χ, of the number of ^{123}Xe atoms to the number of ^{123}I atoms at the end of a bombardment of length T, is

$$\chi = \frac{T_1(1 - e^{-\lambda_1 T})}{T_2(1 - e^{-\lambda_2 T}) + \frac{T_2 T_1}{T_2 - T_1}(e^{-\lambda_1 T} - e^{-\lambda_2 T})} \qquad (2)$$

where λ_1, T_1, and λ_2, T_2 are the decay constants and half-lives for ^{123}Xe and ^{123}I respectively.

At the end of a bombardment of length T there is an optimum waiting period, T_w, at which the ^{123}I in the trap is a maximum. The expression for T_w is

$$T_w = \frac{T_1 T_2}{\ln 2\,(T_2 - T_1)} \ln\left[\frac{T_2}{T_1}\left(\frac{\chi}{\frac{T_2 - T_1}{T_2} + \chi}\right)\right] \qquad (3)$$

This is the time at which the ^{123}I yield is calculated. The yield per unit bombardment time is then given by

$$\frac{N^I}{T} = \frac{fR}{\lambda_1}\frac{(1 - e^{-\lambda_1 T})}{T}\left[\frac{\lambda_1}{\lambda_1 - \lambda_2}(e^{-\lambda_2 T_w} - e^{-\lambda_1 T_w}) + \frac{e^{-\lambda_2 T_w}}{\chi}\right] \qquad (4)$$

The result is plotted in Fig. 3 as a function of the bombardment time. An obvious advantage of the generator is seen to be that the yield per hour of bombardment is nearly independent of the bombardment time. At the end of 10 h bombardment, the yield is 93% of the maximum obtainable. From this result it can be concluded that, were a cyclotron intended for production of large amounts of ^{123}I, the length of the bombardment time could be determined from schedule considerations, or perhaps from an effort to minimise the number of trap changes, rather than from a consideration which optimised the yield.

Cyclic system. The possibility that only a small fraction of ^{123}Xe produced in the $^{122}Te(\alpha, 3n)$ reaction would diffuse and escape from the microcrystal of tellurium in the generator target led to the consideration of another bombardment arrangement. In this, tellurium is bombarded for some period T and is kept cold during this period, so that no ^{123}Xe escapes. After bombardment, the tellurium target is heated to a temperature sufficient to remove all the ^{123}Xe. The same target could be re-used for successive bombardments by repeating the cycle.

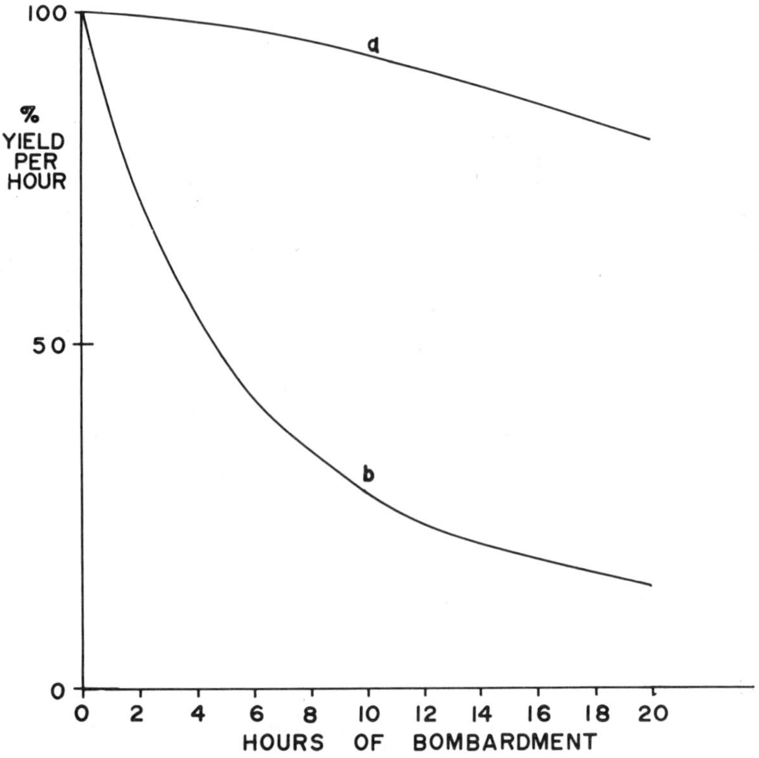

Fig. 3. Per cent of the maximum yield per hour vs bombardment time for curve a, the generator method, and for curve b, the cyclic method

The cyclic method was studied after it was determined that high gas flow rates, sufficient to give cooling of the powder target in the generator arrangement, could not be used without carrying a considerable amount of ^{124}I into the liquid nitrogen trap. The cyclic method permits a controlled heating cycle so that, in principle, all the xenon can be diffused from the tellurium at a temperature low enough to ensure that none of the tellurium is vaporised and lost from the system. One purpose of studying the cyclic system was to determine if the time required to diffuse the xenon from the target was short enough to avoid a large loss of ^{123}Xe by decay.

The arrangement of the cyclic system during bombardment is shown in Fig. 4. The cyclotron beam passes through an open ball valve and a collimator onto the target, which is tellurium powder contained in a small depression in a copper plate. The thickness of the tellurium powder is ~90 mg/cm^2, which is about optimum for the (α, 3n) reaction at 40 MeV. The tellurium powder is enriched (90% ^{122}Te) and is prepared by reduction from solution. The microcrystal size is ~1 μ. The powder is contained in the depression in the copper plate by a 0·002 5 cm aluminium foil. The copper plate forms the end of a canister of liquid nitrogen, which is kept filled during the bombardment by an automatic filling system.

At the end of the bombardment period the ball valve is closed and the target kept under high vacuum while the target assembly is removed from the cyclotron for the ^{123}Xe removal cycle. The arrangement for this is shown in Fig. 5. As soon as possible after removal, a chamber is mounted to the flange that had been connected to the cyclotron beam duct and air is evacuated from the volume in front of the ball valve. The latter is then opened and the foil cutter, a 1·6 cm diam. stainless steel tube, attached to a 6 mm tube serving as a handle, is pushed through a sliding seal to cut the aluminium foil. The enlarged detail in Fig. 5 shows the foil cutter fully inserted, so that its end is in a circular groove in the copper plate. To the end of the 6 mm tube a length of flexible tubing is connected and closed off with a clamp, so the tube can be evacuated. With the cutter tube in the fully-inserted position, helium gas is slowly admitted to the system. When the pressure reaches 1 atm, the clamp on the flexible tube on the end of the foil cutter tube is opened. Helium then flows into the dry ice trap, the liquid nitrogen trap and an oil bubbler.

When the helium flow between the cutting edge of the cutter and the groove in the copper plate is adjusted to ~20 ml/min, power is supplied to a heater inserted in the re-entrant axial tube in the liquid nitrogen container. The heater is in contact with the back of the copper platê and thus heats the tellurium powder. The temperature of the copper plate is measured by a thermocouple

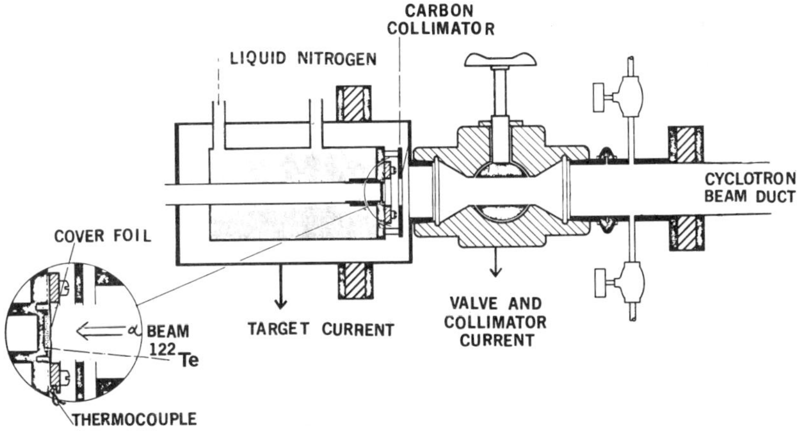

Fig. 4. Cyclic assembly during bombardment cycle

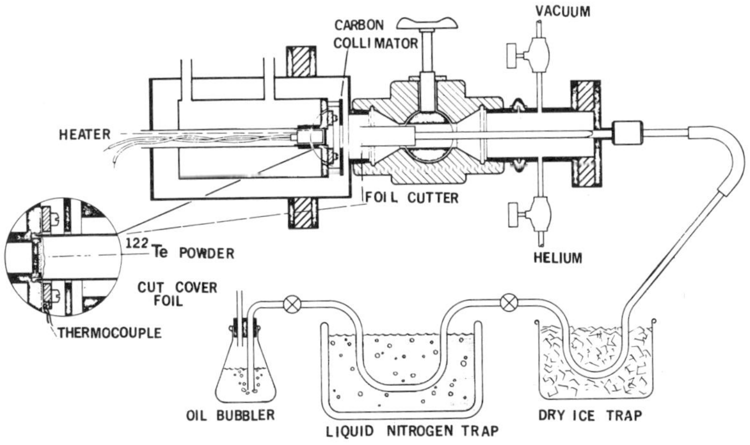

Fig. 5. Cyclic assembly during thermal cycle

held against the copper plate by the aluminium foil hold-down ring. This same thermocouple is used to monitor the temperature during the bombardment.

Bombardment time consideration for the cyclic system. In the cyclic system ^{123}Xe produced during the bombardment is not effective in producing ^{123}I if it decays in the target material before it is removed in the heating cycle. The ^{123}I yield in the trap per unit of bombardment time is given by

$$\frac{N^I}{T} = \frac{fR}{\lambda_1} \cdot \frac{(1-e^{-\lambda_1 T})}{T} \cdot \frac{\lambda_1}{\lambda_1 - \lambda_2} (e^{-\lambda_2 T_W} - e^{-\lambda_1 T_W}) \tag{5}$$

If it is assumed that the diffusion time is short compared to the half-life of ^{123}Xe, i.e. 2·1 h, then essentially all the ^{123}Xe is collected before it decays, with the result that the ratio of ^{123}Xe atoms to ^{123}I atoms at the end of the heating cycle is infinite. The optimum waiting period, T_W, for $\chi = \infty$, is given by

$$T_W = \frac{1}{\lambda_1 - \lambda_2} \ln \frac{\lambda_1}{\lambda_2} = 6{\cdot}4\,\text{h} \tag{6}$$

The ^{123}I yield per unit bombardment time is shown in Fig. 3. It is apparent from the figure and from a comparison of the yield expressions for the generator, Eqn (3), and for the cyclic method, Eqn (5), that the cyclic method has a lower yield. This is because ^{123}I produced by ^{123}Xe decay during the bombardment is collected in the liquid nitrogen trap in the generator method and is lost in the cyclic method.

The more rapid fall-off of the yield in the cyclic system indicates that the length of the irradiation should not be much more than 2 h, since at this time the yield is already 25% less than the theoretical maximum. The maximum cannot in practice be achieved, because it would require very brief irradiation periods and then a heating cycle.

It should be pointed out that the factor f in the generator expression is the fraction of ^{123}Xe escaping from the tellurium powder during the irradiation, whereas in the cyclic yield expression it is the fraction of ^{123}Xe removed during the heating cycle. Furthermore, in the cyclic method there will be a reduction in the ^{123}I yield from that given by Eqn (5), because no allowance was made for the decay of ^{123}Xe during the heating cycle. If the heating cycle is $<$15 min, then the loss of ^{123}Xe by decay will be $<$10%.

RESULTS

Generator method. When operated at low flow rates in the range 20–30 ml/min, the generator method has produced ^{123}I that is essentially free of ^{124}I, i.e. the concentration of ^{124}I was $<$0·005%. Fig. 6 shows a Ge(Li) detector spectrum of ^{123}I after removal from the liquid nitrogen trap. After waiting for ^{123}Xe decay, a trap is warmed to room temperature and flushed with helium gas to remove remaining ^{123}Xe and ^{125}Xe. The trap is rinsed with dilute NaOH and the solution counted. The low ^{124}I is shown by the absence of annihilation radiation and of the 603 keV line. Weak high-energy γ-rays are seen that have not been seen in recent studies[3,4] of ^{123}I, because they were obscured by the high-energy γ-rays

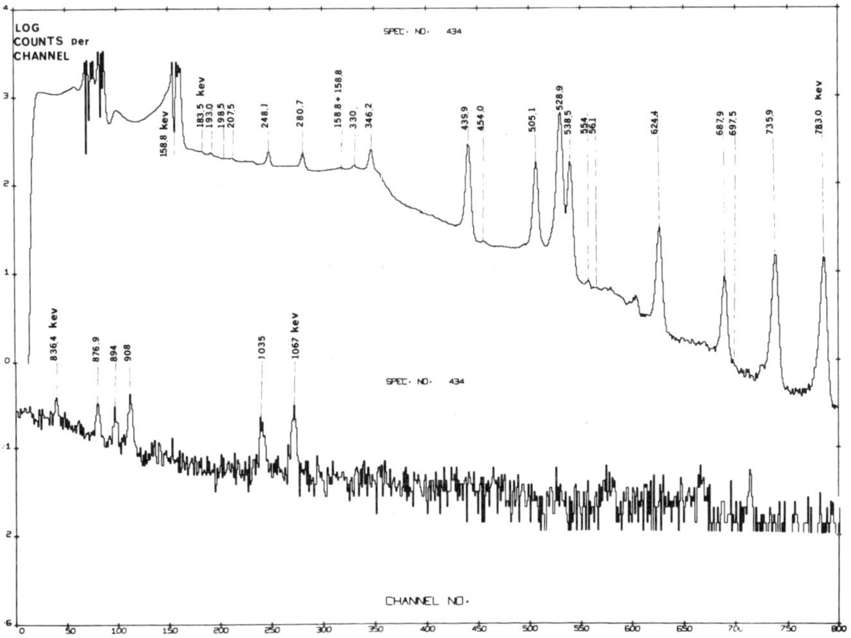

Fig. 6. Ge(Li) spectrum of high purity ^{123}I

of ^{124}I. The γ-rays attributed to ^{123}I decay are labelled in Fig. 6.

The best yields obtained with the generator are 0·2 mCi/μAh at beam current densities of 1 mA/cm^2. It has been difficult to obtain reproducible results, due to the transport of the ^{122}Te target material through the stainless steel screen (Huyck Metals Co., Feltmetal, 30 Rayls). The pore size of the screen is sufficiently small to hold the aggregated clusters of micron-size tellurium particles. After the irradiation, more than half of the tellurium has often passed through the screen, which indicates that the aggregates either break up or vaporise.

Estimate of f for generator method. An accurate determination of *f*, the fraction of the ^{123}Xe released from the tellurium, has not been made. This is due to the uncertainty in the amount of bombardment the target material receives. Small tellurium samples, taken from the target chamber and examined by γ-counting for the relative intensities of the ^{123}Xe 149 keV γ-ray and the ^{123}I 159 keV γ-ray, establish a value for χ. This should be less than the value calculated from Eqn (2) by the amount of ^{123}Xe that has diffused out. This calculation assumes that none of the ^{123}I diffuses out, which is certainly not correct, because ^{124}I is found in the dry ice trap. A comparison of the activity of ^{124}I in the dry ice trap to that in the tellurium in the target chamber does indicate that only a few percent of the radioiodines diffuse from the tellurium. Therefore, neglecting both the loss of ^{123}I from the sample and the ^{123}I from the direct reaction ^{122}Te(α, t), the value of *f* is observed to be between 10 and 20%.

Cyclic method. The cyclic arrangement has been used for only two irradiations. The first was a bombardment of a 30 mg/cm^2 target of virgin ^{122}Te, with a beam current density of about 5 $\mu a/cm^2$. The counting rate of a detector in the proximity of the liquid nitrogen trap vs the temperature of the copper plate is plotted in Fig. 7. The temperature rise was 5°C/min to 200°C, and 2·5°C/min

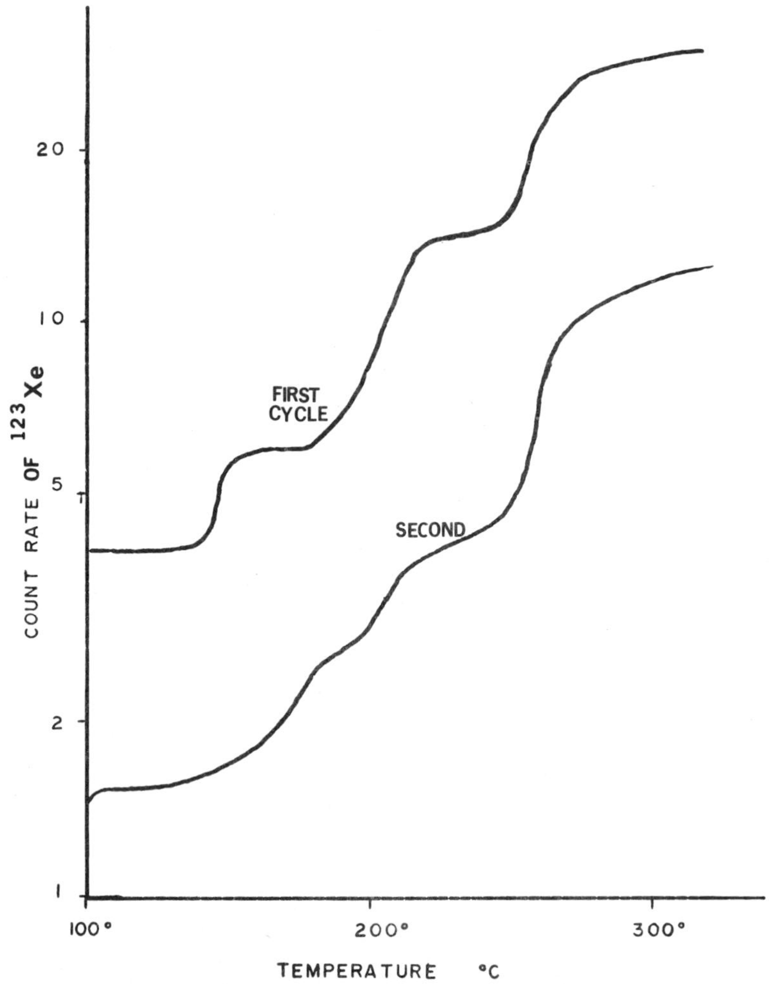

Fig. 7. Thermal release of ^{123}Xe *from tellurium*

between 200° and 350°C. These data show that ^{123}Xe can be removed by a brief heating at 300°C. There are discrete steps in the curve, with onset of ^{123}Xe release at 140°C, 180°C, and 240°C. The steps at 180°C and 240°C were observed again on a second cycle, i.e. after the same target was re-irradiated and then reheated. There was no apparent loss of target material after the first heating cycle, although the particle aggregates did appear to be larger in size.

Estimate of f for the cyclic method. The same method was employed as described earlier, in which χ is observed and compared with a calculated value. In this case the technique is more accurate, since the question as to where the tellurium is during the bombardment does not exist. It is estimated that 80% of the ^{123}Xe had diffused from the sample at the end of the heating cycle.

Performance of the traps. In the studies so far, traps of ¼-in copper tubing, ¼-in Tygon tubing and 3 mm ID glass have been used. The copper and Tygon traps work as expected, with radioiodines sticking in the trap at dry ice temperature and xenon at liquid nitrogen temperature. The glass trap at dry ice temperature retains a significant amount of ^{123}Xe.

The ease of removal of activity from the traps by leaching decreases in the order, glass > copper > Tygon. Most of the activity can be removed from the glass trap with a fraction of a ml of water. About 10 ml of dilute NaOH removes 60–70% of the activity from a copper trap and only 20–30% from the Tygon trap.

Attempts at direct labelling. The ^{123}Xe atom sticks to the wall of the trap and then decays to ^{123}I. This iodine atom is very reactive and undoubtedly undergoes some reaction with the wall of the trap or with some other molecule in the trap, e.g. oxygen gas or water vapour. The resultant species could be sufficiently reactive to chemically combine with an organic molecule. On this basis, macroaggregated serum albumin (MAA) was used to leach traps. The solution was 0·1%, pH 5·5, saline. No labelling was observed on leaching a glass trap, while the same solution used to leach the Tygon tube showed 34% labelling. So far attempts to tag human serum albumin have been unsuccessful.

DISCUSSION

General. The production of pure ^{123}I via ^{123}Xe, produced by any of the reactions ^{122}Te(^{4}He, 3n), ^{122}Te(^{3}He, 2n), or ^{123}Te(^{3}He, 3n), appears to be quite feasible. Yields are satisfactory and the purity of the ^{123}I is superior to that from any other method. In addition, the possibility of non-destructive removal of the ^{123}Xe from the target has been demonstrated. The advantage of this feature for any large-scale production of ^{123}I is obvious.

Generator method. This method was somewhat disappointing, since the fraction of ^{123}Xe diffusing from the tellurium was low and there was difficulty in keeping the tellurium on the screen and in the beam. The advantage of the method is its physical simplicity and potentially higher yield, because of the collection of ^{123}I produced from ^{123}Xe decay during the bombardment. Additional experimentation for a better arrangement and a means of increasing the fraction of ^{123}Xe released from the target is warranted. The use of a thin deposit of tellurium on thin porous carbon sheets will be investigated. The loss of tellurium through the porous mesh and the low flow rate of helium gas, required to keep the radioiodine contaminants in the dry ice trap, indicate that a screen material of smaller pore size could be used.

Cyclic method. This method gives nearly complete ^{123}Xe removal by a post-bombardment thermal cycle at 300°C for about 10 min. The target material is not lost on heating and can be re-irradiated and carried through another thermal cycle. The present arrangement has the disadvantage of requiring a cover foil after each irradiation. This step might be circumvented by the use of a porous screen to contain the target material.

^{123}I *removal from the liquid nitrogen trap.* The success in labelling MAA with ^{123}I by bringing the solution in direct contact with the contents of the trap is a promising technique, and the removal of the activity from a glass trap with a fraction of a ml of solution means that carrier-free solutions of high specific activity can be produced.

There are many possibilities to examine in the radioiodine pickup from the trap. The fact that labelling was possible from a Tygon tube suggests that a glass tube might be coated with the organic compound to be tagged before ^{123}Xe enters the trap. When ^{123}Xe enters and decays to ^{123}I the reactive iodine atom will combine with and label the organic molecule on the trap wall.

REFERENCES

1. Sodd, V. J. and Blue, J. W., previous paper.
2. Sodd, V. J. and Blue, J. W., *J. Nucl. Med.* 9, 349, (1968).
3. Sergolle, H., Albouy, G., Bouloumie, J., Lagrange, J. M., Marcus, L., and Pautrat, M., *J. Physique* 28, 383, (1967).
4. Ragaini, R. C., Walters, W. B., Gordon, G. E., and Baedecker, P. A., *Nucl. Phys.* **A115**, 611, (1968).

Some problems of RbCl targets irradiated with deuterons in the cyclotron

J. J. Hrubý

Nuclear Research Institute, Czechoslovak Academy of Sciences, Řež, Nr. Prague, Czechoslovakia.

ABSTRACT

Deuteron irradiations of molten RbCl without a protecting metallic foil were performed in the extracted beam in the U-120 cyclotron. The target technique used is briefly described. γ-spectra and decay curve analysis were used to identify the following radionuclides in the irradiated target: ^{84}Rb, ^{86}Rb, ^{88}Rb, ^{85m}Sr, ^{85}Sr, ^{87m}Sr, and ^{38}Cl. For some the 'saturation activity per μA' was estimated (^{87m}Sr – 6·4 mCi/μA, ^{85m}Sr – 23·5 mCi/μA, ^{38}Cl – 14·9 mCi/μA) as well as the total thick-target yield for the production of ^{85}Sr (15 μCi/μAh). The losses of the activity produced (^{85}Sr) and of the target material were studied. Attempts were made to find the optimum conditions for irradiation (other Rb salts, especially in mixtures, pretreatment of target, and programmed irradiation). These aspects are discussed from the viewpoint of optimising the production of ^{85}Sr.

INTRODUCTION

^{85}Sr is one of the most popular cyclotron-produced radioisotopes and in our laboratory we have studied the problem of irradiating Rb by deuterons in the U-120 cyclotron.[1] When deciding on the most convenient target compound we rejected metallic rubidium, and its alloys with other metals, for their inconvenient physical and chemical properties. RbCl seems to be the most suitable target compound and we have irradiated RbCl melted on Cu or Ag target mountings without a protecting foil. The target material, as well as the product, may be partially lost during irradiation in the vacuum system of the cyclotron, and we have studied methods by which losses during irradiation can be reduced. We have also tried other target compounds, especially mixtures of RbCl + Rb_2SO_4 and RbCl + RbF. For production purposes we have determined thick-target yields of ^{85}Sr for different ion beam currents. For health physics reasons the short-lived products in the irradiated targets were also determined.

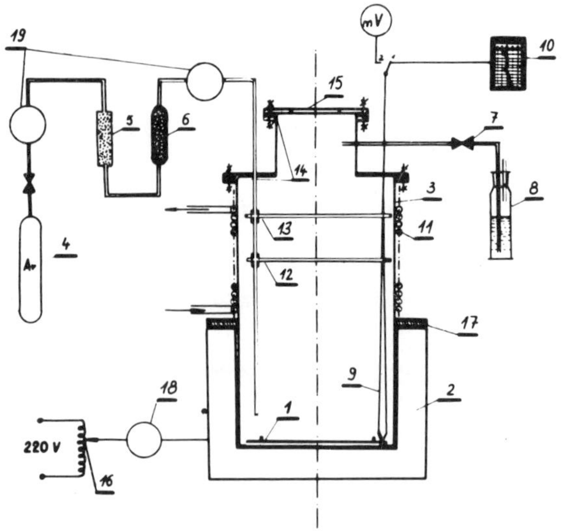

Fig. 1. Device for cyclotron target preparation from molten material: 1. Target; 2. Furnace; 3. Stainless steel vessel; 4. Argon cylinder; 5. $PdCl_2$ *catalyst; 6. Molecular sieve; 7. Gas pressure regulation valve; 8. Outlet vessel (filled with silicone oil); 9. Thermocouple; 10. Temperature recorder; 11. Water cooling; 12.* Cu *heat shield; 13.* Al *heat shield; 14. Rubber gasket; 15. Observation window; 16. Transformer; 17. Asbestos; 18. Wattmeter*

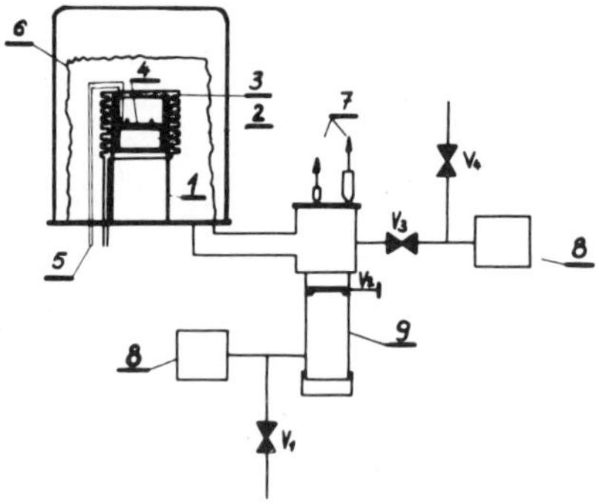

Fig. 2. Device for vacuum annealing of targets: 1. Ceramic crucible; 2. Glass support; 3. Heating wire; 4. Target; 5. Thermocouple; 6. Al *screening; 7. Vacuum gauge; 8. Vacuum pump; 9. Oil diffusion pump;* V_1, V_2, V_3, V_4*—Vacuum valves*

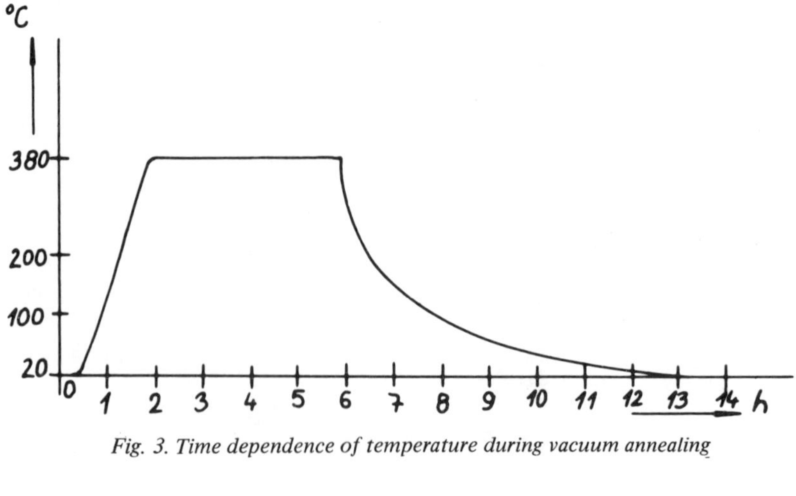

Fig. 3. Time dependence of temperature during vacuum annealing

EXPERIMENTAL

Melting of target compound on target mountings. Fig. 1 shows details of a special device which enabled us to melt the salt in an argon atmosphere. Argon was purified by passing over a $PdCl_2$ catalyst to remove oxygen and through molecular sieve A-4 to remove moisture. The target mountings used were circular discs of Ag or Cu, of outer diam. 48 mm and inner diam. 39 mm, i.e. the inner surface was 12 cm^2. Some of the mountings used were provided with 17 grooves. The covered surface in this case amounts to ~14 cm^2. The amount of target melt attached to the mountings was 1·6–2·2 g. During melting the temperature inside the main body of the device was held at 750°C for 10 min. Two hours were needed to reach this temperature and 0·5–2 h to cool down to 20°C, depending on the rate of cooling. Variation of the latter was possible because the main body of the device could easily be pulled out of the furnace and left to be cooled either by its own water-cooling system alone, or by immersion in cold water. The Ag mountings were degreased before covering with the melt, by washing in CCl_4. They were then washed with water and alcohol and stood for 5 min in 10% HNO_3. After repeated washing with water and alcohol they were annealed at 600°C for 5 min. The Cu mountings were cleaned in the same way, the only difference being that, instead of 10% HNO_3, 10% H_2SO_4 was used. Some were annealed in vacuum for 3–6 h at 300°C.

Pre-irradiation annealing of targets in vacuum. For this purpose a special device was constructed (Fig. 2). Prepared targets were annealed in a vacuum of $1\text{-}3 \times 10^{-5}$ torr for 4–8 h at various temperatures (300, 380 and 450°C). The variation of target temperature as a function of time for an annealing temperature 380°C (annealing time 4 h) is given in Fig. 3.

The programme of heating and cooling illustrated was used for all annealed targets.

Target irradiations. Targets were irradiated in the extracted beam of the U-120 cyclotron with 13 MeV deuterons at a current of 8–40 μA. The targets were located in a special target holder which permitted cooling by distilled water during irradiation. About 10 mm in front of the target holder was a water-cooled oblong screen with a circular slot 32 mm in diam. The deuteron beam passing through the slot bombarded the target. The target was monitored during irradiation through an observation window by a T.V. camera. The intensity of the beam bombarding the target and the screen was measured electrically. The values obtained were checked calorimetrically at a constant flow-rate of cooling water.

Measurement of activity produced and of target material losses. After irradiation the target melt was dissolved from the target mounting with 30 ml of 1 N HNO_3 containing a small amount of inactive Sr carrier. The mounting itself was then washed several times with 1 N HNO_3 and the washings added to the bulk solution. The total solution was diluted to exactly 100 ml, and from this solution aliquots were taken for measurements. The target material deposited during irradiation on the rear side of the screen was also dissolved and the solution used for measuring the irradiation losses. The losses of target material, as well as the change in weight of the target mounting, were checked by weighing the target and the target mounting at every stage of the target treatment. The losses were also checked by estimating the residue of the screen washings after evaporation.

For activity measurements an ionisation chamber and a γ-spectrometer, with a 200-channel or 1000-channel analyser and a semiconductor detector, were used.

RESULTS

Transfer of target mounting material into the melt. For Cu target mountings the amount of Cu transferred into the RbCl melt ranged from 1·5 to 3 mg, the average of 20 determinations being 2·1 mg. For Ag target mountings the amount transferred was 1·8–5·3 mg, the average being 3·2 mg. On Ag target mountings with RbCl + 5 wt% Rb_2SO_4 the amount of Ag transferred was 2·4–2·8 mg, average 2·7 mg, and for the same mounting with RbCl + 0·2 wt% RbF the values were 8·6–9·1 mg, average 8·9 mg Ag.

Losses from melts during pre-irradiation annealing. The results are given in Table 1, which shows that losses of RbCl are larger on Cu mountings than on Ag mountings. Lower losses are observed for melts containing Rb_2SO_4 or RbF.

Radionuclides produced in the RbCl *target bombarded by deuterons.* Fig. 4 gives the time-dependence of the total activity of the irradiated target, together with its resolution into individual components.

As can be seen the following half-lives were determined: 2·8 h, 70 min, 38 min, and 18 min, corresponding to the half-lives of ^{87m}Sr, ^{85m}Sr, ^{38}Cl, ^{88}Rb.

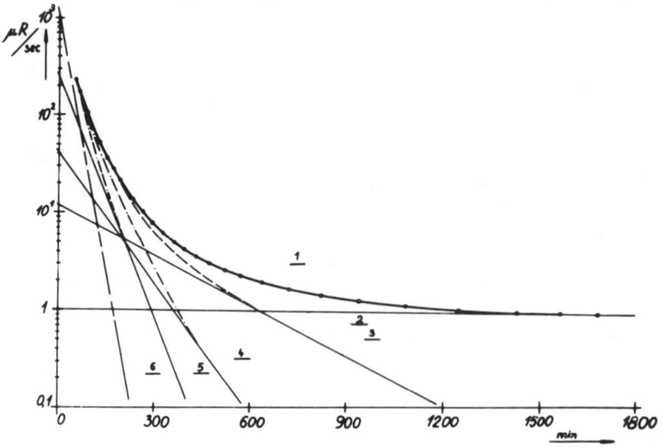

Fig. 4. Total γ radiation of irradiated RbCl *target as a function of time, and its analysis. (Deuteron beam intensity 11·5 μA, energy 13 MeV, irradiation time 6·6 h) 1. Total activity; 2. Activity due to* $^{86}Rb + ^{85}Sr$*; 3. Activity due to* ^{87m}Sr*; 4. Activity due to* ^{85m}Sr*; 5. Activity due to* ^{38}Cl*; 6. Activity due to* ^{88}Rb

By means of γ-spectra the following radionuclides were identified: ^{88}Rb, ^{86}Rb, ^{84}Rb, ^{85m}Sr, ^{85}Sr, ^{87m}Sr, and ^{38}Cl. We did not succeed in identifying ^{34m}Cl, ^{36}Cl, ^{37}Ar, ^{83m}Kr, ^{85}Kr, and ^{87}Kr, or radionuclides with half-lives <1 min, although they might be present in the irradiated target. For ^{87m}Sr, ^{85m}Sr, and ^{38}Cl we determined the values of saturation activity for 1 μA (Table 2).

Thick-target yields were determinated for ^{86}Rb, ^{84}Rb and ^{85}Sr on the basis of γ-spectra measured on the 3rd, 10th and 18th days after the end of the irradiation. The results presented here were obtained for the irradiation of RbCl targets by deuterons at a current of 8·2–9 μA for 1 h. The average values (in μCi/μAh) are: 0·11 for ^{84}Rb, 11·4 for ^{86}Rb, and 14·1 for ^{85}Sr. Table 3 gives data[2,3] for the radionuclides and Table 4 gives possible products of the interaction of deuterons with RbCl.

Table 1. ANNEALING OF MELTS IN VACUUM

Target mounting	*Temperature of annealing* °C	*Annealing period* h	*Type of melt*	*Evaporation of melt* *total, %*	*Evaporation of melt* *during 1 h* % h	*Remarks*
Cu	300	8	RbCl	0·147	0·0184	
Ag	300	8	RbCl	0·109	0·0137	
Cu	380–385	8	RbCl	2·505	0·313	1st annealing
Cu	380–386	4	RbCl	1·13	0·282	2nd annealing
Cu	380–387	4	RbCl	1·05	0·267	3rd annealing
Cu	380–385	4	RbCl	0·848	0·212	4th annealing
Ag	388–395	4	RbCl	1·45	0·362	1st annealing
Ag	382–386	4	RbCl	0·664	0·166	2nd annealing
Ag	455–460	8	RbCl	17·5	2·19	
Ag	480–485	4	RbCl	14·4	3·6	
Ag	405–410	4	RbCl	8·4	2·2	
Ag	400–405	4	RbCl	5·4	1·35	
Ag	380–386	4	RbCl + 5 wt% Rb_2SO_4	0·666	0·166	1st annealing
Ag	380–385	4	RbCl + 5 wt% Rb_2SO_4	0·703	0·175	2nd annealing
Ag	382–386	4	RbCl + 0·2 wt% RbF	0·677	0·169	1st annealing
Ag	382–386	4	RbCl + 0·2 wt% RbF	0·555	0·138	2nd annealing

Table 2. YIELDS OF SOME RADIONUCLIDES IN THE IRRADIATED RbCl TARGET

Radionuclide	*Deuteron beam intensity*	*Irradiation time*	*Activity at the end of irradiation*		*Saturation activity*		*Average of saturation activity determinations*
	μA	h	μR/s	mCi	mCi	mCi/μA	mCi/μA
^{87m}Sr	19·3	1	15	28·5	129	6·7	6·4
	11·5	6·6	12	40·5	50·03	4·3	
	8·5	12·1	12·5	65·4	68·9	8·1	
^{85m}Sr	19·3	1	58·3	171	370	19·2	23·5
	11·5	6·6	41·0	216	220·5	19·2	
	8·5	12·1	34·0	273	273	32·2	
^{38}Cl	11·5	6·6	240	197	197	17·1	14·9
	8·5	12·1	85	109	109	12·8	

Table 3. NUCLEAR DATA FOR RADIONUCLIDES PRODUCED IN THE IRRADIATED TARGET

Radionuclide	*Half-life*	$K\gamma$	*Energy of radiation*, keV, *and percentage abundance*
^{84}Rb	33·0 d	4·55	880 (74%) 1010 (0·5%) 1900 (0·8%)
^{86}Rb	16·66 d	0·546	1078 (8·8%)
^{86m}Rb	1·02 s	3·10	560 (98%)
^{88}Rb	17·8 min	3·37	1863 (21%) 898 (13%) 2680 (2·3%)
^{85}Sr	64 d	2·94	514 (100%)
^{85m}Sr	70 min	1·10	231 (85%) 150 (14%)
^{87m}Sr	2·83 h	1·72	388 (80%)
^{34}Cl	1·56 s	5·90	511 (200%)
^{34m}Cl	32 min		511 (100%) 145 (45%) 2120 (38%) 1170 (12%) 3300 (12%)
^{34m}Cl + ^{34}Cl		14·31	511 (196%) 2130 (51·9%) 145 (26·6%) 1160 (16·6%) 3300 (16·6%) 4100 (0·6%)
^{36}Cl	$3{\cdot}08 \times 10^5$y		511 (0·003%)
^{38}Cl	37·29 min	6·98	2170 (47%) 1600 (38%)
^{38m}Cl	0·74 s	3·73	660 (100%)
^{83m}Kr	1·86 h		9·0 (9%)
^{85}Kr	10·76 y	0·021	514 (0·41%)
^{85m}Kr	4·4 h	0·804	150 (74%) 305 (13%)
^{87}Kr	76 min	5·03	430 (84%) 2570 (35%) 850 (16%)
^{35}Ar	1·83 s	6·38	511 (200%) 1220 (5%) 1760 (2%)

Table 4. PRODUCTS OF INTERACTION OF DEUTERONS WITH RbCl

Nuclear reaction	^{85}Rb	^{87}Rb	^{35}Cl	^{37}Cl
(d, p)	^{86m}Rb ^{86}Rb $Q = 6{\cdot}35$	^{88}Rb $Q = 3{\cdot}97$	^{36}Cl	^{38m}Cl ^{38}Cl $Q = 3{\cdot}88$
(d, n)	^{86}Sr stab.	^{88}Sr stab.	^{36}Ar stab.	^{38}Ar stab.
(d, 2n)	^{85m}Sr ^{85}Sr $E_t = 5{\cdot}16$	^{87m}Sr $E_t = 2{\cdot}79$ ^{87}Sr stab.	^{35}Ar	^{37}Ar $E_t = 4{\cdot}03$
(d, t)	^{84}Rb $E_t = 4{\cdot}37$	^{86m}Rb ^{86}Rb $E_t = 3{\cdot}74$	^{34m}Cl ^{34}Cl $E_t = 6{\cdot}67$	^{36}Cl
(d, α)	^{83m}Kr ^{83}Kr stab.	^{85m}Kr ^{85}Kr	^{33}S stab.	^{35}S
(d, 2p)	^{85m}Kr ^{85}Kr	^{87}Kr	^{35}S	^{37}S
(d, 2pn)	^{84}Kr stab.	^{86}Kr stab.	^{34}S stab.	^{36}S stab.

Isotopic abundance of the target nuclides: ^{85}Rb 72·15%
^{87}Rb 27·85%
^{35}Cl 75·4%
^{37}Cl 24·6%

Irradiation efficiency. The dependence of thick-target yields for ^{85}Sr on the deuteron beam intensity is given in Fig. 5.

In Fig. 6 the loss of RbCl during irradiation is shown to be a function of the deuteron beam intensity and the irradiation period. It can be seen that evaporation of RbCl from the target is greatest at the beginning of an irradiation. In this connection it is interesting to note that the thick-target yields of ^{85}Sr (RbCl on Ag) for different irradiation times (7, 11 and 16 h) and constant beam intensity (25 μA) are practically the same (12·3, 10·7, and 11·3 μCi/μAh). When irradiating the same target (RbCl on Ag) at 8 μA (for 15 h) the loss of ^{85}Sr is 1% and the loss of inactive material 10%. For 22 μA deuteron beam intensity the loss of ^{85}Sr is <18% and the loss of RbCl is 25%. Irradiation with a deuteron beam intensity >35 μA leads to an activity loss higher than the loss of inactive material, both values being very high (>50%). The loss of ^{85}Sr at higher deuteron beam intensities can be substantially reduced by pre-irradiation annealing of the target at 380°C for 4 h. The same result can be reached by a programmed time dependence of the beam intensity: beginning at ~5–10 μA for 1 h, then increasing to 30 μA and lowering it before the end of the irradiation to two thirds of the maximum value. A similar effect is achieved when RbCl + 5 wt% Rb_2SO_4 is used as the target compound instead of pure RbCl: for 21 μA (9·3 h irradiation time) and 23 μA (8·83 h) the yields were 14·5 and 13·2 μCi/μAh. RbCl + 0·2 wt% RbF was not irradiated, as it is unstable and becomes brittle after a time.

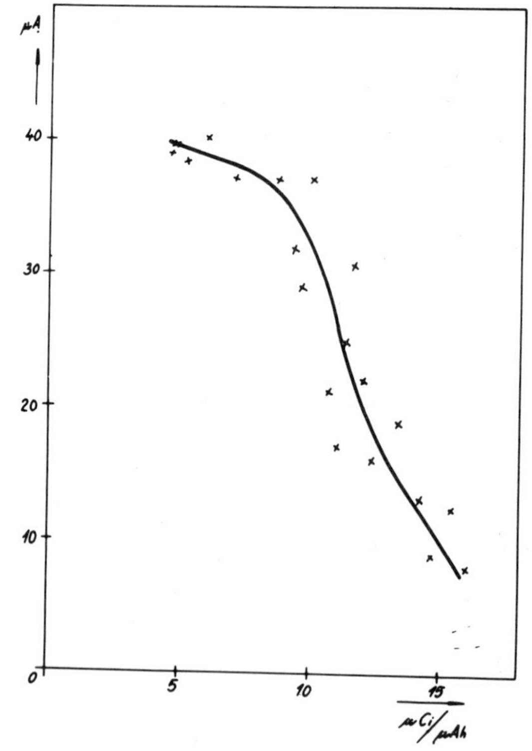

Fig. 5. ^{85}Sr *thick-target yield as a function of deuteron beam intensity (ordinates in* μA, *abscissae in* $\mu Ci/\mu Ah$*)*

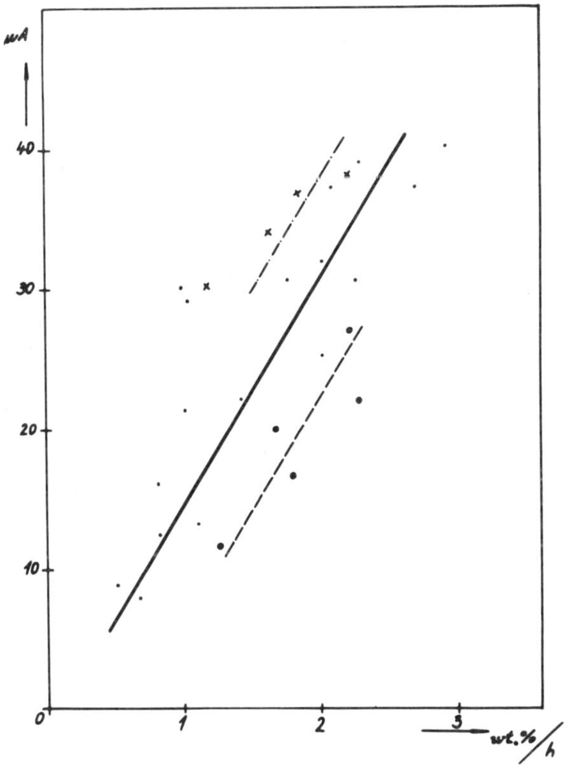

Fig. 6. Weight decrease of RbCl *melt during irradiation as a function of the intensity of the deuteron beam for different total irradiation times:* x – *19–22 h irradiation time;* ● – *14–16 h irradiation time;* ○ – *9–11 h irradiation time; The weight decrease is given in weight % per hour*

DISCUSSION

The loss of active ^{85}Sr during irradiation is not a simple process, and is influenced by: evaporation of target material (due to heating on bombardment), thickness of the target compound layer, cooling of the target (temperature and flow-rate of cooling water), dimensions of the irradiated crystals and defects produced during target preparation and irradiation, energy at which the excitation function of the nuclear reaction is its maximum, etc. From our results it follows that molten RbCl can be irradiated without a protecting foil by a deuteron beam of intensity up to 30 μA; the thick-target yields are $\geqslant$10 μCi/μAh. As can be seen from Fig. 5, a rapid decrease in yield occurs at 32 μA. Most of the ^{85}Sr activity lost from the target is found on the screen (70–80%) and on the tapered surface of the target holder (20–30%). The decreased ^{85}Sr loss after pre-irradiation annealing or on programmed irradiation can be explained by the 'improved' physical properties of the RbCl layer, whether this is achieved by annealing before irradiation or by the irradiation cycle itself. The physical properties of the layer can also be improved by using mixtures of RbCl with other Rb salts. The best results were obtained with a mixture of RbCl and Rb_2SO_4 (2–4%). At Rb_2SO_4 >5% the layer becomes inhomogeneous and its properties deteriorate, the melt becoming brittle. The melt with 5 wt% Rb_2SO_4 is hygroscopic; in air its weight increases by 0·1–0·16% during the first 10 h, and afterwards remains constant. Very good mechanical properties, e.g. strength, adhesion to the target mountings of Cu or Ag, remain even after absorbing the small amount of moisture. Such a melt has a straight and smooth surface. In comparison with the pure RbCl melt these melts contain 20% less transferred Ag, their losses during annealing at 380°C are lower and remain constant when annealing is repeated.

REFERENCES

1. Alexeev, A. V., *et al, Atomn. Energ.* 7, 148, (1959).
2. Gusev, N. G., Mashkovich, V. P., and Verbitskii, B. V., 'Gamma emission of radioactive isotopes' (62, 63), Atomizdat (Moscow) (1964).
3. Lederer, C. M., Hollander, J. M., and Perlman, I., 'Table of Isotopes' 220–225, John Wiley and Sons, Inc. New York, London, Sydney (1968).

Radioisotope production in the I.Ph.P.E. cyclotron

N. N. Krasnov, P. P. Dmitriyev, I. O. Konstantinov, N. A. Konjakin, A. A. Ponomarev, A. A. Ognev, A. E. Romanchenko, and V. M. Tuyev

Institute of Physics and Power Engineering, Obninsk, U.S.S.R.

ABSTRACT

The main problems of radioisotope production on the cyclotron are considered and operational data for isotope production in the cyclotron are presented. Three types of targets used for the bombardment of different materials by the cyclotron beam are described. Data are given on the isotope yields for 40 isotopes produced on the I.Ph.P.E. cyclotron.

INTRODUCTION

Radioactive isotope production in a cyclotron requires the solution of a number of special problems. The three main problems are (1) to obtain beams of different particles of the required energy and intensity, (2) to design target assemblies for irradiation of different materials, and (3) to find suitable nuclear reactions giving maximum isotope yields. This paper describes briefly the solution of these problems in the I.Ph.P.E. 1·5 m cyclotron,[1] which began operation in 1963.

OPERATING CHARACTERISTICS OF THE CYCLOTRON

Consideration of the nuclear reactions leading to the formation of the desired isotopes shows that a cyclotron should be able to accelerate three types of particles, viz. protons, deuterons and He ions. Any radioactive isotope of practical interest can be obtained in a cyclotron and the nuclear reaction which provides the highest isotope yields can be chosen for every isotope.

The main operating characteristics of the I.Ph.P.E. cyclotron are given in Table 1. The cyclotron operates under pulsed conditions, and the modulation frequency is adjusted in the range 25–300 Hz. This cyclotron is a conventional machine in which the required magnetic field form is achieved by means of shims. In changing from one magnetic field (15·5 kilogauss) to another (10·5 kilogauss) the external shims are changed. A comparatively high internal beam intensity

Table 1. MAIN OPERATING CHARACTERISTICS OF THE I.Ph.P.E. CYCLOTRON

Particle type	*Particle energy* MeV	*Beam current* (μA) *Internal* *In pulse*	*Internal* *Average*	*Deflected* *Average*	*Magnetic field* (kilogauss)	*Radiofrequency* (Mc/s)
Protons	22	3000	600	50	10·6	15·9
Protons*	11	3000	600	100	15·5	11·6
Deuterons	22	3000	600	50	15·5	11·6
He ions	44	1500	300	50	15·4	11·6

* In this case protons are accelerated as H_2^+ ions.

was obtained by eliminating vertical beam instability resulting from the lack of coincidence of the magnetic plane with the geometrical median plane of the magnet.[2] For regular control of the vertical position of the beam a three-plate probe is used.

When the targets are irradiated by an internal beam an artificial increase of the beam cross-section is achieved by excitation of radial oscillations of particles. The position of the orbit centre is determined by the two-probe method.[3]

To increase the coefficient of deflection the beam is collimated in the first revolution using a one-slit movable collimator with a 10 mm slit. In this case the coefficient of deflection is ~65%.

TARGET ASSEMBLIES

Three types of target assembly were designed which permitted the irradiation of various materials having different properties. Two targets are used for irradiation by the internal beam and one by the deflected beam. The position of targets in the cyclotron is shown in Fig. 1.

Rotating Target. The head of the rotating target is shown in Fig. 2. The irradiated part (7) is a cylinder with a hemispherical bottom of 90 mm diam. The target head consists of a body (1) on which a movable carriage (3) with sprung catches (4), a rubber sealing ring (5) and a fairing (6) are mounted. Cooling water flows through the clearance between the fairing and the irradiated cylinder. A screw-nut (8) is used to press the sealing ring. The direction of the cooling water flow and that of the particle beam are shown in the figure by arrows. The body of the target head is connected with the rod (2), which is rotated. The target is removed remotely after irradiation. The screw-nut is first relaxed by back rotation. The pusher mounted on the flange then countersinks the catches and pushes off the irradiated target while the head of the target is moving back. The rotation speed of the target is 210 rpm and the cooling water flow-rate 30 l/min. The maximum power dissipation depends on the properties of the irradiated material and varies from 5 to 10 kW. The rotating target is used to irradiate materials from which a cylinder with a hemispherical bottom can be fabricated (e.g. magnesium, copper, iron, silver). This target is also used to irradiate metals plated on to the copper backing (e.g. chromium, nickel, zinc).

Stationary Internal Target. This target (Fig. 3) is intended for irradiation of materials which are not suitable for irradiation on the rotating target.

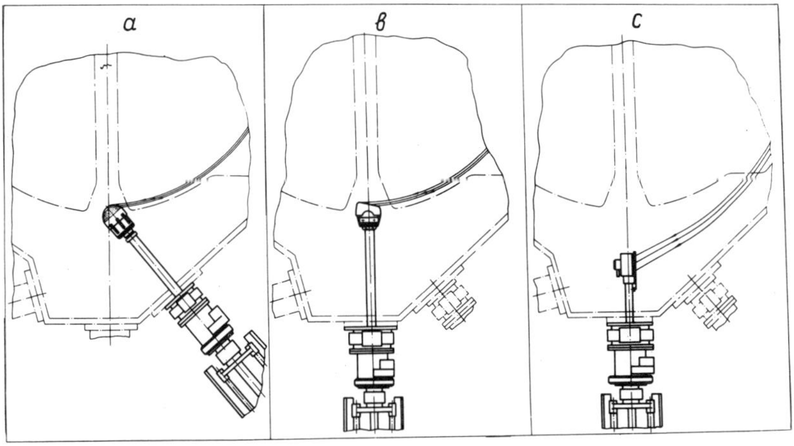

Fig. 1. Position of targets in the cyclotron:
(a) rotating target
(b) stationary internal target
(c) target for deflected beam

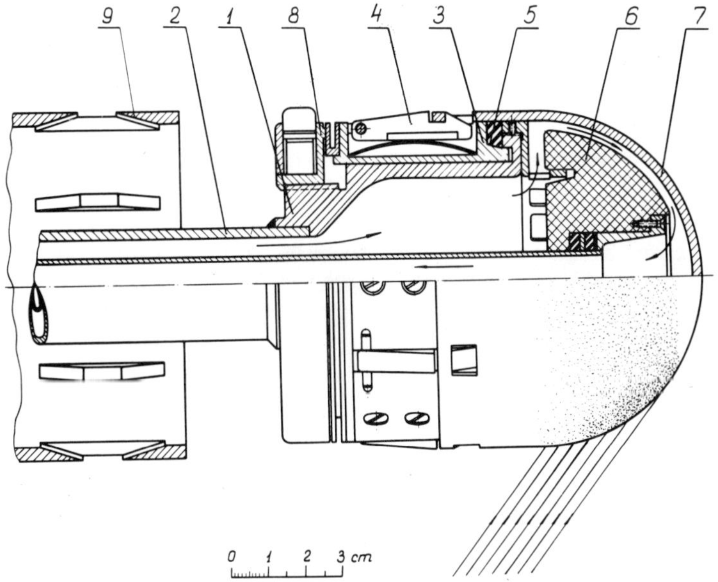

Fig. 2. Head of rotating target: (1) Body; (2) Rod; (3) Movable carriage; (4) Catches; (5) Rubber ring; (6) Fairing; (7) Irradiated cylinder; (8) Screw-nut; (9) Pusher

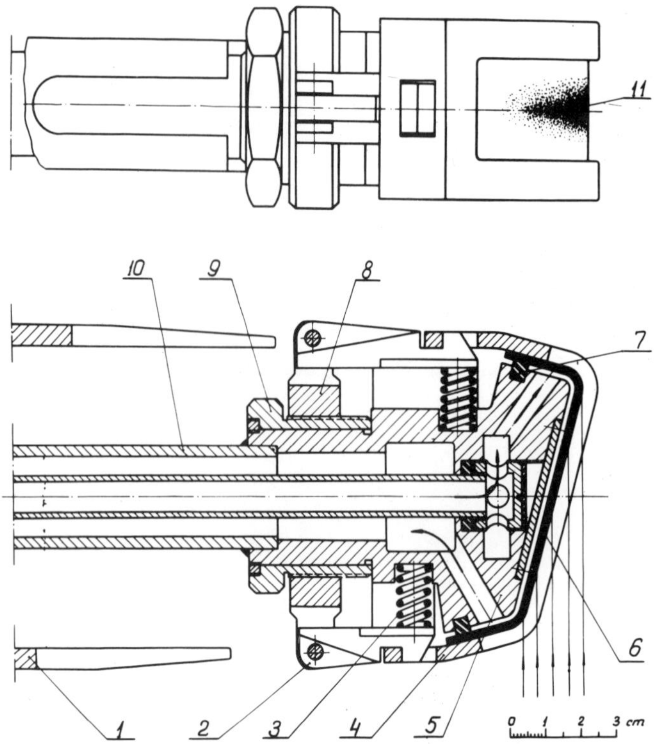

Fig. 3. Head of stationary internal target: (1) Lugs; (2) Catches; (3) Springs; (4) Pressing frame; (5) Body; (6) Irradiated part; (7) Sealing gasket; (8) Screw-nut; (9) Screw-bush; (10) Rod; (11) Irradiated surface of the plate

The irradiated part (6) is a plate with bent ends; the flat part of the plate is 40 × 70 mm, and the size of the irradiated surface is increased by positioning the plate at an angle of 15° to the beam direction. The plate is pressed to the head of the body (5) by means of a pressing frame (4). The latter is held by catches (2) connected with a screw-nut (8), which is mounted on the screw-bush (9). After positioning the plate and the frame, the screw-nut, together with the catches, is moved back by rotating the screw-bush. At the same time the plate is pressed to the rubber sealing gasket (7) in the grooves of the head. Cooling water flows through the clearance between the head and the plate. The direction of the cooling water flow and of the beam are shown by arrows. The target is removed after irradiation by remote control. When the target head is moving back the lugs mounted on the flange countersink the catches and push off the pressing frame, together with the irradiated plate. The cooling water flow-rate is 25 l/m. The power dissipation varies according to the material irradiated from 1–4 kW. This target is used to irradiate materials such as tantalum, platinum, rhodium, titanium, vanadium, antimony, lanthanum, etc. The irradiated material is made in the form of a plate with bent ends or is welded to a backing of this form.

Stationary Target for Deflected Beam. This target (Fig. 4) is intended for irradiation of easily-melted metals such as lithium, gallium, rubidium, etc. The irradiated material is placed into a container (4), the front wall of which is a thin foil. This container is pressed to the head (3) by means of the pressing frame (6). Pressing is done in the following way: when the lever (9) is moved to the left it turns the screw (7). The screw plug (8) therefore rests against the back surface of the body and the pressing frame presses the container on to the rubber sealing gasket (5). During irradiation water cools the back surface of the container at a flow-rate of 25 l/min. The direction of the water flow and of the beam are shown by arrows. The collimating frame (2) is mounted on the

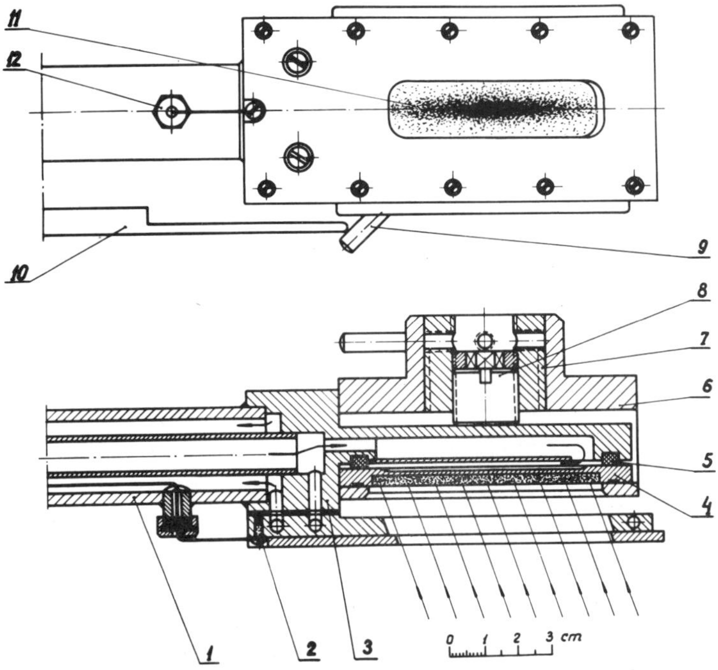

Fig. 4. Head of stationary target for deflected beam: (1) Rod; (2) Collimating frame; (3) Body; (4) Container with irradiated material; (5) Sealing gasket; (6) Pressing frame; (7) Screw; (8) Screw-plug; (9) Lever; (10) Lug; (11) Irradiated surface of the container; (12) Electrical line of the collimating frame

target body. This frame is plated with graphite and is electrically insulated from the target body. The size of the collimating window is 12 × 65 mm. The target is removed remotely after irradiation. As the target head moves to the left the lever runs against the lug (10) mounted on the flange and turns the screw (7); this releases the pressing frame. In the subsequent movement of the target to the left the pressing frame and the container are pushed off by the lug. The power dissipation during irradiation is usually from 0·5 to 1 kW.

ISOTOPE YIELDS

At present more than 40 radioactive isotopes (Table 2) are produced in the I.Ph.P.E. cyclotron. The yields of radioisotopes produced by different methods and the types of targets used are also shown, where ‘a’ is the rotating target, ‘b’ the stationary internal target, and ‘c’ the external target. The isotope yields

Table 2. YIELDS OF RADIOACTIVE ISOTOPES PRODUCED IN THE I.Ph.P.E. CYCLOTRON

Isotope	*Half-life*	*Production mode*	*Yield* $\mu C/\mu Ah$	*Target type*
1	*2*	*3*	*4*	*5*
^{7}Be	53 d	Li + p Li + d	185 60	c
^{22}Na	2·62 y	Mg + p Mg + d	0·5 3·4	a
^{28}Mg	21 h	Mg + α Al + α	1 0·1	a
^{26}Al	$7{\cdot}4 \times 10^{5}$ y	Al + p Mg + p Mg + d Mg + α	$8{\cdot}3 \times 10^{-6}$ $1{\cdot}3 \times 10^{-6}$ $2{\cdot}3 \times 10^{-6}$ $2{\cdot}2 \times 10^{-6}$	a
^{47}Ca	4·7 d	Ca + p	0·65	c
^{44}Ti	47·3 y	Sc + p	0·03	b
^{46}Sc	84 d	Ti + d	7·1	b
^{48}V	16 d	Ti + p Ti + d	510 290	b
^{49}V	330 d	Ti + p Ti + d	3·2 8·3	b
^{51}Cr	27·8 d	V + p V + d	590 460	b
^{52}Mn	5·7 d	Cr + p Cr + d	550 390	a
^{54}Mn	313 d	Cr + p Cr + d Cr + α	0·45 1·9 5·8	a
^{55}Fe	2·7 y	Mn + p Mn + d	9 8	c
^{56}Co	77·3 d	Fe + p	74	a
^{57}Co	267 d	Fe + d Ni + p	3·1* 35	a
^{58}Co	71·3 d	Fe + α	28·5	a
^{65}Zn	245 d	Cu + p Cu + d	16 18·7	a
^{67}Ga	78 h	Zn + p Zn + d Cu + α	550 330 140	a
^{68}Ge	280 d	Ga + p	15	c

Table 2 continued

Isotope	*Half-life*	*Production mode*	*Yield* $\mu C/\mu Ah$	*Target type*
1	*2*	*3*	*4*	*5*
^{73}As	76 d	Ge + p Ge + d	33 26	b
^{74}As	17·5 d	Ge + p Ge + d	170 215	b
^{72}Se	8·6 d	Ge + α	19·5	b
^{75}Se	127 d	Ge + α	6·2	b
^{85}Sr	64 d	Rb + p Rb + d	165 115	c
^{87}Y†	80 h	Sr + p Sr + d Rb + α	2340 510 420	c
^{88}Y	105 d	Sr + p Sr + d	83 72	c
^{93}Mo	3000 y	Nb + p Nb + d	41×10^{-3} $5{\cdot}3 \times 10^{-3}$	b
^{103}Pd	17 d	Rh + p Rh + d	245 340	b
^{105}Ag	40 d	Rh + α	17	b
^{109}Cd	416 d	Ag + p Ag + d	5·1 6·6	a
^{111}In	2·8 d	Ag + α	270	b
^{113}Sn	119 d	Cd + α	2·5	b
^{121m}Te	154 d	Sb + d	12	b
^{123m}Te	104 d	Sb + d	10	b
^{139}Ce	140 d	La + p La + d	9·5 41	b
^{151}Gd	120 d	Eu + p Eu + d	1·6 8	c
^{181}W	140 d	Ta + p Ta + d	3·3 12·4	b
^{195}Au	183 d	Pt + p Pt + d	4 5·8	b

Table 2 continued

Isotope	*Half-life*	*Production mode*	*Yield* $\mu C/\mu Ah$	*Target type*
1	*2*	*3*	*4*	*5*
^{203}Pb	52 h	Tl + p Tl + d	680 410	c
^{205}Bi	15·3 d	Pb + p Pb + d	90 40	b
^{208}Bi	6·4 d	Pb + p Pb + d	203 225	b
^{207}Bi	28 y	Pb + p Pb + d	0·2 0·22	b

* At E_d 7·5 MeV (without ^{56}Co) † After decay $^{87m}Y \longrightarrow {}^{87}Y$

shown in Table 2 are physical yields, i.e. those measured in the absence of isotope losses during irradiation. The error in the isotope yields, ±15%, results from errors in measuring the isotope activity and the exposure quantity. The technological isotope yields may depend on the irradiation conditions if partial dispersion or evaporation of the bombarded material occurs. To determine the optimum method of producing several isotopes their isotope yields were measured from different nuclear reactions Results of these investigations have been published separately: ^{22}Na,[4] ^{26}Al,[5] ^{52}Mn,[6] ^{54}Mn,[7] ^{57}Co,[8] ^{65}Zn,[9,10] ^{93}Mo,[11] ^{139}Ce,[13] ^{181}W,[14] ^{195}Au,[15] ^{109}Cd.[12]

The availability of three types of target which enable us to irradiate different materials and the possibility of using three types of charged particles allow us to produce in the I.Ph.P.E. cyclotron almost any desired isotope which is necessary for practical application and scientific investigations.

REFERENCES

1. Krasnov, N. N., Ognev, A. A., and Ponomarev, A. A., *Pribory i tehnika eksperimenta* **4,** 219, (1965).
2. Krasnov, N. N., Ognev, A. A., and Ponomarev, A. A., *Pribory i tehnika eksperimenta* **4,** 23, (1965).
3. Krasnov, N. N., Ognev, A. A., and Ponomarev, A. A., *Pribory i tehnika eksperimenta* **2,** 16, (1968).
4. Krasnov, N. N. and Dmitriyev, P. P., *Atomnaya energiya* **21,** 400, (1966).
5. Krasnov, N. N., Dmitriyev, P. P., Sevastyanov, Yu. G., and Bezmaternych, A. S., *Atomnaya energiya* **19,** 62, (1965).

6. Dmitriyev, P. P., Konstantinov, I. O., and Krasnov, N. N., *Atomnaya energiya* **26,** 467, (1969).
7. Krasnov, N. N. and Dmitriyev, P. P., *Atomnaya energiya* **21,** 52, (1966).
8. Krasnov, N. N. and Dmitriyev, P. P., *Atomnaya energiya* **20,** 57, (1966).
9. Dmitriyev, P. P. and Krasnov, N. N., *Atomnaya energiya* **18,** 184, (1965).
10. Dmitriyev, P. P., Konstantinov, I. O., and Krasnov, N. N., *Atomnaya energiya* **27,** 279, (1968).
11. Dmitriyev, P. P., Konstantinov, I. O., and Krasnov, N. N., *Yadernaya fizika* **6,** 209, (1967).
12. Dmitriyev, P. P., Konstantinov, I. O., Krasnov, N. N., *Atomnaya energiya* **22,** 310, (1967).
13. Dmitriyev, P. P., Konstantinov, I. O., and Krasnov, N. N. *Atomnaya energiya* **24,** 278, (1968).
14. Krasnov, N. N. and Dmitriyev, P. P., *Atomnaya energiya* **20,** 154, (1966).
15. Dmitriyev, P. P., Konstantinov, I. O., and Krasnov, N. N., *Atomnaya energiya* **23,** 61, (1967).

DISCUSSION

B. D. Pate (Simon Fraser University): How are the hemispherical targets fabricated?
N. N. Krasnov: They are pressed from sheet metal.
G. Burton (MRC, Hammersmith): What is the beam size striking the rotating target, and what is the maximum power which the latter can dissipate?
N. N. Krasnov: The beam size is $\sim$1 cm^2, and the maximum power which can be used without loss of isotopes by evaporation is 10 kW.

Isotope production in the Harwell variable energy cyclotron (V.E.C.)

I. L. Jenkins and A. G. Wain

Applied Chemistry Division, A.E.R.E., Harwell, Berks., England

ABSTRACT

The V.E.C. is primarily a research machine but it has been used for a limited isotope production programme. Production is restricted to those isotopes which the U.K.A.E.A. needs for research purposes and which are not easily available from other sources. This paper describes the irradiation and processing facilities provided by the V.E.C. and some of the irradiations that have been carried out.

Yields of 60 μCi of ^{236}Pu and 72 μCi of ^{235}Np have been obtained by irradiating metallic targets of natural uranium with 50 MeV protons. Proton beams of lower energy (<15 MeV) have been used to produce millicurie amounts of ^{56}Co by the ^{56}Fe(p, n)^{56}Co reaction, and excitation functions for some of the proton-induced reactions of ^{56}Fe have been determined.

Proton and deuteron bombardments of ^{151}Eu have been used to prepare ^{151}Gd, and millicurie quantities were made by the ^{151}Eu(d, 2n)^{151}Gd reaction using an oxide target. Small amounts of isotopes such as ^{140}Nd, ^{87}Y, and ^{66}Ga have also been produced.

INTRODUCTION

The important design features of the Harwell Variable Energy Cyclotron (V.E.C.) have been described by Lawson[1], and Amphlett[2] has discussed its main uses. These are for studies in chemistry, metallurgy and radiation damage. Broadly speaking, nuclear reactors only allow the preparation of stable or neutron-rich isotopes, but a charged-particle accelerator enables us to produce neutron-deficient isotopes. This paper describes the facilities available and some of the work carried out to produce radioisotopes in the V.E.C.

EXPERIMENTAL

Irradiation facilities. Targets may be irradiated in either the Chemists' Irradiation Rig or in the Users' Probe. The former allows the irradiation of foil targets cooled by air or inert gas, while the latter permits the irradiation of thick

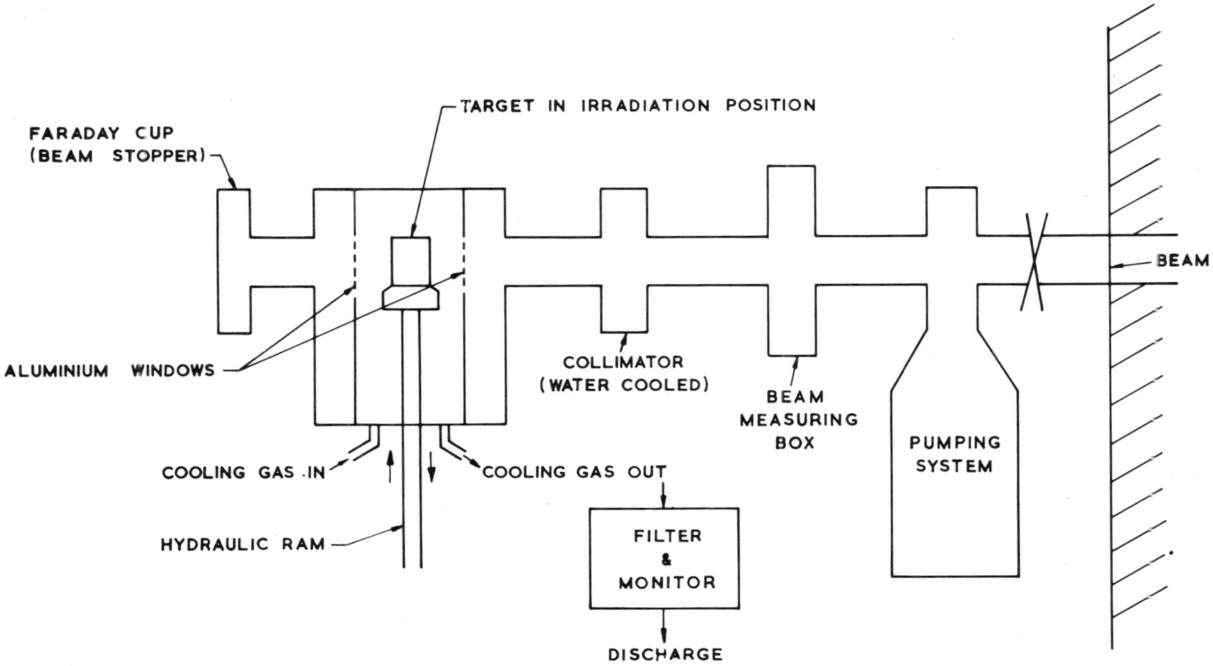

Fig. 1. Diagram of Chemists' Irradiation Rig

water-cooled targets which may be metallic, non-metallic (e.g. oxide) or liquid. The layout of the Chemists' Irradiation Rig is shown diagrammatically in Fig. 1. The irradiation chamber consists of a vacuum-tight steel cylinder fitted with aluminium windows (~0·003 in thick) through which the beam passes. The target is introduced from below and can be parked in an intermediate position while setting up the beam. For irradiation the target is raised hydraulically into position between the two aluminium windows, and is cooled by gas under pressure so that the beam is slightly degraded by passing through the gas and window before hitting the target. The coolant gas is activated by the beam and is passed through a filter before discharge. Monitoring facilities are provided so that any airborne contamination due to damage to the target can be detected. The target holder is removed from the assembly by means of remote handling tongs and the target is dismantled in a fume-hood or glove-box. A pneumatically-operated target transporter system is also available so that the irradiated target can be rapidly transferred from the target room into a fume-hood in the radiochemical laboratory.

The beam current can either be measured using a Faraday cup as a beam stopper or the actual current on the target may be monitored. The total integrated current is measured with a beam-current integrator accurate to ±3%.

Two types of target holders are shown in Fig. 2; these are made of aluminium and will accept targets up to 1½ in (3·8 cm) diam. Several foil targets, separated by aluminium spacers to allow efficient cooling, can be irradiated simultaneously using the stacked-foil technique. Heat dissipation of ~100 W total can be achieved in the Chemists' Irradiation Rig.

The Users' Probe allows much greater energy dissipation than the Chemists' Irradiation Rig and, to date, heat loads of up to 6·5 kW/in^2 (1 kW/cm^2) at 79°C have been achieved. Details of the target systems are given by Ball.[3] A schematic diagram of the Users' Probe assembly is given in Fig. 3. In this equipment, irradiations are carried out in vacuo and the target assembly can be presented normally or inclined to the beam. The target is water-cooled and irradiations can be carried out either with a static target or the specimen can be rotated

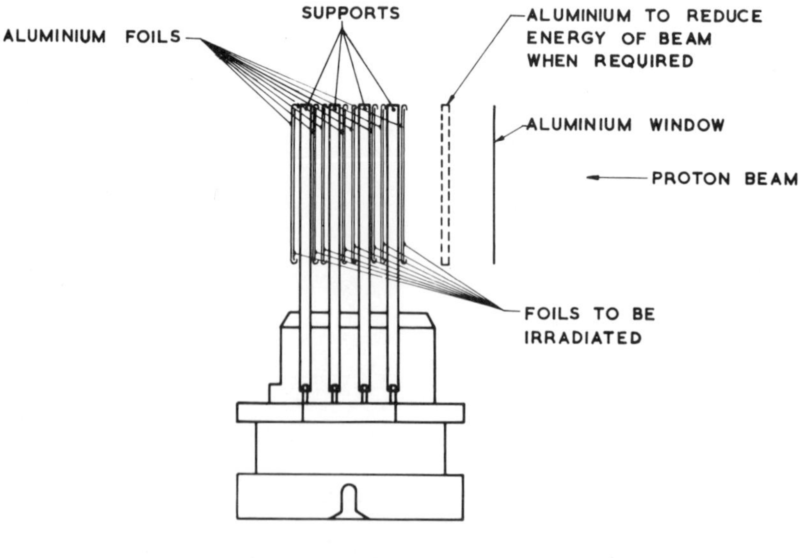

Fig. 2(a). Target assembly for Chemists' Irradiation Rig

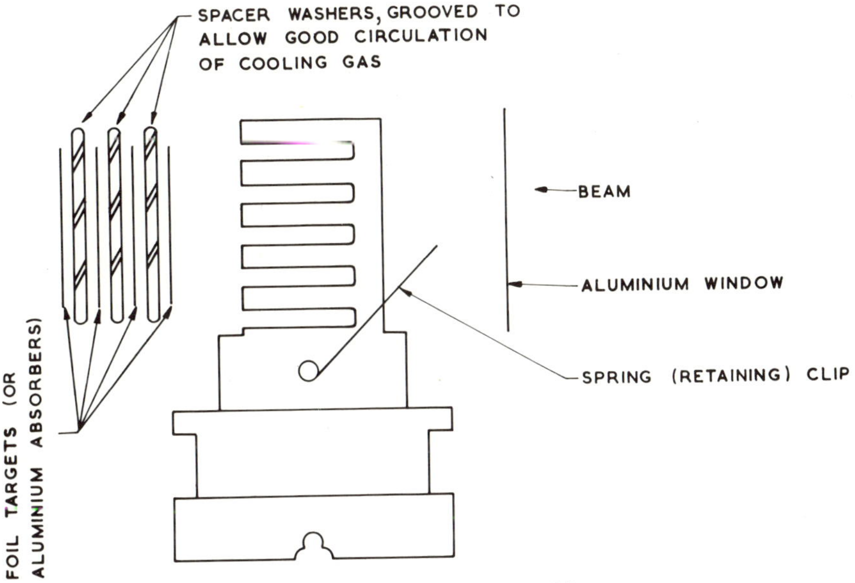

Fig. 2(b). Alternative target assembly

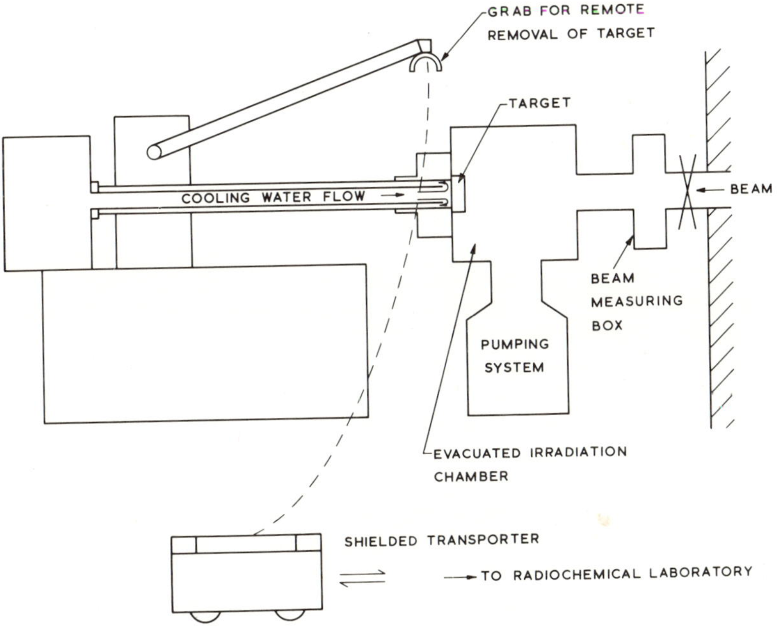

Fig. 3. Diagram of Users' Probe Irradiation Facilities

about the beam axis at 180 rpm. A mechanical grab enables the target assembly to be removed remotely and loaded into a lead-shielded, remotely-controlled, transporter which delivers the irradiated assembly into a shielded cave in the radiochemical laboratory. A plan of the target room and radiochemical laboratory is given in Fig. 4. The transporter carries the irradiated target assembly into a

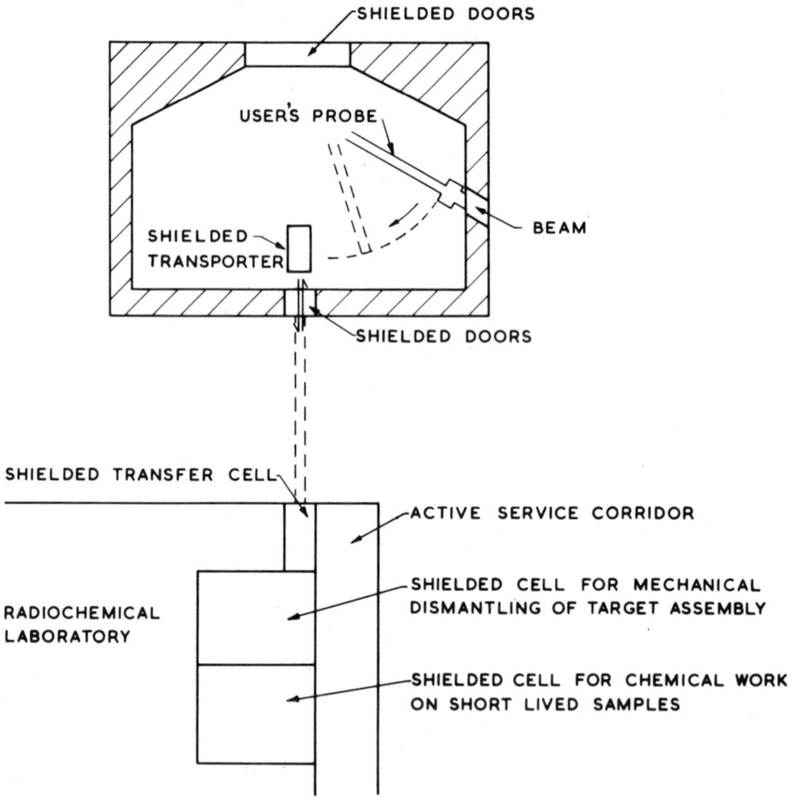

Fig. 4. Lay-out of Users' Probe and shielded cells

small shielded transfer cell from which the target assembly can be posted into a shielded cell equipped with over-the-wall master-slave manipulators and in which mechanical dismantling of the target assembly is carried out. Usually, the target is then transferred in a shielded 'coffin' to the laboratory where it is to be processed, but if the half-life of the product is short chemical processing can be carried out in an adjoining shielded cell also fitted with manipulators.

The beam current falling on the target is monitored by leads which are broken when the target assembly is removed and the thermocouple leads are also broken at the end of a run. As with the Chemists' Irradiation Rig the total integrated current is measured by the beam-current integrator, while the temperature is measured by means of the e.m.f. generated in a thermocouple inserted into the target assembly.

Isotope production ^{236}Pu. This is a useful isotope for the analysis of plutonium by the isotope dilution method, as neither it nor its precursor, ^{240}Cm, is produced to any appreciable extent in nuclear reactors. Excitation functions for the production of the 22 h isomer of ^{236}Np (which decays by β-emission to ^{236}Pu) have been given by Vandenbosch, Thomas, Vandenbosch, Glass, and Seaborg[4] and by Wing, Ramler, Harkness, and Huizenga.[5] The thick-target yield for the $^{238}U(p, 3n)^{236}Np$ reaction using 50 MeV protons is ~500 μCi/mAh;[6] ^{235}Np is also produced by the corresponding (p, 4n) reaction. We have prepared[7] 60 μCi of ^{236}Pu and 72 μCi of ^{235}Np by a 100 μAh irradiation of uranium of natural isotopic composition with 50 MeV protons.

The target was a uranium disc of 3·56 cm diam. 0·38 cm thick, and its back was directly cooled by a stream of water. The disc was thick enough to prevent the diffusion of fission products into the cooling water and its front was covered by a tantalum foil 0·002 in (0·005 cm) thick. Further protection from contamination to the V.E.C. was given by a foil of 'Havar' alloy (0·000 5 in, 0·012 7 cm thick). The target assembly is shown diagrammatically in Fig. 5.

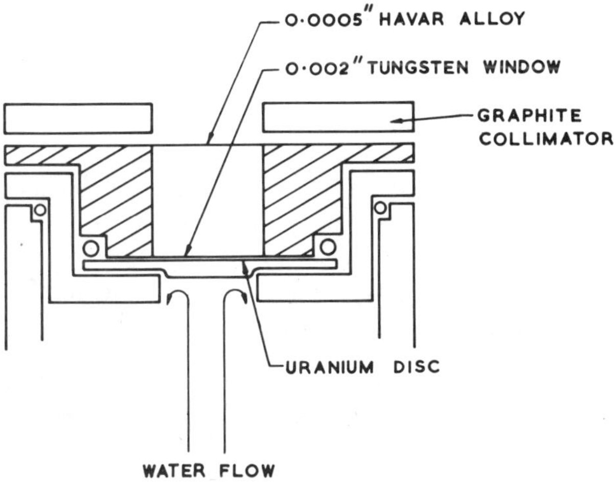

Fig. 5. Target assembly for ^{236}Pu production

The irradiated target assembly was dismantled remotely using manipulators and the uranium disc dissolved in 12M HNO_3. Plutonium and neptunium fluorides were coprecipitated with lanthanum trifluoride and after dissolving the precipitate in a mixture of nitric acid and aluminium nitrate plutonium and neptunium were separated from one another and purified by ion exchange.[7] The decontamination factors (DF = amount of fission product in initial material/amount of fission product in sample) are given in Table 1.

^{56}Co: This isotope has a half-life of 77·3 days and decays by electron capture and positron emission.[8] The γ-rays associated with the decay scheme have a broad spectrum and in particular several have energies >1 MeV, e.g. 1·04 MeV (15%), 1·24 MeV (66%), 1·76 MeV (15%), 2·02 MeV (11%), 2·60 MeV (17%), and 3·26 MeV (13%). As a consequence of its comparatively long half-life and the range of energetic γ-rays ^{56}Co is a very useful isotope for the calibration of γ-ray detectors at high energies. Millicurie amounts of ^{56}Co have been made by irradiating thick foils (91 mg/cm^2 of the separated isotope ^{56}Fe) with protons: $^{56}Fe(p, n)^{56}Co$. Excitation functions for the reaction are given in

Table 1. DF's OF ^{235}Np AND ^{236}Pu FROM FISSION PRODUCTS

	^{235}Np	^{236}Pu
^{95}Zr	$>10^4$	$2{\cdot}6 \times 10^3$
^{103}Ru	$>10^5$	$>5 \times 10^4$
^{124}Sb		$>3 \times 10^4$
^{137}Cs		$>6 \times 10^3$
^{140}Ba		$>10^4$
^{141}Ce		$4{\cdot}7 \times 10^3$
^{144}Ce		$3{\cdot}6 \times 10^3$

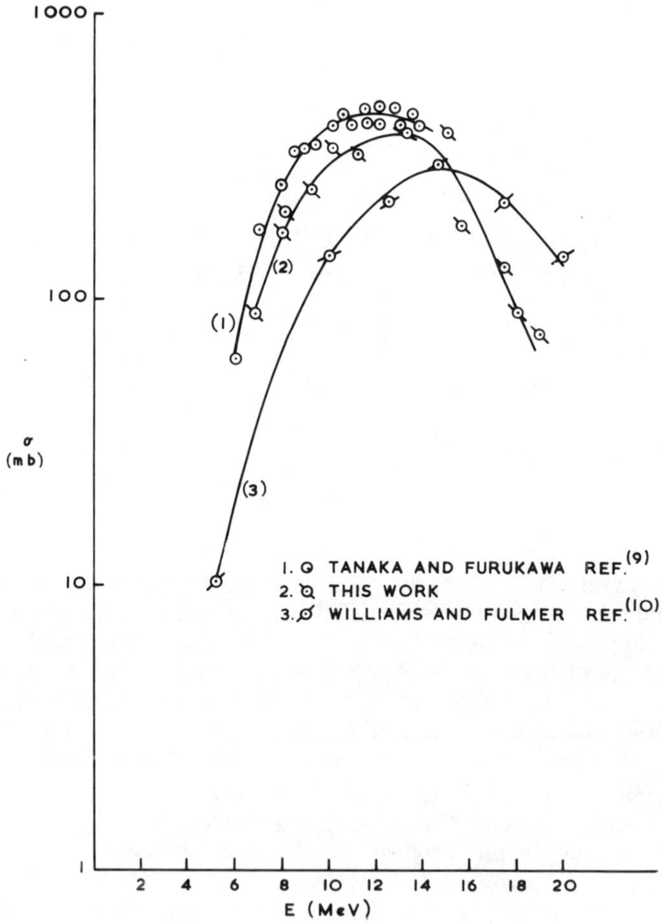

Fig. 6. Excitation functions for the $^{56}Fe(p, n)^{56}Co$ *reaction*

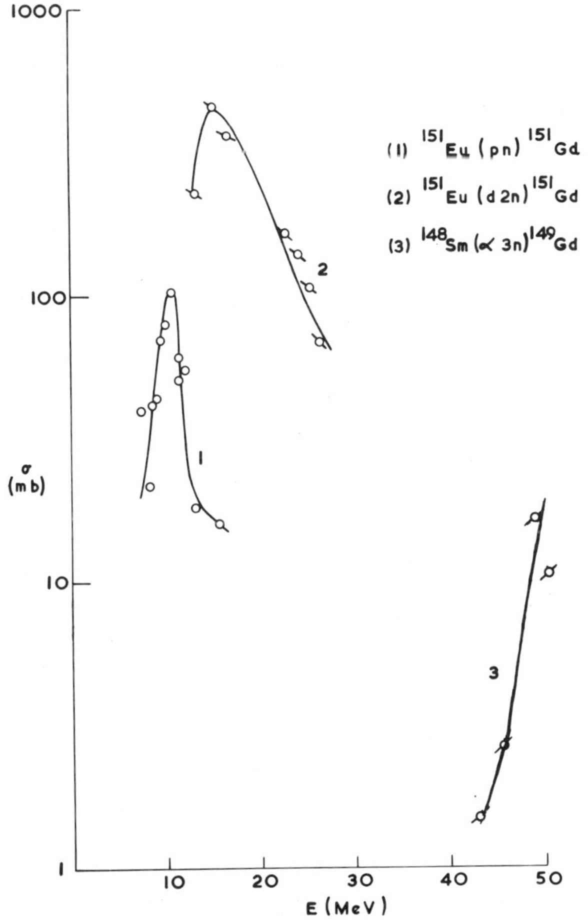

Fig. 7. Excitation functions for the production of ^{151}Gd *and* ^{149}Gd

Fig. 6 where the results of Tanaka and Furukawa[9] (points taken from compilation by F. K. McGowan, W. T. Milner, and H. J. Kim) and of Williams and Fulmer[10] (points interpolated from graphs) are also given. Excitation functions for other proton-induced reactions of ^{56}Fe, viz. ^{56}Fe(p, 2n)^{55}Co; ^{56}Fe(p, 2pn)^{54}Mn; ^{56}Fe(p, αn)^{52}Mn; and ^{56}Fe(p, pn)^{55}Fe have also been measured in the V.E.C.[11] The thick-target yield of ^{56}Co calculated from our excitation functions for the irradiation of a target of iron of natural isotopic composition is 51 mCi/mAh with 14 MeV protons; the observed yield[12] is ~60 mCi/mAh.

^{151}Gd: This isotope is a suitable source for the study of the Mössbauer spectra of ^{151}Eu. No excitation functions for reactions leading to the production of ^{151}Gd have been published, but some measurements have been made using the V.E.C.; these are given in Fig. 7. Martin, Livingston, Murray, and Rankin[13] have calculated the yield of ^{151}Gd produced by the ^{151}Eu(p, n)^{151}Gd reaction using 22 MeV protons. They estimated 10 mCi/mAh, but in practice our irradiations with protons have given much lower yields. We have produced ^{151}Gd by the $^{151}Eu(p, n)^{151}Gd$ and $^{151}Eu(d, 2n)^{151}Gd$ reactions and the preparations have been described in detail elsewhere.[14] The target consisted of the separated isotope ^{151}Eu as the oxide Eu_2O_3. This was contained in an aluminium capsule which fitted into a target head of the type described by Ball.[3] The assembly is shown in Fig. 8, and the yields are given in Table 2.

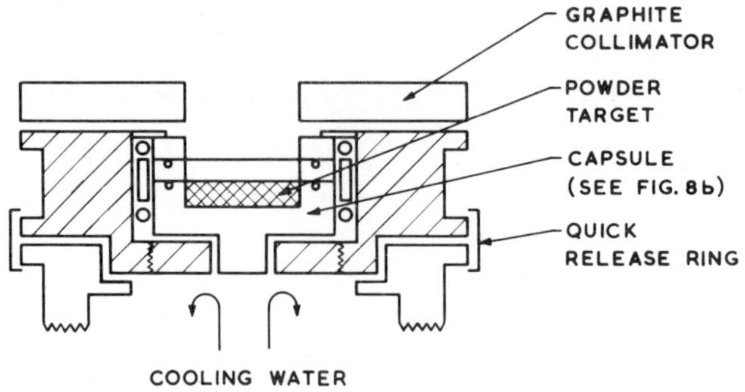

Fig. 8(a). Schematic diagram of target assembly for deuteron irradiations of Eu_2O_3

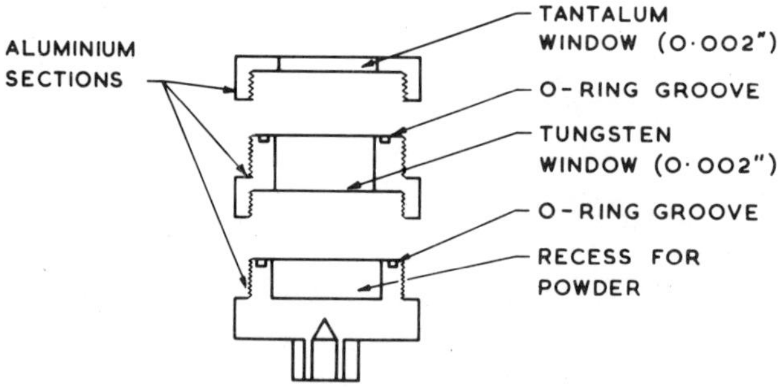

Fig. 8(b). Exploded diagram of capsule

Table 2. IRRADIATION CONDITIONS AND YIELDS OF ^{151}Gd

Beam	*Energy degradation in assembly*	*Packed density of* Eu_2O_3	*Yield of* ^{151}Gd
H^+	17·5–12 MeV	0·48 g/ml	$<$ 0·07 μCi/μAh
H^+	17·5–0 MeV	1·35	$\leqslant$ 0·2
D^+	16·0–0 MeV	0·71	5·1
D^+	16·0–0 MeV	0·78	$\nless$ 4

Irradiations of ^{148}Sm with α-particles using the stacked-foil technique (~300 μg/cm² $^{148}Sm_2O_3$ on aluminium) over the energy range 50–43 MeV did not yield ^{151}Gd, but excitation functions for the reaction $^{148}Sm(\alpha, 3n)^{149}Gd$ are given in Fig. 7.

^{87}Y: 100 μCi amounts of ^{87}Y have been made by irradiating 0·5 mg/cm^2 targets of $^{87}SrO_2$ on aluminium.

^{66}Ga: This was made by the $^{66}Zn(p, n)^{66}Ga$ reaction using 1·8 mg/cm^2 Zn metal foils of normal isotopic composition and a 17 MeV proton beam. ^{66}Ga has a half-life of 9·45 h, and although this is very much shorter than that of ^{56}Co it is also useful for calibrating γ-ray detectors, as it has several energetic γ-rays, viz. 1·039 MeV (37%); 1·91 MeV (3%); 2·183 MeV (5%); 2·748 MeV (25%); and 4·30 MeV (5%). The yield was about 15 μCi/μAh.

^{140}Nd: Metallic foils of ^{140}Pr (16·5 mg/cm^2) were sealed between aluminium covers using Araldite. Since finely-divided praseodymium is pyrophoric the target preparation was carried out in an inert (Ar) atmosphere in a glove box. The initial 48 MeV proton beam was attenuated to ~22 MeV with tantalum foil and at the end of the irradiation the tantalum was found to be severely oxidised. The yield of ^{140}Nd was ~6 mCi/μAh.

ACKNOWLEDGEMENTS

The authors would like to thank the operating staff of the V.E.C. for help with the irradiations, Mr. R. J. Bullock for help with the γ-counting and computing of the excitation functions.

We would also like to thank Mr. M. A. A. Stewart for permission to publish the $^{151}Eu(p, n)^{151}Gd$ excitation function curve shown in Fig. 6, which was measured while he was a member of the V.E.C. Operations Group.

REFERENCES

1. Lawson, J. D., NIRL/R/85 (1965).
2. Amphlett, C. B., Paper presented to the European Colloquium on AVF Cyclotrons, Eindhoven, Holland, (1965).
3. Ball, A. R., U.K.A.E.A. Document P. & E. Memo. No. 76, (1968).
4. Vandenbosch, R., Thomas, T. D., Vandenbosch, S. E., Glass, R. A., and Seaborg, G. T., *Phys. Rev.* **111,** 1358, (1958).
5. Wing, J., Ramler, W. J., Harkness, A. L., and Huizenga, J. R., *Phys. Rev.* **114,** 163, (1959).
6. Scargill, D., private communication.
7. Jenkins, I. L. and Wain, A. G., AERE-R 5790 (1968).
8. Lederer, C. M., Hollander, J. M., and Perlman, I., 'Table of Isotopes', Sixth Edition, John Wiley & Sons Inc., New York, (1967).
9. Tanaka, S. and Furukawa, M., *J. Phys. Soc. Japan* **14,** 1269, (1959).
10. Williams, I. R. and Fulmer, C. B., *Phys. Rev.* **162,** 1055, (1967).
11. Jenkins, I. L. and Wain, A. G., unpublished work.
12. Finlan, M., private communication.
13. Martin, J. A., Livingston, R. S., Murray, R. L., and Rankin, M., *Nucleonics* **13,** 28, (1955).
14. Bullock, R. J., Large, N. R., Jenkins, I. L., Wain, A. G., Glentworth, P., and Newton, D. A., *J. Inorg. Nucl. Chem.* **31,** 1929, (1969).

DISCUSSION

M. A. Chaudhri (M.R.C., Hammersmith): We are under the impression that available time on the Harwell V.E.C. is very limited, and I would like to know what is the cost of producing isotopes with such a versatile machine. Secondly, at high incident energies multiple neutron emission will take place, and I wonder why you did not produce impurities from such reactions, e.g. impurities arising from (p, 2n) and (p, 3n) reactions when studying a (p, n) reaction.
I. L. Jenkins: The V.E.C. is always fully booked, but some time can be made available for isotope production; normally isotopes are made to meet specific requirements within the U.K.A.E.A. The charge for use of the V.E.C. is £160/h.

In reply to your second question, it is sometimes possible to work at a sufficiently low proton energy to minimise the formation of products from other reactions such as (p, xn), (p, pn), etc. When irradiating foils of separated ^{56}Fe, if the incident beam energy is <15 MeV very few other products are formed, e.g. at 15 MeV the (p, n) cross-section is ~350 mb, that for (p, 2n) is <3 mb, that for (p, pn) is ~100 mb and those for other reactions are very small indeed (Jenkins and Wain, unpublished work). The product of the (p, pn) reaction is ^{55}Fe, which can easily be separated from the (p, n) product ^{56}Co. In many cases where (p, xn) and other reactions interfere, the products can be separated chemically.
H. A. C. McKay (A.E.R.E.): I would like to hear comments from those producing isotopes as to the suitability of their machines, and what improvements they would like to see.

R. S. Tilbury (Sloan-Kettering Institute): I would like to have a machine of slightly higher energy, e.g. 30 MeV ^{3}He ions and 20 MeV protons. It is very desirable to have a method of measuring the shape and distribution of the beam, so as to spread it uniformly over a large area.

D. J. Silvester (M.R.C., Hammersmith): Our use of the high beam current machines already available is strictly limited by the kind of targets we have to irradiate. I would like to see developed beam-sharing systems such as those proposed in the preceding Conference, so that two or three targets could be irradiated simultaneously.
J. L. Need (New England Nuclear Corp.): With the ORNL 86 in cyclotron, which was designed specifically for production and has a 2 in vertical aperture, we could routinely put 2·5 mA on silver or copper targets. The present AVF cyclotrons have a ¼-in vertical amplitude, which presents a difficult problem in target design. There is a limit to the heat flux that can be handled, and any increase in the vertical extent of the beam will increase the total power-handling capacity of the target. I would like to see steps taken in this direction.

B. Hoop (Massachusetts General Hospital, Boston): The justification of a cyclotron within a medical centre hinges critically on its capability to produce rapidly and efficiently a wide variety of clinically-useful radioisotopes, particularly short-lived ones. For this reason it is essential to develop efficient beam handling and versatile target systems.
M. A. Chaudhri (M.R.C., Hammersmith): It would be of great interest and value if the compact cyclotrons could accelerate heavier ions and could be made energy-variable. Axial ion injection would be required for heavy-ion sources, which are too large to be located internally, and polarised sources could also then be used. Such a machine would be of interest to nuclear physicists and radiobiologists. High-intensity beams would not be required as in isotope and

neutron production, since internal sources of p, d, ^3He, and ^4He ions would perform this task satisfactorily.

I was disturbed to see in the results presented by Dr. Hrubý that the yield did not increase linearly with increasing intensity, and would be interested to hear any comments on this point.

J. L. Need (New England Nuclear Corp.): We found that silver-plated targets gave lower yields than solid silver ones. We used full thick plating at 1750 μA and solid silver at 2500 μA, and the yield on the former was down by ~50%. The material being made, viz. ^{109}Cd, has a low vapour pressure and the plated target a more porous structure than the solid one; possibly there is loss to the cyclotron interior. The yield vs current curves also showed definite maxima for lead and lithium targets, although the current at the maximum was well below that at which the target melted.

B. D. Pate (Simon Fraser University): Many workers face the problem of using facilities which were designed for other purposes. At TRIUMF 500 MeV and 100 μA will be used to produce nuclides which cannot be produced in other ways. 50 kW of beam energy may be handled by 'heat-pipe' targets, for example.

G. Burton (M.R.C., Hammersmith): We should seriously consider developing the separated-sector cyclotron in the medium and lower energy ranges for production of radioisotopes and fast neutron beams. Intense external beams would then be more readily available, since extraction is easier and septum/deflector problems less severe. These beams could be shared simultaneously between several targets. External ion sources may require development in order to produce high beam currents of ^3He or ^4He ions for isotope production.

N. Hazewindus (Philips Research Laboratories): An axial injection system would be useful for ions heavier than He, since ^3He and ^4He ions can be produced in sufficient intensity in a normal internal source. For heavier ions the external source has its advantages. Furthermore, axial injection can be used in compact machines. In fact, our prototype Compact Cyclotron is only equipped with an axial injection system, and at present has no internal source.

SESSION 4

Radiation Biology and Medicine

Chairman: Dr. H. A. C. McKay

Energy and L.E.T. spectra of cyclotron beams—prediction and measurement

Howard D. Maccabee*

Medical Research Council, Radiobiological Research Unit, Harwell, Berks., England

ABSTRACT

In many experiments in radiation biology, chemistry and physics, we need to know the particle energy spectrum of an accelerator beam after slowing down in a thin or thick absorber. It is often desirable also to know the distribution in L.E.T. (linear energy transfer, analogous to dE/dx, stopping power, or specific energy loss) of the beam particles. The straggling theories involved in predicting these spectra are reviewed briefly.

Semiconductor detectors have been used to measure energy loss distribution and residual energy distribution for many accelerator beams. The measured energy loss spectrum of a 45·3 MeV proton beam in a thin silicon absorber, and the residual energy spectra of beams of 19·68 MeV protons and 79·8 MeV He^{+2} ions slowing down in thick aluminium absorbers are presented. In these and other cases, the measured distributions are in good agreement with the theoretical predictions of Vavilov for thin absorbers, and with those of Tschalär for thick absorbers.

We have developed, with C. Tschalär of the Rutherford Laboratory, a computer code for predicting L.E.T. distributions of heavy charged-particle beams slowing down in matter, and the calculated L.E.T. spectrum is given for an experiment in which 14·5 MeV protons were slowed to 12% of their initial energy. The implications are discussed for experiments in which degrading absorbers are used to increase the L.E.T. of an accelerator beam by reducing its energy. In extreme cases the usefulness of the concept of an 'average' L.E.T. is vitiated by the severe skewness of the L.E.T. distribution due to straggling.

INTRODUCTION

Straggling. Cyclotron beams are slowed down in matter by a series of random collisions with the atoms of the material, causing energy deposition by electronic excitation and ionisation. The stochastic nature of these processes results in

*Address after 1 October: Physics Research Laboratory, Massachusetts General Hospital, Boston, Mass. 02114, U.S.A.

statistical fluctuations in the energy lost by the charged particles in the beam after it has passed through any given thickness of material. This phenomenon is often called 'straggling'. Thus, at any point in an absorber, there will be a statistical *distribution* of energy-values in the beam, as well as a distribution of energy-losses expected in the next portion of the beam path.

Since many experiments in biology, chemistry and metallurgy depend on the deposition of energy by cyclotron beams, the energy spectrum of the beam particles is of practical importance. There are several theoretical formulations of energy straggling, each of which has a limited region of validity.

Collision spectrum. The fundamental physical event which is basic to each of the theories is the single discrete coulombic collision between an energetic charged particle and an atomic electron. The distribution of energy losses expected for a beam of particles undergoing one collision each is called the collision spectrum. In the approximation that the atomic electrons are free (i.e. not bound to atoms of the material), the simplified single-collision spectrum is:

$$P(\epsilon, T) = (k/2T)\epsilon^{-2} \tag{1}$$

where $P(\epsilon, T)$ is the probability of an energy loss ϵ by a particle with kinetic energy T in a single collision.

$$k \equiv 2\pi e^4 z^2 NZ/\mu \tag{2}$$

where $e \equiv$ electron charge, $z \equiv$ particle charge number, $N \equiv$ number of atoms/cm^3 of absorber, $z \equiv$ atomic number of absorber, and μ is the ratio of the electron mass to the particle mass.

Thin absorbers. We define a thin absorber as one in which the energy loss is small compared to the initial kinetic energy T_0, so that we can neglect the variation of the collision spectrum and the stopping power with T. A 'rule of thumb' useful for high-energy beams is that the energy loss in a *thin* absorber is less than 10% of the initial energy. Landau[1] solved the thin absorber case under the condition that the number of collisions in the highest collision-energy interval is small, or equivalently,

$$\xi/\epsilon_{\max} \ll 1 \tag{3}$$

where $\xi \equiv kx/2T_0$, $x \equiv$ absorber thickness, and $\epsilon_{\max} = 4\mu T_0/(1-\beta^2)$, the maximum possible energy transfer in a non-relativistic collision between a heavy particle and an electron. In such cases the distribution of total energy losses is extremely asymmetric, with a broad peak around the most probable energy loss, and a long tail corresponding to higher-energy losses. The full width of the peak at half-maximum $(FWHM)_L = 3{\cdot}98\xi$, and the most probable energy loss Δ_{mp} is significantly less than the average energy loss $\overline{\Delta}$:

$$\Delta_{mp} = \xi\left\{\ln[4\mu T_0\xi/I^2(1-\beta^2)] - \beta^2 + 0{\cdot}20\right\} \tag{4}$$

where $I \equiv$ mean excitation potential of material.

Bohr[2] solved the thin absorber case for the opposite condition, i.e. when there are many collisions in every collision-energy interval; this is equivalent to

$$\xi/\epsilon_{\max} \gg 1 \tag{5}$$

Bohr's energy-loss distribution is Gaussian, with variance given by

$$\sigma_B^2 = 4\pi e^4 z^2 NZ x = 2\mu k x \tag{6}$$

which leads to a width given by

$$FWHM_B = 3{\cdot}33 \sqrt{(\mu k x)} \tag{7}$$

The problem of cases intermediate between those of Landau and Bohr was treated by Vavilov,[3] who gave an exact general solution of the thin absorber problem. Numerical evaluation of Vavilov's equations yields a family of curves with dimensionless parameters β^2 and κ, where

$$\kappa \equiv \xi/\epsilon_{max}, \tag{8}$$

which effect a smooth transition between the Bohr and Landau distributions and include them as special cases. Vavilov's solution has been tabulated by Seltzer and Berger.[4] Note that up to this point we have spoken in terms of the energy loss for a particle at any point in the absorber, defined by

$$\Delta \equiv T_0 - T, \tag{9}$$

where T is the residual kinetic energy of the particle; in other words, the distribution of energy losses for a beam of particles is the exact complement of the distribution of residual energies of the beam. Thus, for the properties of the distributions,

$$\begin{aligned} \bar{\Delta} &\equiv T_0 - \bar{T} \\ \Delta_{mp} &\equiv T_0 - T_{mp}, \text{ etc.} \end{aligned} \tag{10}$$

If the initial beam is relatively monoenergetic, the width of the energy loss distribution is the same as that of the residual energy distribution. In other words, a high-energy-loss 'tail' corresponds to a low-energy 'tail' in the residual energy distribution.

Thick absorbers. A thick absorber is defined as one in which the energy loss is *not* small compared to the initial kinetic energy T_0, so that we may not neglect the changes in the stopping power and the collision spectrum with decreasing T. Thus, as a beam penetrates deeper into an absorber, its theoretical description must change from a thin absorber regime to a thick absorber regime. Actually the Bohr solution described above remains valid part of the way into the thick region, e.g. up to 20% energy loss for 20 MeV protons and even further for higher-energy particles.

At greater depths of penetration, however, the width (and variance) of the energy distribution continues to increase, and a 'bulk' effect begins to become important compared to the stochastic fluctuation effects we have considered up to now. The lower-energy particles in the distribution have a higher stopping power and therefore lose, on average, more energy in the succeeding thickness element than the higher-energy particles. This effect causes an additional broadening of the energy spectrum.

Symon[5] improved on Bohr's solution for these cases, and gave an approximate solution to the thin absorber case. Where Bohr's variance increases directly with absorber thickness Symon adds a term to account for the variation of the stopping power over the distribution, leading to a variance which increases faster than the absorber thickness.

Tschalär[6] pointed out that Symon had neglected a term which contributed strongly to the skewness of the distribution, and thus Symon's distributions erroneously remain practically Gaussian up to very large energy losses. Tschalär[6] gave the numerical solution to the corrected equations and tabulated the distributions and their properties (e.g. variance, skewness, higher moments, etc.). In general, as the penetration increases, a non-Gaussian low-energy 'tail' begins to reappear. In addition, Tschalär[7] showed that when a heavy-particle beam has penetrated to such a depth that particles start to be lost by complete stopping, the previous theories break down, and energy distributions must be propagated numerically to account for the cutting off of the low-energy tail at $T = 0$. Recently Payne[8] has given a solution to the problem of predicting the energy distributions in thick absorbers by working 'backwards' from the distribution of residual ranges, which remains practically Gaussian throughout the whole depth of penetration. Payne's distributions agree with Tschalär's for the cases considered until now.

Resonance broadening. It is well known[9] that the collision spectrum given by Eqn (1) is imperfect, due to its failure to deal properly with resonance (or 'glancing') collisions at small values of ϵ. In other words, the ϵ^{-2} dependence for free electrons is violated when the collision energy is small enough for atomic binding effects to become important. The net result of these quantum-mechanical effects is a broadening of the energy-loss distributions; this subject has been treated by several authors.[10] Blunck[10] gave a rough estimate for the variance of the resonance collision spectrum,

$$\sigma^2_{\text{res}} \approx \bar{\Delta} Z^{4/3} \times 10\ \text{eV} \tag{11}$$

which may be added to the variance of the previously-described energy spectra to yield an approximate resonance-corrected variance. Note that the resonance correction is most important for very thin absorbers of high-Z material.

Other effects. Up to this point we have neglected several other effects which tend to perturb and broaden the energy distributions. We have assumed the incident beam to be mono-energetic. In real cases, there will be a distribution of initial kinetic energies; if the width of this distribution is small but not negligible compared to the width due to straggling, an approximate correction may be made by 'folding' them together in quadrature. Slit scattering effects due to collimators can introduce a long low-energy 'tail' in the incident beam, which may severely perturb the resulting energy distributions in the absorber. In addition, multiple-scattering effects tend to increase the pathlengths associated with a given depth of penetration, and therefore decrease the calculated energies of the particles at that depth. In addition there are more exotic effects such as blocking and channelling of narrow beams in single crystals, etc.

L.E.T. spectrum. The Linear Energy Transfer (L.E.T.) of a particle is defined as the amount of energy transferred to the absorber medium per unit path length along the particle track. If we neglect effects of disequilibrium of δ-ray entry and escape, etc., the energy deposited per differential pathlength is the same as the energy lost per differential pathlength, i.e. the stopping power, specific energy loss, or dT/dx. This quantity is properly called L.E.T._∞ Since L.E.T. is believed to be an important parameter of radiation quality, influencing many experimental effects in radiation biology and chemistry, it is often important to know the L.E.T. distribution of a particle beam at a depth in an absorber. The L.E.T. distribution is derived simply by assigning an L.E.T. to each energy in the particle energy distribution, using a stopping-power relation such as that tabulated by Janni.[11] Note that the L.E.T. distribution, which incorporates the *mean* value of energy loss per *differential* pathlength, is *not* the same as the energy-loss distribution, which accounts for *fluctuations* from the mean energy loss in a *finite* pathlength.

In order to simplify calculations of radiation effects involving L.E.T. distributions, several different types of averages are taken over the L.E.T. distribution. If we denote the energy distribution by $f(T)$ and the corresponding L.E.T. by $L(T)$, the 'track-average' L.E.T. is given by

$$\bar{L} \equiv \frac{\int L(T)\, f(T)\, dT}{\int f(T)\, dT} \tag{12}$$

and the 'dose-average' or 'energy-average' L.E.T. is given by

$$\bar{\bar{L}} \equiv \frac{\int L^2(T)\, f(T)\, dT}{\int L(T)\, f(T)\, dT} \tag{13}$$

EXPERIMENTAL

Semiconductor detectors for cyclotron beam spectrometry. In the past seven years, semiconductor particle detectors have proved extremely valuable for measuring energies and energy losses of heavy charged-particle beams. In general, they have superior energy resolution due to their higher charge output per given ionisation energy input, as well as linear response and high stopping power, when compared to other detectors. Perhaps their main disadvantage is the fact that their response is degraded due to radiation damage after absorbing a large number of particles (e.g. $\sim 10^{10}$ protons).

We have used planar silicon detectors to measure the energy loss distribution in silicon, by analysing the charge pulse spectrum due to cyclotron beams passing through the detectors. We have also measured the residual energy distribution after aluminium absorbers, by analysing the charge pulse spectrum on stopping the emergent beam in the detector. We have used lithium-drifted 'p-i-n junction' detectors, as well as surface barrier and 'p-n' junction detectors, to measure beams at the 88 in and 184 in cyclotrons in Berkeley, the Variable Energy Cyclotron, the Van de Graaff and Tandem Generators at Harwell, the Proton Linear Accelerator at Rutherford Laboratory, and the Proton Generator at the M.R.C. Radiobiology Unit. The data we present here represent only a small sample of these results.

Method. The measuring system consists of the detector (in its holder) and a charge-sensitive preamplifier in the irradiation cave, with the detector face normal to the axis of the incident beam, following closely on any aluminium absorber sets, if used. Charge pulses from the preamplifier are transmitted out of the cave by a long coaxial cable to a linear amplifier and pulse-shaping system, and thence to a multi-channel pulse-height analyser which sorts them individually as to size; the data, in the form of a set of counts-per-channel versus channel number, are then taken from a digital printer. Assuming that the charge pulse height is proportional to the energy deposited in the detector, and that the channel number is linear re the pulse height, the data may be analysed to yield a plot of the frequency distribution of particle energies, or energy losses in the detector. Energy calibration is done with standard radioactive sources of α-particles and electrons, using a linear test pulse generator, by the usual method. Details of the experimental method and associated systems may be found in previous publications.[12]

RESULTS

Thin absorbers. Fig. 1 shows an example of an energy-loss distribution measured in a thin absorber, for a 45·3 MeV proton beam passing through 0·070 g/cm^2 of silicon (~0·3 mm thick). The value of the parameter κ is 0·590, between the regions of validity of the Landau and Bohr theories, and hence the Vavilov[3] theory must be used to predict the energy loss distribution. The solid line is the distribution predicted by Vavilov; the experimental points are shown

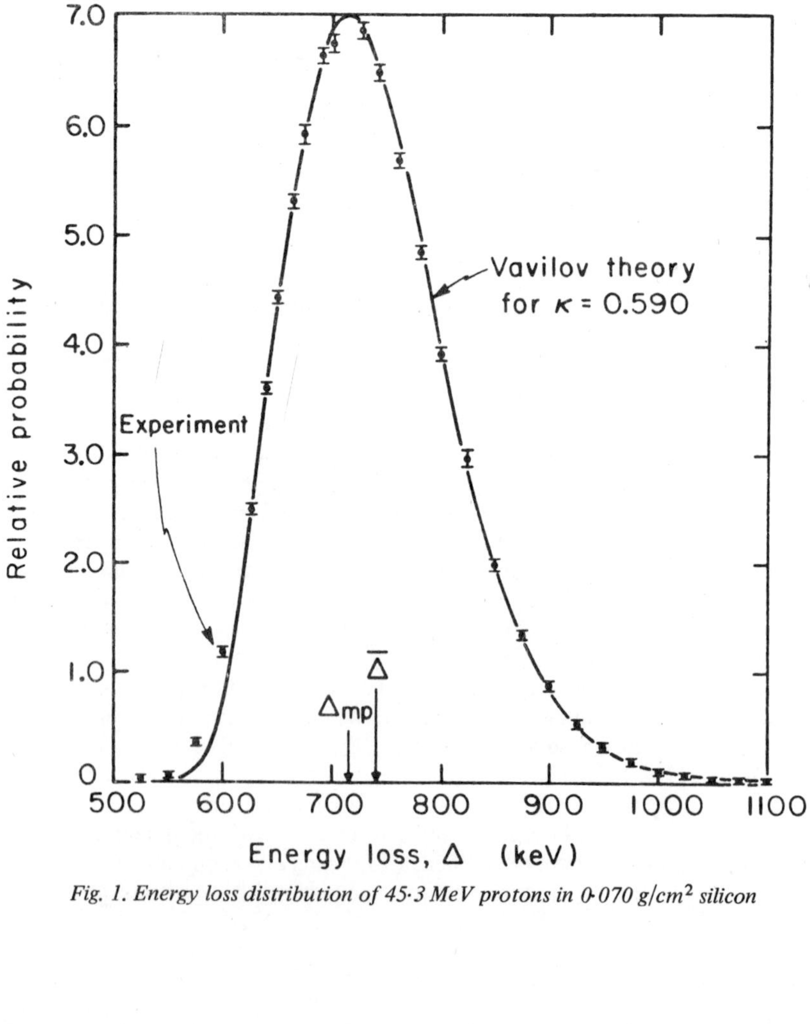

Fig. 1. Energy loss distribution of 45·3 MeV protons in 0·070 g/cm² silicon

with statistical error bars whose heights are calculated as the square root of the numbers of counts in a channel. Note the asymmetry of the distribution and the presence of a high-energy-loss 'tail'; the most probable energy loss Δ_{mp} is significantly smaller than the mean energy loss $\bar{\Delta}$. There is very good agreement between experiment and theory, except for a small unexplained deviation at the low-energy-loss end, which could be due to resonance effects or to δ ray escape from the detector. The resolution of the experimental system is ~30 keV *FWHM*; this correction is negligible compared to the *FWHM* peak width due to straggling. The error in the energy scale has been estimated as $< \pm 2\%$.

This curve, and many others similar to it, confirm our belief that the Vavilov theory and its special cases, the Landau and Bohr distributions, can be used to predict accurately the energy loss spectrum due to straggling of cyclotron beams in thin absorbers.

Thick absorbers. Fig. 2 shows the results for a 19·68 MeV proton beam slowing down in various thick absorbers of aluminium.

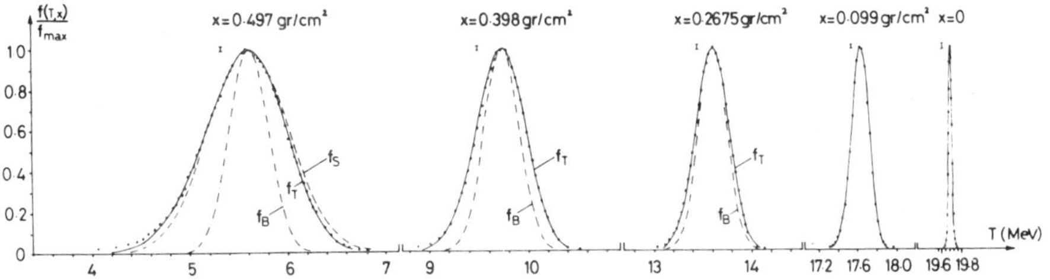

Fig. 2. Residual energy distributions of a 19·68 MeV proton beam straggling in various thick aluminium absorbers. f_T, f_S and f_B *represent the predictions of Tschalär, Symon and Bohr respectively*

The ordinate shows the relative frequency, normalised to unity at the maximum peak height. The abscissa shows the residual kinetic energy (MeV) of the beam particles; the scale is broken for compactness of presentation. The distributions are plotted with absorber thickness increasing from right to left. In these cases the thickness x has already been multiplied by ρ, the density of the absorber, to yield the conventionally-used 'mass thickness'. The plot for $x = 0$ corresponds to the measurement of the energy distribution of the incident 'virgin' beam (40 keV *FWHM*), including the system resolution (~25 keV *FWHM*). The experimental points are shown as solid dots, with a single error bar near the peak to represent the statistical error, calculated as above. The plotted curves represent the predictions of various theories for $f(T, x)$, the distribution of energies T at thickness x. The solid line f_T is the prediction of Tschalär,[6,7] the dashed-dotted line f_B that of Bohr,[2] and the dashed line f_S that of Symon.[5] The system resolution is negligible compared to the widths due to straggling.

Note that the first absorber ($x = 0{\cdot}099$ g/cm^2), with an energy loss of about 10%, gives a distribution that is essentially Gaussian, well described by all three theories. After 30% energy loss ($x = 0{\cdot}2675$ g/cm^2) the distribution begins to become asymmetric, and the Bohr distribution is too narrow; these tendencies are accentuated after about 50% energy loss ($x = 0{\cdot}398$ g/cm^2). After about 70% energy loss (0·497 g/cm^2), the low-energy 'tail' is developed, as predicted by Tschalär, and Symon's curve is not sufficiently skewed.

To show the effect of a different particle beam at approximately the same velocity, Fig. 3 gives the results of a 79·8 MeV beam of He^{2+} ions slowing down in the same set of thick aluminium absorbers.

The notation is the same as in Fig. 2. Again the extreme right plot represents the 'virgin' beam measurement (~150 keV *FWHM*). At ~10% energy loss the distribution is approximately normal, as predicted by all the theories. The Bohr prediction becomes too narrow at 30% energy loss (x = 0·267 5 g/cm²) and the

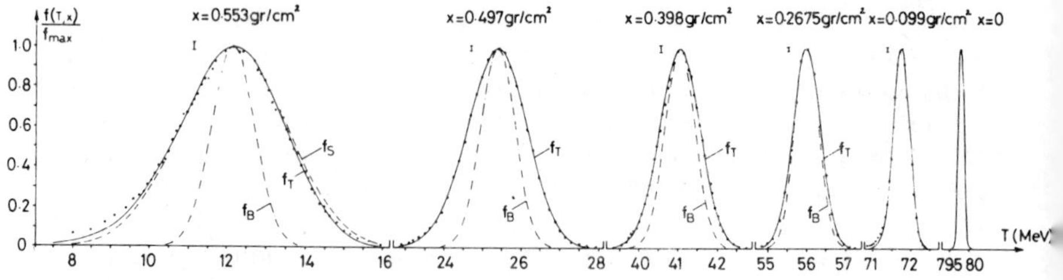

Fig. 3. Residual energy distributions of a 79·8 MeV beam of He^{+2} *ions slowing down in thick aluminium absorbers*

tendency is accentuated at about 50% and 70% energy loss (x = 0·497 g/cm²), where Tschalär's and Symon's models remain adequate. At about 85% energy loss (x = 0·553 g/cm²), however, the experimental results are more highly skewed than Symon's prediction, but Tschalär's model still holds. Here, as in Fig. 2, there is a small deviation at the low-energy end, which remains unexplained, but which is probably an experimental effect. Note that all of the theoretical curves f_T and f_S in Figs. 2 and 3 include a 3% correction in the width to account for resonance broadening, as calculated from the tables of Golovin *et al.*[10] Absolute calibration errors in the energy scale for these data are estimated as < ± 50 keV; relative errors between plotted points are much smaller.

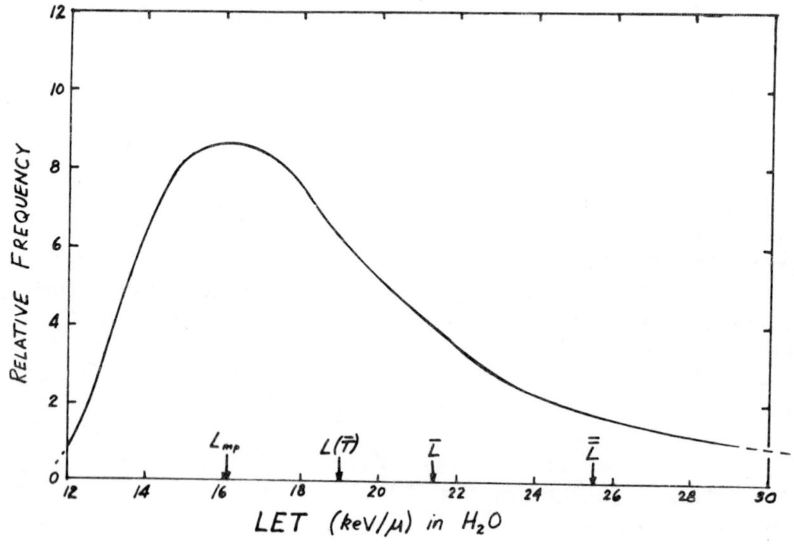

Fig. 4. Calculated L.E.T. distribution of a 14·50 MeV proton beam slowed down in 317·68 mg/cm² Al, so that the residual mean energy is 1·60 MeV and the most probable energy 1·74 MeV. Initial beam spread was 150 keV (FWHM)

L.E.T. spectrum. Fig. 4 shows the calculated L.E.T. distribution corresponding to an experiment at the proton generator of the M.R.C. Radiobiology Unit, Harwell. In this experiment the measured energy spread of the initial beam was 150 keV (*FWHM*), and the proton beam was slowed down in 317·68 g/cm^2 aluminium, so that the measured most probable energy (T_{mp}) of the residual energy distribution was 1·74 MeV.

The mean energy ($\overline{T}$) was 1·60 MeV, corresponding to an L.E.T. of 19·0 keV/μ. The most probable L.E.T. in this case is 16·1 keV/μ, the track-averaged L.E.T. ($\overline{L}$) is 21·4 keV/μ, and the dose-averaged L.E.T. ($\overline{\overline{L}}$) is 25·5 keV/μ. The L.E.T. distribution calculations were carried out using a computer code developed by C. Tschalär of the Rutherford Laboratory. In this case, the beam has lost ~88% of its initial energy, and is nearly at the Bragg maximum in ionisation; a significant percentage of particles has already been lost by complete stopping, and the residual energy distribution has been severely broadened. The result of this is that the L.E.T. distribution is extremely skew, with a long high-L.E.T. tail corresponding to the low-energy tail of the energy spectrum. These effects are manifest in the wide variation of the different averages of the distribution, e.g. ($\overline{\overline{L}}$) is half as large again as L_{mp}.

DISCUSSION AND SUMMARY

In general there is good agreement between the experimental measurements and the theoretical predictions of Vavilov for thin absorbers, and of Tschalär for thick absorbers, provided that small resonance broadening corrections are taken into account. The theory of energy straggling is well developed, and can be satisfactorily used to provide an accurate prediction of energy and energy-loss distributions for almost any irradiation case, but there is some mathematical labour involved, and one must be careful to account for other physical effects which disturb the spectra. These effects include energy spread and angular divergence in the 'virgin' beam, multiple scattering, slit scattering, contamination and 'satellite' beams, nuclear interactions, etc. In many cases it is easier simply to measure the relevant spectrum using a semiconductor detector or similar system, rather than try to account quantitatively for all the possible physical effects.

Summary. When a charged particle beam leaves the homely near-vacuum environment inside the accelerator, it is subject to a myriad of collisions, mostly gentle, some jarring, before it straggles to a complete stop in some absorbing material. As it slows down, its path in 'energy-space' may be described by several different theoretical regimes.

In the beginning of its penetration, the Landau[1] distribution holds, when the parameter $\kappa \lesssim 0{\cdot}01$, where

$$\kappa = 0{\cdot}150 \cdot \frac{\rho\, x\, Z\, z^2}{A} \cdot \frac{1-\beta^2}{\beta^4} \tag{14}$$

and ρx is the absorber mass-thickness in g/cm^2, A the atomic number of absorber material, and β the particle velocity compared to light. In such cases, the most probable energy loss is given by

$$\Delta_{mp} = \xi[\ln(\xi\epsilon_{max}/I^2) - \beta^2 + 0{\cdot}20] \tag{15}$$

and the width of the distribution is given by

$$FWHM_L = 3{\cdot}98\,\xi \tag{16}$$

where

$$\xi = 0{\cdot}1536\ \rho x Z z^2/A\beta^2 \text{ (in MeV)} \tag{17}$$

When the thickness penetrated increases, so that $0{\cdot}01 \lesssim \kappa \lesssim 10$, the Vavilov solution[3] should be used; unfortunately, there are no short formulae in this case, but Seltzer and Berger[4] have tabulated the distributions.

When the thickness penetrated is sufficiently large so that $\kappa \gg I$, Bohr's solution[2] becomes valid; it holds through the end of the thin absorber region and into the beginning of the thick absorber region, e.g. up to ~20% energy loss for 20 MeV protons. In such cases the energy-loss spectrum (and the energy spectrum) are Gaussian, with

$$f_B(\Delta) = (2\pi\sigma^2)^{-1/2} \exp[-(\Delta - \bar{\Delta})^2/2\sigma^2] \tag{18}$$

so that Δ_{mp} (Bohr) = $\bar{\Delta}$. The variance of the distribution is given by

$$\sigma_B^2 \cong 0{\cdot}157\ \rho x Z z^2/A \text{ (in MeV}^2\text{)} \tag{19}$$

The *FWHM* width can be obtained by the relation which is valid for Gaussian distributions:

$$FWHM = 2{\cdot}356\sigma. \tag{20}$$

Resonance corrections to these straggling distributions can be obtained from references 9 and 10.

When the particle penetration is so great that there is a considerable spread in the stopping powers of the particles in the beam, the Bohr solution is too narrow and the Tschalär solution should be used. Unfortunately there are no short formulae for these cases, but the distribution shapes, widths, etc., have been tabulated graphically.[6,7]

If a simple approximation to thick absorber energy straggling is necessary, we may remember that the range straggling distribution remains nearly Gaussian throughout the whole range of the particles.[8] The residual range distribution can be obtained from Lewis[13] or Sternheimer,[14] and by assigning an energy to each residual range, using range-energy tables, the energy distribution can be assembled.

L.E.T. averages. In many experiments, degrading absorbers are used to decrease the energy of accelerator beams, thereby increasing the L.E.T. This process, when carried to extremes, results in significant broadening of the energy spectrum and severe skewness in the L.E.T. spectrum. In such cases, the usefulness of the concept of a single 'average' L.E.T. is vitiated, and the whole L.E.T. distribution should be taken into account when considering L.E.T. dependent radiation effects.

ACKNOWLEDGEMENTS

I wish to thank Chris Tschalär for his collaboration and for preparing calculated energy distribution curves and the L.E.T. distribution code. Bob McGrath and Gordon Wozniak helped with runs at the Berkeley 88 in cyclotron. I also express my appreciation to G. J. Neary, D. G. Martin, D. A. Bance, A. Stretch and R.E. Wilkinson for their co-operation in experiments at Harwell accelerators.

REFERENCES

1. Landau, L. D., *J. Phys. U.S.S.R.* 8, 201, (1944).
2. Bohr, N., *Phil. Mag.* **30**, 581, (1915).
3. Vavilov, P. V., *Soviet Phys. JETP* **5**, 749, (1957).
4. Seltzer, S. and Berger, M. J., *Natl. Acad. Sci.-Natl. Res. Council*, Publ. **1133**, 187, (1964).
5. Symon, K. R., Ph.D. Thesis, Harvard University (1948), unpublished.
6. Tschalär, C., *Nucl. Instr. Methods* **61**, 141, (1968).
7. Tschalär, C., *Nucl. Instr. Methods* **64**, 237, (1968).
8. Payne, M. G., *Phys. Rev.* (to be published).
9. Livingston, M. S. and Bethe, H. A., *Rev. Mod. Phys.* 9, 261, (1937).
10. Blunck, O. and ~~Leisegan~~, S., *Z. Physik* **128**, ~~(500)~~ (1950); Shulek, P., Golovin, B. M., ~~Kylyukina~~, L. A., Medved, S. V., and Pavlovich, P., *Soviet J. Nucl. Phys.* **4**, 400, (1967)/ Golovin, B. M., Kulyukina, L. A., Medved, S. V., Pavlovich, P., and Shulek, P., preprint P1–3190, Joint Inst. for Nucl. Research, Dubna (1967); see also refs 9 and 14.
11. Janni, J. F., U.S. Air Force Weapons Laboratory Report AFWL-tr-65-150 (1966).
12. Maccabee, H. D., Raju, M. R., and Tobias, C. A., *Phys. Rev.* **165**, 469, (~~1967~~)/ Tschalär, C. and Maccabee, H. D., *Phys. Rev.* (in press).
13. Lewis, H. W., *Phys. Rev.* **85**, 20, (1952).
14. Sternheimer, R. M., *Phys. Rev.* **117**, 485, (1960).

DISCUSSION

W. Wild (A.E.R.E.): How precisely is it necessary to know variations in L.E.T.? In Figs. 2 and 3, three methods of computing an 'L' value only differed over a range of 1·5 to 1. I know of no case where, in chemical studies, the observed values vary so rapidly with L.E.T. as to require knowledge of the latter to a precision finer than this. Is the situation different in radiobiological studies?

H. D. Maccabee: It does seem that many L.E.T. effects vary only slowly, e.g. as the logarithm of L. There are, however, biological effects where the damage appears to vary directly with L.E.T., e.g. in Dr. Neary's paper in this session. In such cases it is desirable to have a detailed knowledge of the characteristics of the L.E.T. spectrum.

D. K. Bewley (M.R.C.): Does your theory and computer program allow for charge exchange, particularly for ions heavier than He ions?

H. D. Maccabee: The program uses as its input a table of stopping power vs energy. Insofar as the latter includes charge-exchange effects, these are accounted for in the resulting energy distributions. We have not, however, studied any cases where the energy is low enough for charge exchange effects to become important. The straggling theory itself takes into account multiply-charged heavy ions through the z^2 dependence of the collision spectrum, which is valid only within the limits of the Born approximation. I believe that Lindhard and his associates have treated straggling in the lowest energy cases.

Irradiation of simple biological specimens by charged particles

G. J. Neary

M.R.C. Radiobiology Unit, Harwell, Didcot, Berks., England.

ABSTRACT

In order to study basic mechanisms of the biological effects of radiation the L.E.T. should be defined as closely as possible, since the effects depend on the quality of the radiation as well as on the dose. Consideration of stopping power data shows that the biological specimens must be of cellular dimensions and the minimum range of the particles $\sim 50\,\mu$. Furthermore, any particular value of L.E.T. requires the use of particles of some minimum mass and speed, and to cover the whole region of L.E.T. of biological interest a variety of particle types is needed.

The use of a foil of heavy element to spread the beam by multiple scattering for ensuring satisfactory uniformity of dose over the specimen, the determination of dose by a small air-filled ionisation chamber, the monitoring of the beam by secondary emission from a thin foil *in vacuo*, and the method of containment of the specimens are described.

Some examples of the findings in this type of radiobiological investigation are given, including strand breakage of DNA *in vitro*, cell killing and chromosome aberration production. Differences in the radiation quality dependence of cellular effects and effects on cellular components *in vitro* show that the repair mechanisms of cells probably act in a selective way on the various types of lesion initially produced in a cell, and the results also show that L.E.T. alone is not a complete specification of radiation quality.

INTRODUCTION

The chemical and biological effects of ionising radiation depend not only on the dose (energy absorbed per unit mass) but also on the quality of the radiation. In this context, the quality of radiation relates primarily to the spatial concentration of the primary collision processes along the tracks of charged particles. Since the spectrum of energy losses in primary collisions is to a good approximation the same for various types and speeds of charged particles, a useful simple index of radiation quality is the amount of energy lost per unit length of track, the linear energy transfer (L.E.T.). Therefore, in studying the basic mechanisms of the biological effects of radiation, the L.E.T. should be defined as closely as possible. Since L.E.T. varies along the track of a particle as its energy and speed decrease, it is desirable to use only a small segment of a track. Consequently, the thickness of the biological specimen must be small compared with the total range of the particle, so that in practice such experiments must be restricted to irradiation of monolayers of cells. Furthermore, the layer of cells must be exposed to directly-ionising particles all having nearly the same energy. This concept of 'track-segment' irradiation by beams of charged particles was first introduced by Zirkle and Tobias.[1]

Stopping power and range data. In planning track-segment experiments the first need is for comprehensive stopping power and range data. The compilation of Barkas and Berger[2] is the most extensive, and in principle provides data for calculating the stopping power and range of any heavy ion from protons upwards for any energy above 1 MeV/amu and for any medium. A very useful general presentation is also given by Steward;[3] both sources may be usefully supplemented by the direct tabulation of experimental data given by Northcliffe,[4] particularly for the heavier ions, and by the tables of Janni[5] for protons in various substances, which appear to be in best agreement with latest experimental data.

Track-segment considerations. Some representative data for the stopping power in water (unrestricted energy loss, L.E.T.$_\infty$) of various particles at various energies are shown in Fig. 1; in most of the energy region covered in the curves the ions are nearly fully stripped. Since the total range, R, of the particles used in a track-segment must be large compared with the segment itself, there is a lower limit to the range which may acceptably be used. For planning purposes, a very approximate form of the stopping power relation may be used, neglecting the log term:

$$-\frac{dE}{dx} = z^2 \frac{M}{E} k \qquad (1)$$

where the symbols have their usual meaning and k is a constant. It follows from Eqn (1) that R is approximately equal to $E/2L_\infty$, writing L_∞ for dE/dx; therefore for a given L.E.T. (L_∞) and minimum R, there is a minimum value of particle energy E which is acceptable. In addition, R is also approximately equal to $kz^2 M/2L_\infty^2$, so that for a given L_∞ and R there is a minimum value of particle mass, M, which is acceptable. These considerations are illustrated more precisely in Fig. 1, where the energy and L.E.T.$_\infty$ of a particle of range either 50 μm or 20 μm are indicated; it may be seen that for any one particle type there is an upper limit to the value of L.E.T.$_\infty$ which can be obtained for a given minimum range. Therefore a variety of particle types is required for track-segment experiments to cover all values of L.E.T.$_\infty$ of biological interest. Of course, in principle, the heaviest particle required for the highest L.E.T.$_\infty$ would provide for the lower L.E.T.$_\infty$ values also if its energy could be high enough. Two considerations rule out this possibility: (a) it would normally be impracticable to accelerate the particle to a high enough energy, and (b) the maximum initial energy of the δ-rays is proportional to the energy of the primary particle per atomic mass unit, and it is generally undesirable to exceed values of 10 MeV/amu because the δ-ray slowing-down spectrum becomes rather complex. Thus the two criteria of a maximum particle energy of 10 MeV/amu on the one hand and a minimum range on the other serve to define the region of L.E.T.$_\infty$ which can be covered by a particular particle. Such regions for various particles, based on a minimum range of 50 μm, are illustrated in Fig. 2.

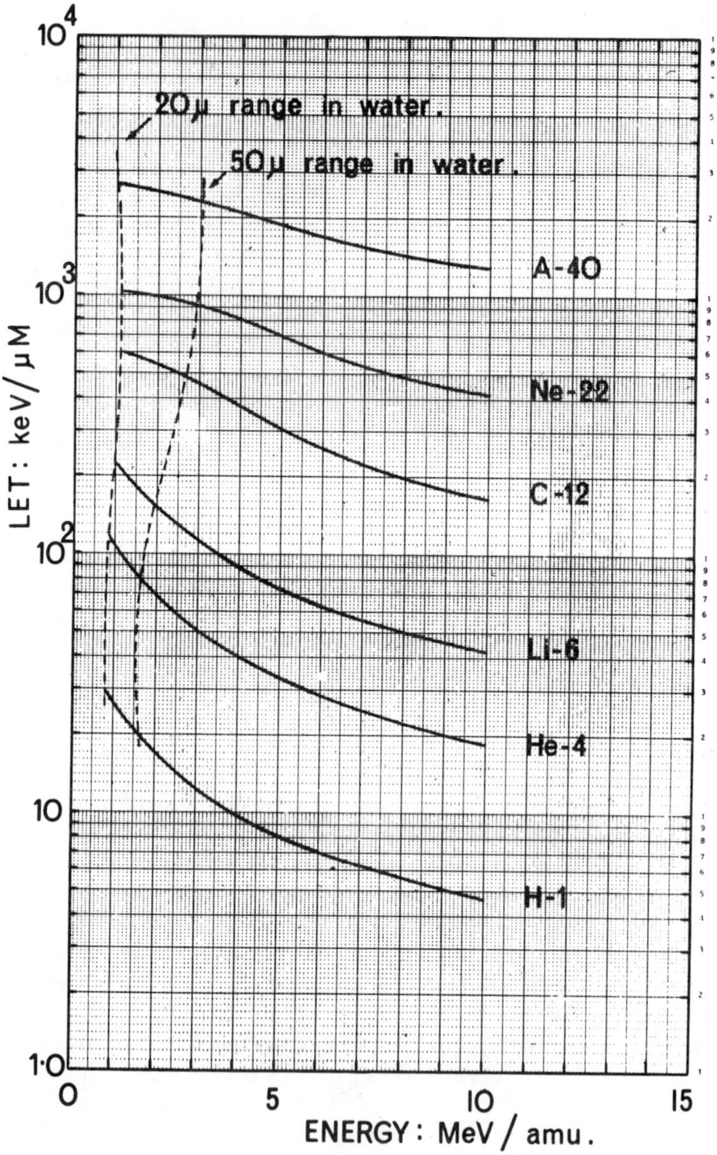

Fig. 1. L.E.T.$_{\infty}$–Energy relationships for various charged particles

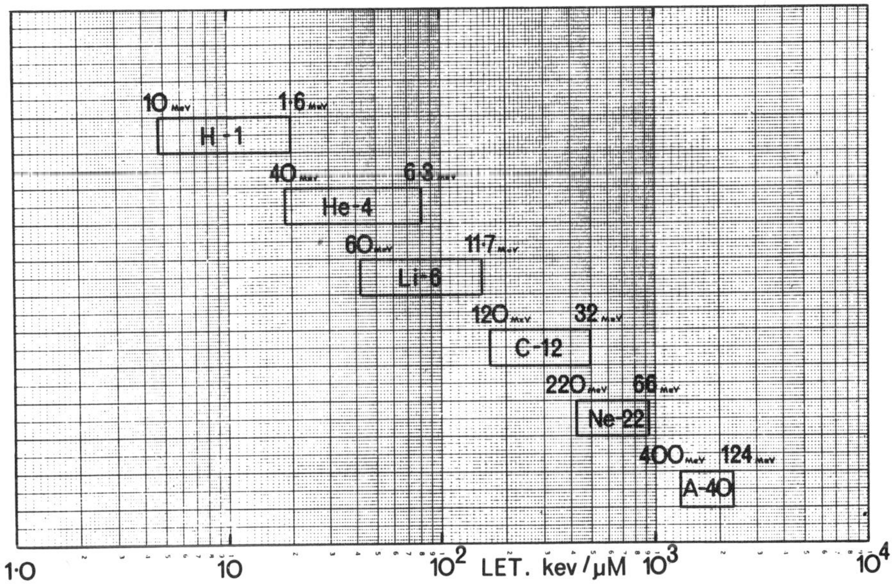

Fig. 2. Track-Segment Irradiations: L.E.T.$_\infty$ regions for various charged particles

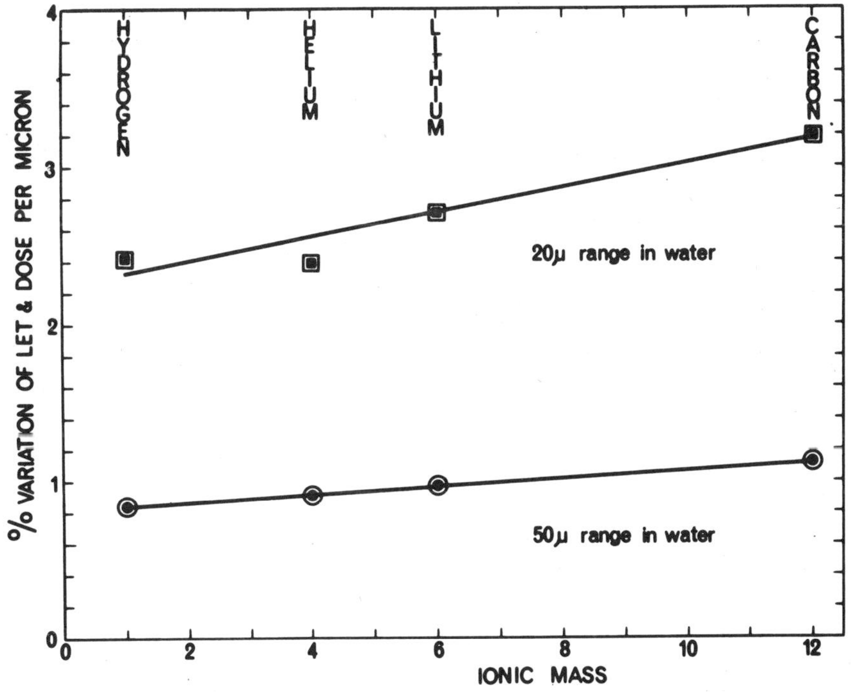

Fig. 3. Percentage variation of L.E.T.$_\infty$ and dose per micron of track-segment for various charged particles of range 50 or 20 μm in water

The variation of dose and L.E.T.$_\infty$ per micron of track-segment for various ions of range 50 μm or 20 μm is shown in Fig. 3; for example, the variation for a yeast cell of diam. 5 μm irradiated by lithium-6 particles of L.E.T.$_\infty$ 155 keV/μm and range 50 μm is about 5% from front to back. The slight upward trend in percentage variation with increasing particle mass displayed in Fig. 3 is not maintained for heavier particles; for argon, the rate of variation appears to be only about half that for the lighter particles.

EXPERIMENTAL

The general arrangement used for exposure of biological specimens to particles from a cyclotron or other accelerators is shown schematically in Fig. 4. It is usual,

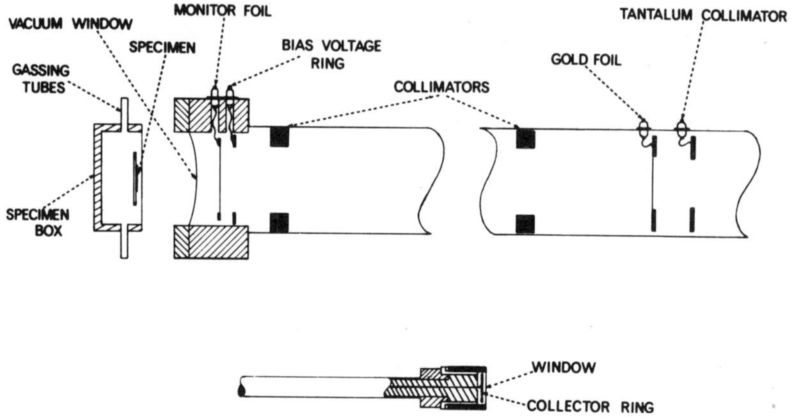

Fig. 4. Experimental arrangement for irradiation of biological specimens by accelerated particles. Below: Ionisation chamber for dose determination

because of shielding considerations, for the particles from the accelerator to pass down a long evacuated tube of diam. ~4 in, with appropriate collimators to minimise scattered particles of reduced energy. Since most biological specimens must be kept at atmospheric pressure and at controlled humidity, the particles must emerge from the evacuated beam-line through a thin window. Melinex is a convenient material for a window; typically, for a 1¾ in diam. window, 300-gauge Melinex (~10 mg/cm^2) is used, and the deflection at the centre is ~3 mm. A thin layer of aluminium (a few μg/cm^2) is deposited by evaporation on both sides of the foil to prevent build-up of charge.

The specimens, deposited directly on a microscope slide or membrane filter, are contained within a Perspex box, with a front window of 25-gauge Melinex (~1 mg/cm^2). Water-saturated air or pure nitrogen is circulated through the specimen box, which is positioned accurately in a jig so that the total absorption path for the particles reaching the specimen is known precisely.

Beam uniformity. The specimen may cover an area up to ~3 cm diam. and it is desirable that the beam should be uniform to within a few per cent, depending on the particular biological effect under investigation. This requirement can sometimes be met by defocusing the beam with a magnetic lens. An alternative is to use a thin foil of gold situated several metres from the exit window to produce sufficient divergence of the beam by multiple scattering;[6] the beam is focused on to the scattering foil. For a total current i concentrated on a scattering foil producing a mean square scattering angle $\overline{\theta^2}$, the current density per cm^2 at a point r cm from the axis of the specimen, itself at d cm from the scattering foil, is

$$(i/\pi d^2 \overline{\theta^2}) \exp(-r^2/d^2 \overline{\theta^2}) \tag{2}$$

As an example, for 4 MeV protons incident on a gold foil 10^{-4} in thick, the energy loss is 0·16 MeV and the root mean square scattering angle is $4{\cdot}8 \times 10^{-2}$ rad; at the specimen 6 m distant, the 99% contour has a radius of 2·9 cm and the 90% contour 9·2 cm.

Beam energy. The beam energy at the position of the specimen can be measured by the standard procedure using a solid state detector. An alternative and quite satisfactory method requiring less equipment is to measure the Bragg curve with a shallow ionisation chamber (see below) and to deduce the mean range from the extrapolated ionisation range. The precise relation between these two ranges must be calculated for the particular particles and energy from a consideration of the expected straggling.

If more than one particle energy is required in an experiment it will frequently be more convenient to obtain the lower energies by adding absorbers in the path of the beam rather than by altering the accelerator settings which determine initial energy. The effect of the use of absorbers on the spread of particle energies and L.E.T. values is analysed and discussed by Maccabee.[7]

Dosimetry. The relations between current density, j, in pA/cm^2, particle fluence rate, F, in particles/cm^2 s, and dose rate in rad/s for particles of charge z and L.E.T.$_\infty$, L_∞, in keV/μm are:

$$F = (j/z)\, 6{\cdot}245 \times 10^6 \tag{3}$$

and

$$\text{Dose rate} = FL_\infty\, 1{\cdot}602 \times 10^{-7} \tag{4}$$

$$= jL_\infty/z \tag{5}$$

In practice, it is usually most convenient to determine the dose rate by measuring the ionisation in a small air-filled ionisation chamber placed at the position to be occupied by the specimen. The ionisation chamber shown schematically in Fig. 4 has an air volume 8 mm diam. X 2 mm deep. The voltage applied to the ion chamber is 240 V, and in view of the high field strength no difficulty about saturation of the current was expected or found at the highest dose rates used ($\sim 10^5$ rad/min). The side walls and base of the chamber are of 'air-wall' bakelite-graphite mixture and the front foil is aluminium leaf 1 μm thick. The effective volume of the chamber is determined from the current produced by exposure to a known dose rate from a standard radium source after adding a cap of bakelite-graphite to ensure electronic equilibrium.

The value of W, the energy per ion pair in air in track-segments of charged particles of the energies used, is probably close to the value for fast electrons. If this value is taken as 34·0 eV/ion pair, a collected charge of 1 esu/cc under standard conditions (0°C, 760 torr) represents 87·7 ergs/g of air. The energy which would be absorbed in any other medium, e.g. water, DNA, etc., can be deduced from the ratio of the stopping powers of the medium and of air for the particles. For example, for protons of a few MeV, 1 esu/cc corresponds to 1·054 rad in water; for carbon-12 particles at 58 MeV, 1 esu/cc corresponds to 1·041 rad in water.

Monitoring of the beam during exposure of a specimen is achieved by measuring the secondary emission current from a thin foil (25-gauge Melinex coated with aluminium) just inside the end of the beam line (Fig. 4). For a given beam, the absolute magnitude of the current collected can be chosen at a convenient level by selecting an appropriate voltage for the bias electrode nearby. This secondary emission system is very reproducible (~1 %) if the pressure in the beam line is low, $\leqslant 10^{-5}$ torr. During an experiment the secondary emission monitor is calibrated by placing the ion chamber in position and observing the currents from it and from the monitor. The currents are amplified by a d.c. electrometer type 1388B or a Vibron, and the outputs are integrated by two electromechanical clocks. Thus a given charge collected from the monitor represents a definite dose at the specimen, independent of beam fluctuations.

This same system is used to survey the uniformity of the beam distribution for a given set of beam settings, observing the ratio of readings of the ion chamber at various positions to the readings of the monitor.

RESULTS AND DISCUSSION

Some examples of the kind of biological investigations which have been carried out at the cellular level using charged particles from accelerators, and a brief indication of their implications for fundamental mechanisms of radiobiological action, are given below. The examples concern physicochemical changes in DNA, cell killing and chromosome aberrations.

Effects on DNA. Ionising radiation produces a variety of molecular lesions in DNA, such as base damage, single-strand breaks and double-strand breaks, all of which would be potentially important in an irradiated cell. In many cases, cellular effects are produced more effectively at high than at low L.E.T., and an explanation frequently suggested for this increase in relative biological effectiveness (RBE) has been a supposed change in the relative numbers of the various types of molecular damage to DNA, particularly an increase in the production of double-strand breaks, a supposedly irreparable type of lesion. In fact, however, the only direct evidence for DNA *in vitro* showed that double-strand breakage by α-particles of $L.E.T._{\infty}$ between ~105 and 145 keV/μm was very little more than by fast electrons.[8] We have recently obtained data for dry films of DNA extracted from T7 bacteriophage, and for freeze-dried whole phage, in track-segment experiments over a wide range of $L.E.T._{\infty}$[9]. The results show that efficiency of production of single-strand breakage, as expected, did not increase with $L.E.T._{\infty}$, but double-strand breakage did not show the increase in proportion to $L.E.T._{\infty}$ which would be expected if this lesion were the consequence of two single-strand breaks independently produced in the two strands by the same particle track; there was very little increase up to at least 100 keV/μm, in agreement with the earlier work,[8] but at 250 keV/μm, the efficiency was about doubled. This variation of double-strand breakage per rad with $L.E.T._{\infty}$ for DNA in an *in vitro* system is unlike that for cellular effects (see below). Furthermore, there was no enhancing effect of oxygen on either double-strand or single-strand breakage at any L.E.T. value, in contrast to the usually marked effect of oxygen on radiation sensitivity for most cellular effects.

Cell killing. We have examined cell killing of bacteria and yeasts by charged particles of various L.E.T.$_\infty$ values and also by monochromatic X-rays of quantum energy of the order of 1 keV.[10,11,12] The full analysis of the L.E.T. dependence of radiosensitivity to charged particles requires detailed consideration of effects of δ-rays,[10] but in any case there is a considerable variation of sensitivity in the region up to an L.E.T.$_\infty$ value of ~100 keV/μm in which there is little variation for DNA double-strand breakage *in vitro*; furthermore, the radiosensitivity of the cells has fallen considerably at an L.E.T. value of 250 keV/μm, where DNA double-strand breakage has doubled. Taken in isolation, such facts might suggest that effects on DNA are not the determining factor in cell killing. However, in view of the very considerable body of evidence from various directions that effects on DNA are very important, the experimental findings point to a difference in the L.E.T. dependence of the initial lesions in DNA and the residual lesions after the operation of cellular repair processes. The same conclusion is suggested by the data for low-energy X-rays, which are twice as effective in killing haploid yeast (SC7)[12] and sensitive bacteria[11] (*E. coli* B_{s-1} and K12 AB 2483 rec$^-$) as are heavy charged particles with L.E.T. equal to the average L.E.T. of the electrons produced. But the efficiency of production of single-strand and double-strand breakage of DNA *in vitro* by the low-energy X-rays is no greater than that of protons of corresponding L.E.T. These findings with low-energy X-rays also show that L.E.T. alone is not a complete specification of radiation quality. Two qualities of radiation may be equally effective in producing a lesion, e.g. strand-breakage of DNA detected in an *in vitro* system, but there might be qualitative differences in the nature of the lesions in the two cases, if examined for their detailed chemical characteristics. Multiple scattering of the low-energy electrons produced by the low-energy X-rays might perhaps lead to a greater local concentration of damage to a DNA molecule than would be produced by the straight track of a heavy particle. Thus a strand break produced by a slow electron might be less likely to be repaired by the cellular repair mechanisms than a break by a heavy particle of comparable L.E.T.

Chromosome aberrations. We have shown that the efficiency of production of chromatid aberrations in pollen tubes of Tradescantia increases at first in direct proportion to L.E.T.$_\infty$, reaches a maximum at an L.E.T.$_\infty$ of ~100 keV/μm and falls to about half the maximum at an L.E.T.$_\infty$ of 250 keV/μm;[13] low-energy X-rays are considerably *less* effective than protons of equivalent L.E.T.[14] These data are compatible with a model in which a chromosome aberration is the consequence of the interaction of two damaged chromosomes, each primary lesion being produced by a separate energy-loss event; the two primary lesions may be produced at a separation of a considerable fraction of a micron, so that the electrons of short track-length produced by low-energy X-rays are relatively inefficient. Similar conclusions have been reached from experiments on chromosome-type aberrations in synchronised Chinese hamster cells irradiated in the G_1 phase.[15] These experiments show, in addition, that killing of these cells is not principally due to the production of chromosome aberrations, at least in the L.E.T. region around 20 keV/μm corresponding to the particular quality of low-energy X-rays used in the investigation.

In conclusion, it seems obvious in view of the complexities of the processes even in experiments when the factor of radiation quality has been simplified as far as possible, that the use of radiations with a broad distribution of L.E.T. and track-length such as hard X- or δ-radiation or fast neutrons must increase greatly the difficulties in unravelling the basic mechanisms of radiation action. For this reason, the use of a comprehensive range of charged particles for track-segment experiments on simple biological entities is essential.

ACKNOWLEDGEMENTS

The writer's thanks are due to D. A. Bance, D. G. Martin, A. Stretch, S. Weal, and R. Wilkinson for expert technical assistance, and to his scientific collaborators for permission to refer to unpublished work.

REFERENCES

1. Zirkle, R. E. and Tobias, C. A., *Arch. Biochem. Biophys.* **47**, 282, (1953).
2. Barkas, W. H. and Berger, M. J., in 'Studies in Penetration of Charged Particles in Matter', 103, National Academy of Sciences–National Research Council, Washington D.C. (1964).
3. Steward, P. G. and Wallace, R., report UCRL-17314, University of California, Berkeley (1966).
4. Northcliffe, L. C., in 'Studies in Penetration of Charged Particles in Matter', 173, National Academy of Sciences–National Research Council, Washington. D.C. (1964).
5. Janni, J. F., report AFWL-TR-65-150, Air Force Weapons Laboratory, Kirtland Air Force Base, New Mexico (1966).
6. Bethe, H. A. and Ashkin, J., in 'Experimental Nuclear Physics', ed. E. Segrè, 282, John Wiley and Sons, Inc., New York (1953).
7. Maccabee, H. D., these proceedings.
8. Alexander, P., Lett, J. T., Kopp, P., and Itazhaki, R., *Radiat. Res.* **14**, 363, (1961).
9. Neary, G. J., Simpson-Gildemeister, H., and Peacocke, A. R., to be published.
10. Munson, R. J., Neary, G. J., Bridges, B. A., and Preston, R. J., *Int. J. Radiat. Biol.* **13**, 205, (1967).
11. Neary, G. J., Preston, R. J., and Munson, R. J., to be published.
12. Neary, G. J., Savage, J. R. K., Preston, R. J., and Hunnable, E. G., *Studia Biophysica* **12**, 198, (1968).
13. Neary, G. J. and Savage, J. R. K., *Int. J. Radiat. Biol.* **11**, 209, (1966) and unpublished work.
14. Neary, G. J., Preston, R. J., and Savage, J. R. K., *Int. J. Radiat. Biol.* **12**, 317, (1967).
15. Preston, R. J. and Neary, G. J., *Studia Biophysica*, in press.

DISCUSSION

W. G. Burns (A.E.R.E.): You say that the L.E.T. distribution of radiation is not a sufficient specification for the correlation of biological effects. Would an L.E.T. distribution which includes all generations of secondary radiation be sufficient?
G. J. Neary: I think we can be fairly sure that detailed consideration of the total track distribution from the viewpoint of L.E.T. only will not provide a complete specification. For example, we have cases where low-energy X-rays are less effective than protons of corresponding L.E.T. and others where the X-rays are more effective. It therefore seems unlikely that analysis of L.E.T. distribution alone could account for the two opposite results. In the production of

chromosome aberrations there is good reason to think that track-length itself is an important parameter. In cell killing, multiple scattering of a low energy electron could give more concentrated, less-reparable damage to DNA than the straight track of a heavy particle.

H. D. Maccabee (M.R.C., Harwell): I would like to comment on Dr. Neary's paper. It is likely that biological systems are more sensitive indicators of L.E.T. effects than chemical ones, because a biological inactivation can result from the breaking of a single macromolecule. Indeed, a genetic mutation can result from a single-atom change in a DNA base. I know of no chemical system with such resolution. In view of this, and of the fact that some biological effects are directly proportional to L.E.T., perhaps our attention to accuracy in L.E.T. spectrum calculations is not misplaced.

Irradiation of phages with fast protons

B. Candoni

Istituto di Scienze Fisiche-Università degli Studi di Milano, I.N.F.N., via Celori 16, Milan, Italy

and C. Corno

Istituto di Microbiologia-Università degli Studi di Milano, via Mangiagalli 31, Milan, Italy

ABSTRACT

Dry preparations of bacteriophages both with single-strand and with double-strand nucleic acid have been irradiated with fast protons from the Milan AVF Cyclotron. The inactivation cross-section was determined by measuring the plaque-forming ability of the phages. For single-strand DNA and RNA phages, experimental data seem to confirm the lethal effect of all hits on the nucleic acid; for double-strand phages we observe that the sensitive volume is much smaller than the DNA volume. Some possible processes of inactivation are discussed.

INTRODUCTION

The behaviour of phages irradiated with different L.E.T. and types of radiations has proved a useful tool for investigating the biological action of ionising radiations.[1-3] The use of protons of ⩾20 MeV enables us to understand the mechanism of radiation damage. It is found that the damage is sufficiently localised because of the rather low L.E.T. and the well-defined track of the incident proton. The possibility of irradiating the samples in a dry state and *in vacuo* reduces significantly the effects of indirect processes due to radicals and oxygen. In this work we present the results obtained by measuring the plaque-forming ability of RNA as well as of single- and double-strand DNA phages, irradiated by 25 MeV protons extracted from the Milan AVF Cyclotron. A few samples have been already studied by other authors but comparison of the results obtained is difficult because of the different environments and radiations used.

EXPERIMENTAL

Sample preparation. The phages irradiated were: $\mu 2$[4] and 8c[5] (RNA), ϕx174, S13 and Ec9[6] (single-strand DNA), T1, T2, T5 and λ (double-strand DNA).

Survival curves were obtained by mixing two or three phages and by plating on separate indicator strains. The mixed suspensions were obtained by diluting stock preparations of the phages in nutrient broth. Samples were prepared alternatively by pipetting 0·02 ml aliquots of a suspension onto sterile Mylar and drying in a vacuum desiccator at ~13·3 Nm^{-2} (0·1 torr) or by lyophilising 0·2 ml portions. The last procedure was followed for phages which were more sensitive to dryness. After irradiation the samples were suspended again overnight in 10 ml of trypton broth at 4°C. Phages T1, Ec9, μ2 and 8c remained unaffected by the drying process and by vacuum, while 40–90% of the plaque-forming units of the other phages were inactivated. Several irradiated samples and unirradiated controls were required to reduce the errors arising from this cause. Plating bacteria were: *E.coli* K12W1895T1^r for 8c, μ2 and Ec9, *E.coli* CC_2T1^r for ϕx174 and S13, *E.coli* B for T1, T2 and T5, and C600 for λ.

Sample irradiation. The dry samples were transferred to an irradiation chamber at a pressure of 0·133 Nm^{-2} (10^{-3} torr). This chamber was equipped with an aluminium scattering foil, 1 mm thick, to distribute the protons uniformly on the sample area, with an aperture to reduce the scattered beam to a measurable area. The beam current, maintained at ~5–10 nA in order to avoid thermal effects, was collected on an aluminium target block and integrated by a current integrator. Control samples were placed in the chamber away from the beam. Before irradiation the beam was monitored to ensure uniform irradiation of the samples.

RESULTS

Typical measurements of the dose-survival relationship are shown in Fig. 1. The experimental points fit one-hit survival curves over at least three logarithmic decades. The dose in krad and the experimental cross-section S_{exp} in units of 10^{-12} m^2 are given in Tables 1 and 2 for the single- and double-strand nucleic acid phages respectively. The mean geometric thickness t and cross-section S_0 of the nucleic acid content are also reported together with the theoretical cross-section $S = S_0\ (1 - e^{-it})$. The value of i is $0{\cdot}47 \times 10^8 \cdot m^{-1}$ for protons of 25 MeV, assuming a value of 75 eV per hit.[2] The error in the cross-section was estimated at ~10% for all phages except T2, T5 and λ, which amounted to 20%.

DISCUSSION

From a careful examination of the experimental data we infer that the theoretical cross-sections are in good agreement with the experimental values for all phages with single-strand nucleic acid. As the inactivation process follows an exponential law we conclude that every hit is almost certainly lethal to these phages; that is, the lethal efficiency per ion-pair produced in the nucleic acids is nearly unity. On the other hand, the absence of hypersensitivity phenomena (experimental cross-sections greater than geometrical ones) and the irradiating environment confirm the assumption that ignoring indirect effects due to the presence of radicals and to the presence of oxygen and δ-rays is valid. In fact, the phages with double-strand DNA are more resistant to radiation, as has been proved by many authors.[7,8]

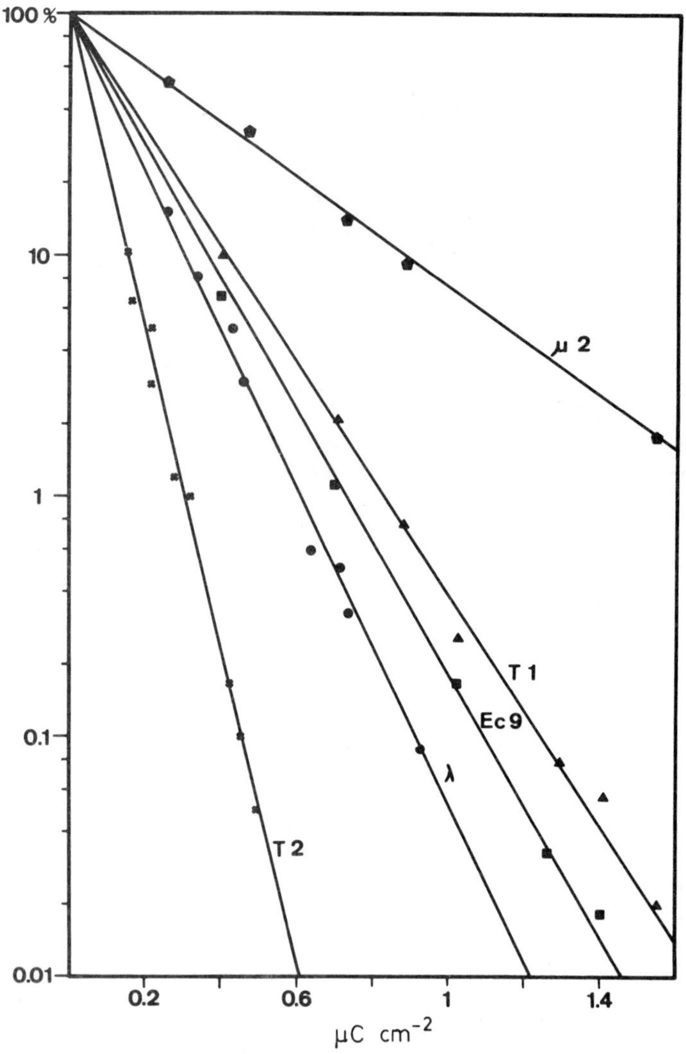

Fig. 1. Typical survival measurements plotted against radiation dose in microcoulombs cm^{-2}

Table 1. TARGET THICKNESS AND AREA, THEORETICAL AND EXPERIMENTAL CROSS-SECTION AND D_{37} DOSE OF RNA AND SINGLE-STRAND DNA PHAGES

Phage	*t* (Å)	S_0	S	S_{exp}	D_{37} (krad)
μ2	80	1·25	0·38	0·40	881
8c	80	1·25	0·38	0·38	928
S13	100	1·7	0·61	0·71	496
φx174	100	1·7	0·61	0·68	518
Ec9	15	11·0	0·75	1·03	342

S_0, S and S_{exp} are in units of 10^{-12} cm^2.

Table 2. TARGET THICKNESS AND AREA, THEORETICAL AND EXPERIMENTAL CROSS-SECTION AND D_{37} DOSE OF DOUBLE-STRAND DNA PHAGES

Phage	t (Å)	S_0	S	S_{exp}	D_{37} (krad)
T1	270	14·4	10·5	0·88	105
λ	330	21	16	1·2	294
T2	400	32	27	2·4	147
T5	345	22	17	4·7	75

S_0, S and S_{exp} are in units of 10^{-12}cm^2.

The theoretical cross-sections (Table 2) are much larger than the experimental ones: this means that the hits are not always lethal. Ignoring the possibility of a multi-hit inactivation process, since the dose-survival relationship is linear on a semilogarithmic plot, we may consider the following possibilities:

(1) existence of a repair mechanism;
(2) existence of a critical target which is responsible for some early enzymatic function, whose complete physicochemical integrity is required for DNA replication;[9]
(3) that in a double-strand structure only some particular lesions (double-strand breaks, crosslinks, etc.) are lethal and generally a single hit is not lethal.

The three hypotheses may, of course, coexist in an inactivation process; this means that a given type of lesion may be lethal to one phage and unable to be repaired, and yet not lethal to another one which possesses the repair mechanism for that particular lesion. Following Szybalski[10] we will discuss the results by assuming that:

(1) the double-strand break is always lethal, and
(2) the direct damage produced by radiation cannot be repaired, the repair mechanism being available only for some type of lesions produced by indirect effects.

On the basis of such assumptions we conclude that in our case, because of the absence of indirect effects, the damage cannot be repaired, sot that either a 'critical target' exists or particular types of lesions are lethal. Such lesions should then be lethal for all phages with double-strand DNA, and their frequency shou should depend only on the kind of radiation used, especially on L.E.T. If a critical target exists we can evaluate its volume $V_{c.t.}$ (Table 3).

Table 3. SENSITIVE VOLUME $V_{c.t.}$ AND PROBABILITY OF INACTIVATION α OF DOUBLE-STRAND DNA PHAGES

Phage	$V_{c.t.}$ (10^{-18}cm^3)	α
T1	1·87	0·048
λ	2·55	0·037
T2	5·1	0·039
T5	10·0	0·131

The values obtained are much lower than the whole DNA content and are comparable with Freifelder's[11] and McCorquodale's[12] data if their results are interpreted in terms of a critical target. If, on the contrary, some type of lethal lesions exist, we can calculate the frequency with which they occur. This frequency will be equal to the proportionality constant α between the critical target volume and the total DNA volume.[8, 11] The values are reported in Table 3; they are almost the same for T1, T2 and λ, while for T5 the frequency of occurrence of lethal lesions is higher. If we assume that the direct lethal lesion is essentially a double-strand break, then the presence of single-strand breaks on the DNA strains of T5[13] could justify the higher value of α. We may thus attribute to α the meaning of the probability that an incoming particle breaks simultaneously both DNA strands.[14]

Physicochemical studies on radiation damage to the nucleic acid are in progress.

ACKNOWLEDGMENTS

Our experimental programme has been carried out with the helpful cooperation of Professor C. Succi and the Milan cyclotron crew, and of Professor R. Deotto. We also acknowledge the expert technical assistance of Miss P. Barassi. This work was supported by a grant from the C.N.R., No. 115.1849.0.1217, by I.N.F.N., and by Istituto di Scienze Fisiche dell' Università di Milano.

REFERENCES

1. Fluke, D. J., Brustad, T., and Birge, A. C., *Radn. Res.*, **13**, 788, (1960).
2. Schambra, P. E. and Hutchinson, F., *Radn. Res.* **23**, 514, (1964).
3. Butts, J. J. and Katz, R., *Radn. Res.* **30**, 855, (1967).
4. Dettori, R., Maccacaro, G. A., and Piccinin, G. L., *Giornale di Microbiologia* **9**, 141, (1961).
5. Dettori, R. and Neri, M. G., *Giornale di Microbiologia* (in press).
6. Dettori, R. and Neri, M. G., *Giornale di Microbiologia*, **13**, 111, (1965).
7. Terzi, M., *Nature* **191**, 461, (1961).
8. Kaplan, H. S. and Moses, L. E., *Science* **145**, 21, (1964).
9. Ginoza, W. and Miller, R. C., *Proc. N.Y. Acad. Sci.* **54**, 551, (1965).
10. Szybalski, W., *J. Ark. Med. Soc.* **66**, 488, (1966).
11. Freifelder, D., *Virology* **36**, 613, (1968).
12. Colter-Paranchych, 'The molecular biology of viruses', 39–40.
13. Burgi, E., Hershey, A. D., and Ingraham, L., *Virology* **28**, 11, (1966).
14. Munson, R. J., Neary, G. L., Bridges, B. A., and Preston, R. J., *Int. J. Radn., Biol.* **13**, 205, (1967).

Minimum cyclotron size for radiation therapy: a study of relevant physical parameters

Leon J. Goodman and Stephen A. Marino

Radiological Research Accelerator Facility, Radiological Research Laboratory, Department of Radiology, Columbia University, 17 Cornell Avenue, Upton, New York, 11973, U.S.A.

James T. Brennan
Matthew J. Wilson *Professor of Research Radiology, Department of Radiology, Hospital of the University of Pennsylvania, 3400 Spruce Street, Philadelphia, Pennsylvania, 19104, U.S.A.*

ABSTRACT

The use of the reaction $^{9}Be(^{2}H, n)^{10}B$ with cyclotron-generated deuterons has been examined with respect to radiotherapy requirements. Attainable dose rates and depth dose are the most important physical characteristics for this application. Sufficient dose rate has hitherto been available, but improvement in depth dose was sought.

For the reaction considered, the depth dose could be improved only by increasing the energy of the deuteron beam. Since this implies an increase in the size and cost of the cyclotron, it became important to know the cyclotron magnet diameter required to produce neutrons having satisfactory depth dose characteristics.

Experiments simulating realistic therapy conditions were performed within the last year at the Texas A and M Variable Energy Cyclotron (TAMVEC). Collimated beams of neutrons were generated by deuterons of various energies bombarding thick and thin beryllium targets. Dose rate, depth dose and secondary particle build-up measurements were made using a tissue-equivalent ionisation chamber in a large tissue phantom.

These studies show the minimum deuteron energy required to be ~30 MeV to satisfy the criterion of 50% depth dose at a minimum depth of 10 cm of tissue. The thin-target investigations demonstrated that depth dose can be increased somewhat if thin targets are used.

*Based on work performed under (1) Contract AT-(30-1)-2740 for the U.S. Atomic Energy Commission and (2) U.S. Public Health Service grant CA-10059 for the National Cancer Institute.

INTRODUCTION

A useful neutron-producing reaction for use with cyclotrons is $^{9}Be(^{2}H, n)^{10}B$. Some spectra for the neutrons produced are available,[1] as is some information concerning dose rates and depth dose.[2, 3] However, complete information on dose rates and depth dose has not been available for deuteron energies above 16 MeV.

In considering the kinds of neutron sources that might be most useful for purposes of radiotherapy, the two physical characteristics of most importance are dose rate and depth dose. A third, less important, but nevertheless significant parameter is skin sparing, that is, the dose reduction to the skin resulting from lack of secondary particle equilibrium near the beam entrance surface.

With respect to cyclotrons, there has never been any difficulty with dose rate. As early as 1939, when the 60-in cyclotron became available at Berkeley, California, dose rates of 10 rad/min at an SSD (skin-to-source distance) of 100 cm have been available. With the most recent AVF cyclotrons dose rates several times as great are readily obtained. On the other hand, some difficulty has been experienced with cyclotrons in obtaining depth dose curves which are satisfactory from the clinical point of view.

A ^{60}Co clinical machine, when used with a SSD of 80 cm, produces a 50% depth dose at 10 cm when the field size is small (e.g. 5 × 5 cm). To be competitive, neutron sources must have depth dose characteristics which compare favourably with those of ^{60}Co machines. If the value of 50% at 10 cm depth in tissue is accepted as the criterion for acceptable depth dose,[4] then it becomes important to know how large a cyclotron must be to meet this requirement. The MRC Cyclotron at Hammersmith Hospital, which is now being used in a clinical trial in London, employs 16 MeV deuterons on a beryllium target. The depth dose characteristics of this cyclotron are satisfactory for most purposes.[3] However, given a choice, one would not at present start a new clinical trial without trying to obtain a machine having somewhat better depth dose characteristics. If the deuteron-beryllium reaction is used, the only way to improve the depth dose of the neutron spectrum produced is to increase the energy of the incident deuterons. Early work by Tochilin[4] and Bond[5] indicated that depth dose would probably improve slowly with increasing deuteron energy, which can only be achieved by increasing the diameter of the cyclotron magnet. Furthermore, the cost of a cyclotron rises sharply with increasing magnet diameter. For these reasons it is important to know the cyclotron diameter required to produce neutrons having satisfactory depth dose characteristics.

Dose rate, depth dose, and secondary particle build-up have been measured for deuteron beam energies from 16 to 30 MeV using thick beryllium targets and also for thin beryllium targets. Some data were also obtained at 35 MeV. This work was carried out in September, 1968 and January, 1969 at the Texas A and M University's Variable Energy Cyclotron (TAMVEC) at College Station, Texas.

Earlier work was confined to the use of a thick beryllium target at deuteron energies up to 30 MeV. 'Thick' is taken to mean a target of thickness equal to or greater than the range of the bombarding particle. Targets thinner than this range are referred to as 'thin'.

In accordance with a suggestion by C. J. Parnell, the latter work investigated the possibility of increasing the dose at depth through the use of thin beryllium targets. Such targets could be expected to yield higher doses at depth due to the increase in the mean neutron energy resulting from utilisation in the beryllium target of only the high-energy bombarding deuterons (i.e. only the initial portion of each deuteron track).

The application of thin targets is complicated by the necessity of stopping the charged particle beam after its passage through the target, in order to prevent its emergence into the air and the severe local heating which this could produce on the collimator, test equipment or patient. It is well known that deuteron bombardment readily generates neutrons in all materials. To minimise the number of unwanted neutrons generated in the beam-stop, a high-Z material must be used. A further complication is the requirement that the beam-stop material have good thermal conductivity and a high melting-point, to facilitate the dissipation of the high beam power expended in the target assembly. Most of the high-Z elements are rare, expensive, and/or unstable. A familiar exception is lead, which is unsuitable for a high-power beam stop due to its low melting-point. Two acceptable metals are tungsten and tantalum. Tantalum was chosen since it is readily obtainable in thick slabs, and is also much easier to fabricate than tungsten.

APPARATUS AND EXPERIMENTAL TECHNIQUE

All of the measurements were made using a disc-shaped, tissue-equivalent ionisation chamber[7] through which a steady flow of tissue-equivalent gas was maintained (Fig. 1.) One wall of the chamber was 0·14 mm thick, which permitted measurements of secondary particle build-up to be made by the external addition of tissue-equivalent plastic discs of various thicknesses, as shown in the figure.

The depth dose measurements were performed along the central axis of a cubic phantom, 35 cm on a side, filled with tissue-equivalent liquid.[8] The chamber was made to traverse the central axis slowly using a remotely-controlled motor. This gave a continuous strip chart record of dose as a function of depth.

The skin sparing measurements were made with the chamber outside the phantom and directly in front of it. Previous neutron measurements at other sites had indicated that the presence or absence of the tissue phantom behind the chamber during the skin sparing measurements had essentially no effect on the curve obtained, except for an overall increase in dose rate of several per cent due to back scatter.

All measurements were monitored and normalised to the response of an air-filled spherical ionisation chamber, 1·2 cm in diam. and with a 4 mm thick tissue-equivalent plastic wall. This instrument was located within the inner edge of the neutron beam at the downstream end of the collimator.

In order to obtain realistic simulation of a radiotherapy arrangement, a neutron beam was collimated using Benelex collimators in a cubic Benelex shield, 60 cm on a side, as shown in Fig. 2. Benelex is an inexpensive, hydrogenous, pressed hardboard which is easy to fabricate and which has good neutron shielding properties. The square cross-section, tapered collimator gave a 5 × 5 cm field size at the front of the phantom for an SSD of 125 cm. This field size was chosen for these measurements since the problems of obtaining adequate dose rate and adequate depth dose are maximised when field size is at a minimum.

The target assembly is shown in Fig. 3. No target backing disc, which serves

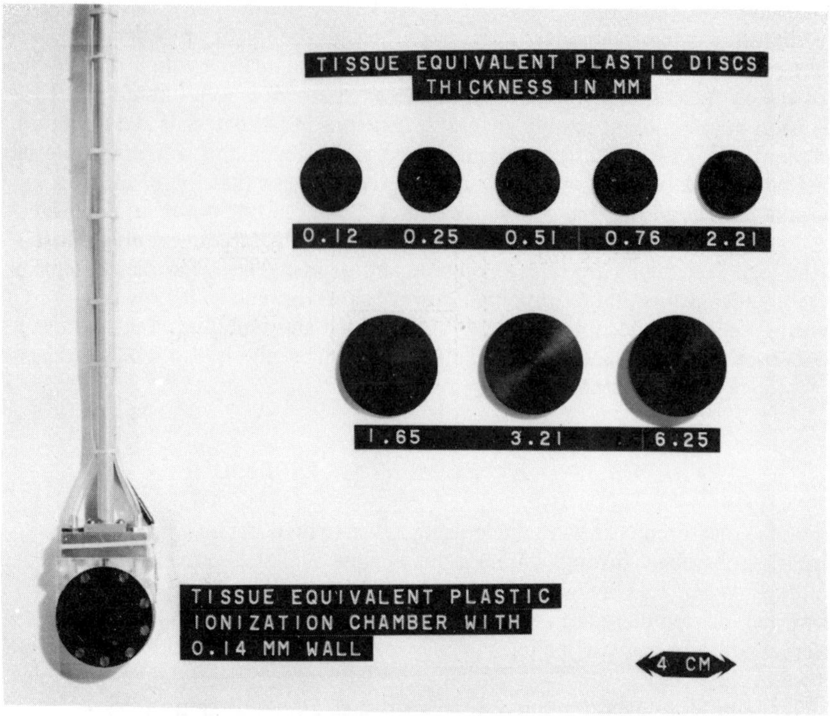

Fig. 1. Disc-shaped ionisation chamber and discs used for skin sparing measurements

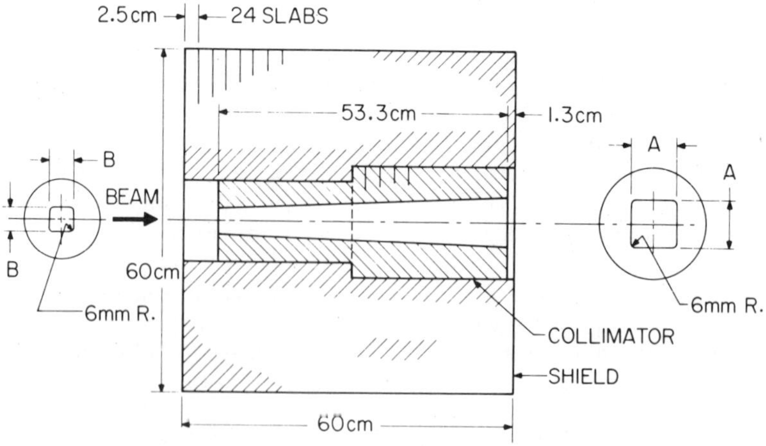

Fig. 2. Diagram of Benelex shield and collimator. A = *4·2 cm and* B = *2·3 cm*

as the beam-stop, was used for the thick-target measurements. A tantalum disc was used behind the thin targets, and for these targets the target holder was changed to water-cooled copper in place of the uncooled aluminium shown in the figure.

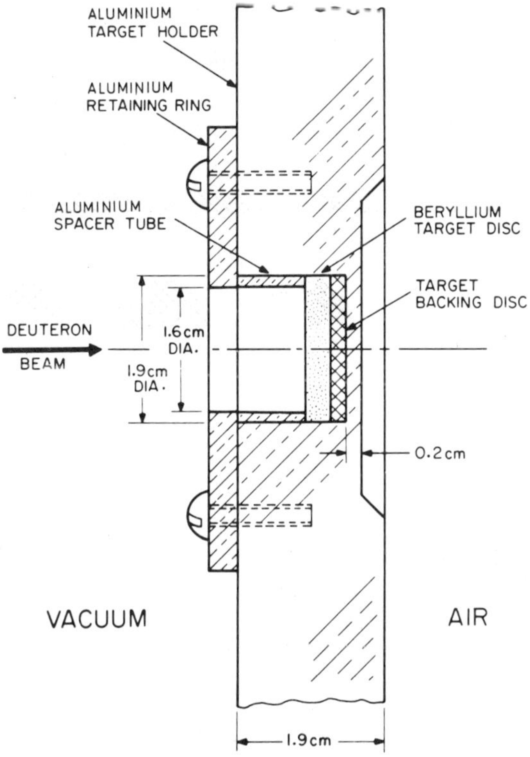

Fig. 3. Target assembly

To verify that the deuteron beam struck only the beryllium target, and not other surrounding target holder materials, measurements were made across two perpendicular axes on the external face of the target holder using thermo-luminescent dosimeters. In addition, visual inspection of the target and holder, subsequent to the irradiations, was used to verify the beam location by observing the discolorations resulting from the intense local heat generated by the beam.

RESULTS

Dose rates. It is advantageous to examine the attainable dose rates before presenting the depth dose data. Curve 1 of Fig. 4 shows the maximum dose rate near the front of the phantom, per unit of beam current, as a function of deuteron energy for a 6·35 mm thick beryllium target. In the energy range from 16 to 30 MeV the dose rate per unit beam current increases rapidly with energy. Curve 2 shows the data for a 0·81 mm thin beryllium target backed with 1·55 mm of tantalum, and again there is a strong increase in dose rate per unit beam current with energy. The thickness of the tantalum beam-stop is equal to the range of a 46 MeV deuteron.[9] Curve 3 is the result of measurements made with a very thin beryllium target, 0·37 mm, and shows a much less marked increase in dose per unit current with energy. This very thin target was similarly backed with 1·55 mm of tantalum. Curve 4 is a line drawn between two data points obtained at deuteron energies of 16 and 30 MeV with 1·55 mm of tantalum alone serving as a thick target. These two data points indicate the relative significance of the dose due to neutrons generated in the tantalum beam-stop.

Decreases in dose rates resulting from the use of thin targets could be compensated by the use of higher beam currents which, for a specific machine,

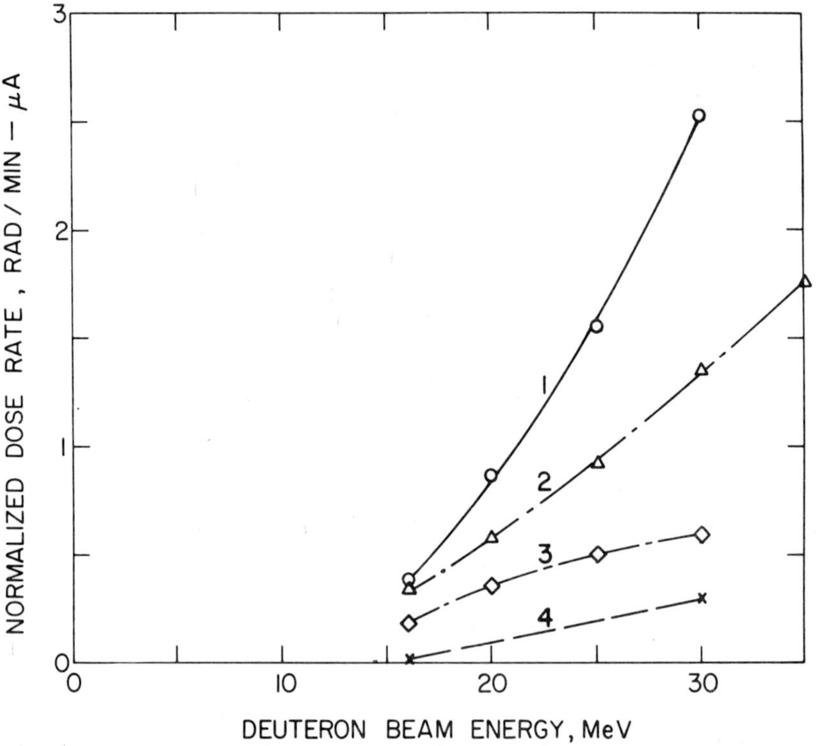

Fig. 4. Normalised dose rates at an SSD of 125 cm as a function of deuteron bombarding energy for various target thicknesses
Curve 1: 6·35 mm thick Be *target*
Curve 2: 0·81 mm thin Be *target with* Ta *beam-stop*
Curve 3: 0·37 mm thin Be *target with* Ta *beam-stop*
Curve 4: 1·55 mm thick Ta *target*

may require modification or re-design of the ion source. At 30 MeV, deuteron beam currents as high as 15 μA were obtainable with the TAMVEC machine. For the thick target and an energy of 30 MeV the data indicate that a minimum beam current of 4 μA would be required to achieve the minimum acceptable dose rate of 10 rad/min. For the 0·81 and 0·37 mm thin targets beam currents of 7·5 and 17 μA respectively would be required at 30 MeV.

An additional problem resulting from the use of higher currents to compensate for the reduced dose rate of a thin target would be the increased power dissipation required in the target assembly.

Depth dose. Central axis depth dose curves were obtained for bombarding energies of 16, 20, 25 and 30 MeV with the thick beryllium target, with the two thin beryllium targets backed with tantalum, and with the thick tantalum target. A curve was also obtained at 35 MeV with one of the thin targets. These data are summarised in Fig. 5, which shows the depth required to halve the dose at 0·5 cm depth as a function of energy.

Curve 1 shows the thick-target data and demonstrates the increased neutron penetration obtainable with higher deuteron beam energy. The increase was less than had been hoped. This may be due to the increase in cross-section for reactions yielding more than one neutron. The deuteron energy increase thus failed to produce proportionately higher mean neutron energies.

Curve 2 is for the 0·81 mm thin beryllium target backed with tantalum, and shows an increase in neutron depth dose with increasing deuteron energy, as well

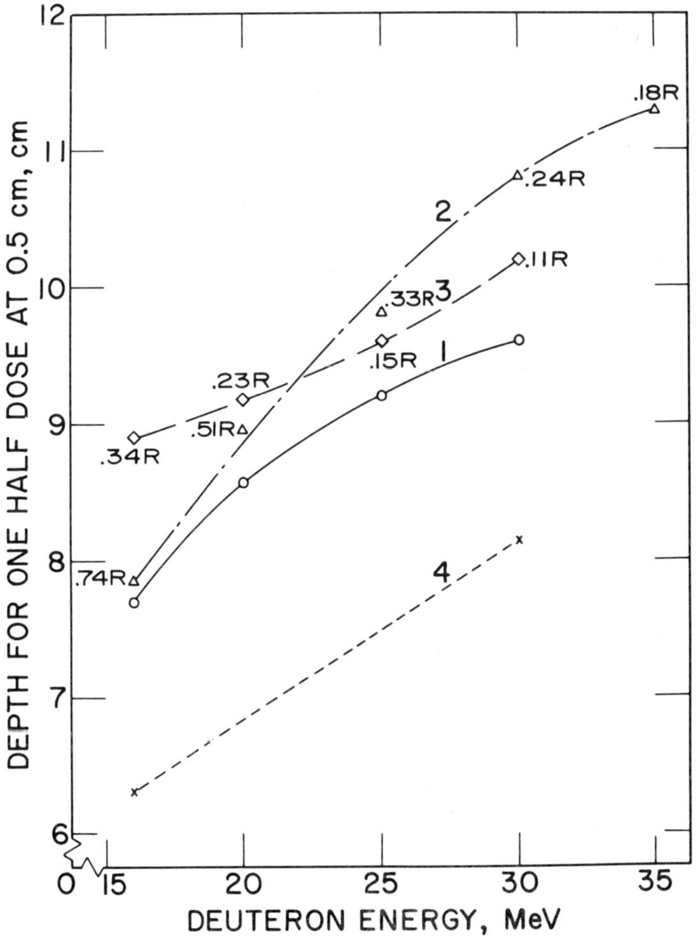

Fig. 5. Half-dose depth at an SSD of 125 cm for a 5 × 5 cm field size as a function of deuteron bombarding energy for various target thicknesses. R = *deutron range in* Be
Curve 1: 6·35 mm thick Be *target*
Curve 2: 0·81 mm thin Be *target with* Ta *beam-stop*
Curve 3: 0·37 mm thin Be *target with* Ta *beam-stop*
Curve 4: 1·55 mm thick Ta *target*

as an overall increase compared to the thick target curve. Curve 3 is for the thinnest target, 0·37 mm thin beryllium backed with tantalum. The data points at 16 and 20 MeV again demonstrate increased neutron penetration with deuteron energy, and increased penetration for this very thin target compared to the previous two curves. Although the data points at 25 and 30 MeV continue to show increased penetration with deuteron energy, they are below the values obtained for the 0·81 mm thick target. That is, for a given energy, these results indicate that there is an optimum target thinness for achieving maximum dose at depth. This can be clarified by noting that as target thickness is reduced, a greater portion of the deuteron energy must be expended in the beam-stop material. Although the beam-stop was chosen to yield relatively few neutrons when bombarded, nevertheless, these neutrons will be much lower in energy than those generated in the beryllium. These low-energy neutrons are apparently effective in decreasing neutron penetration for very thin targets. The numbers shown in the Figure next to the thin target data points (e.g. 0·34R) indicate the fraction of the deuteron range in beryllium represented by the beryllium target.[9] Curve 4 is a straight line drawn between the data points at 16 and 30 MeV obtained for the thick tantalum target. The penetration of these tantalum-generated neutrons is poor compared to those from the beryllium targets.

The data of Fig. 5 have been replotted in Fig. 6 to show the depth required to halve the dose as a function of the ratio of the beryllium thickness to the deuteron range in beryllium. Although the data available gave only three or four points per energy, approximate curves have been drawn to indicate the trend. The dashed portions of the curves are extrapolations that intercept the ordinate at the values obtained for the bare tantalum, that is, for zero beryllium thickness. It is evident that for each curve there is a broad maximum for

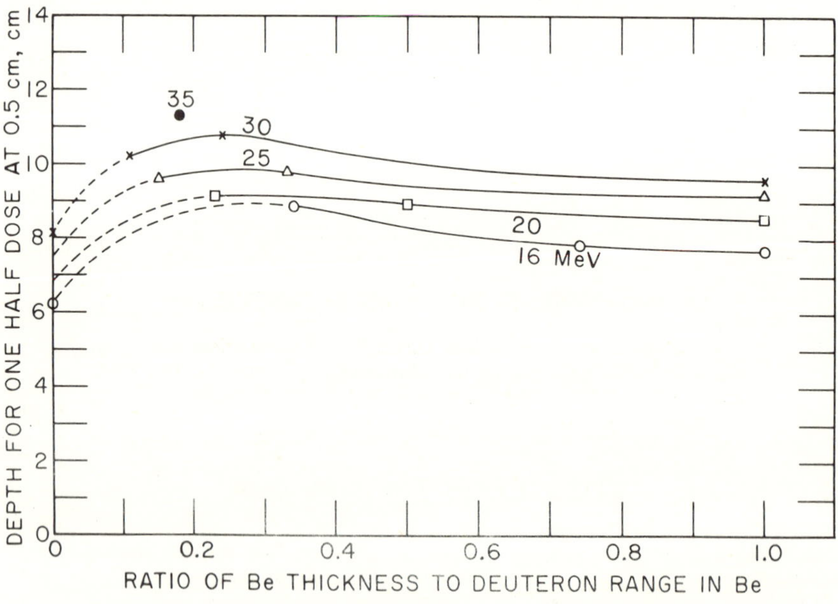

Fig. 6. Half-dose depth at an SSD of 125 cm for a 5 X 5 cm field size as a function of Be *target thickness for several deuteron bombarding energies. SSD = 125 cm*

optimum beam penetration which is at a target thinness of ~0·2–0·3 of the deuteron range in beryllium.

A comparison of the 30 MeV data for the thick target and the 0·81 mm thin target with the data obtained using other sources is shown in the depth dose

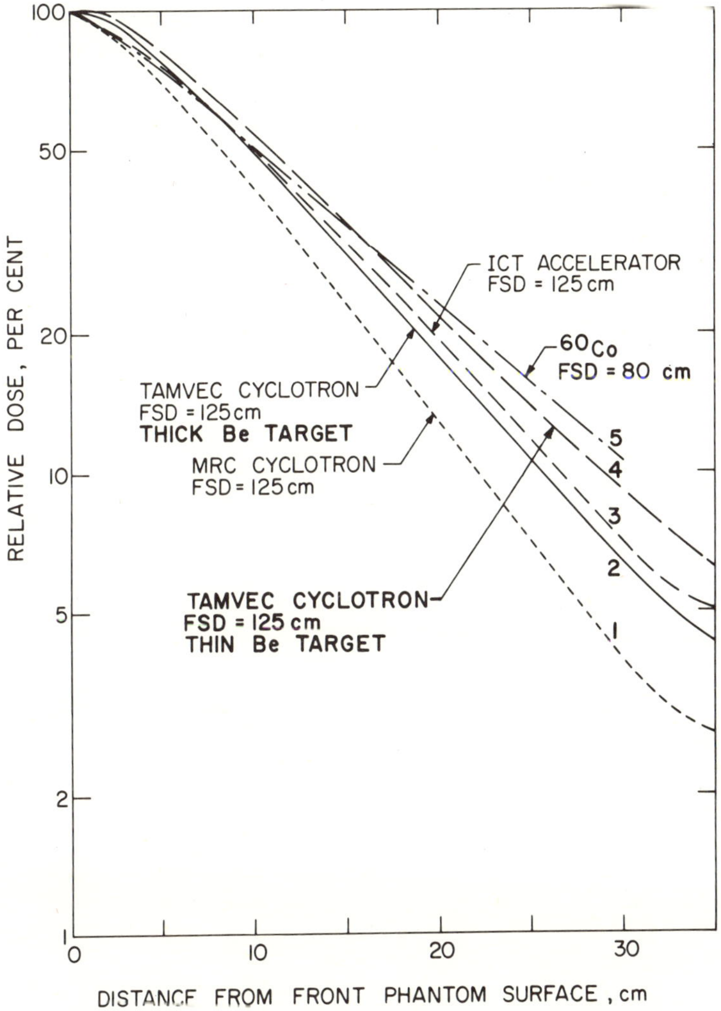

Fig. 7. Comparison of depth dose curves for a 5 × 5 cm field size
Curve 1: MRC Cyclotron; 16 MeV deuterons on 0·8 mm thin Be *target*
Curve 2: TAMVEC; 30 MeV deuterons on 6·35 mm thick Be *target*
Curve 3: ICT Accelerator; 15 MeV neutrons
Curve 4: TAMVEC; 30 MeV deuterons on 0·81 mm thin Be *target*
Curve 5: ^{60}Co *at an SSD of 80 cm*

curves of Fig. 7. Neutron energy is given only for the 15 MeV monoenergetic neutrons of Curve 3. The other four curves are for broad spectra which were measured only crudely (using threshold detectors). Consequently, for these four spectra a value is given for deuteron energy only.

Curve 1 was obtained at the MRC Cyclotron by bombarding a beryllium target, of thickness equal to 0·7 of the deuteron range, with 16 MeV deuterons.

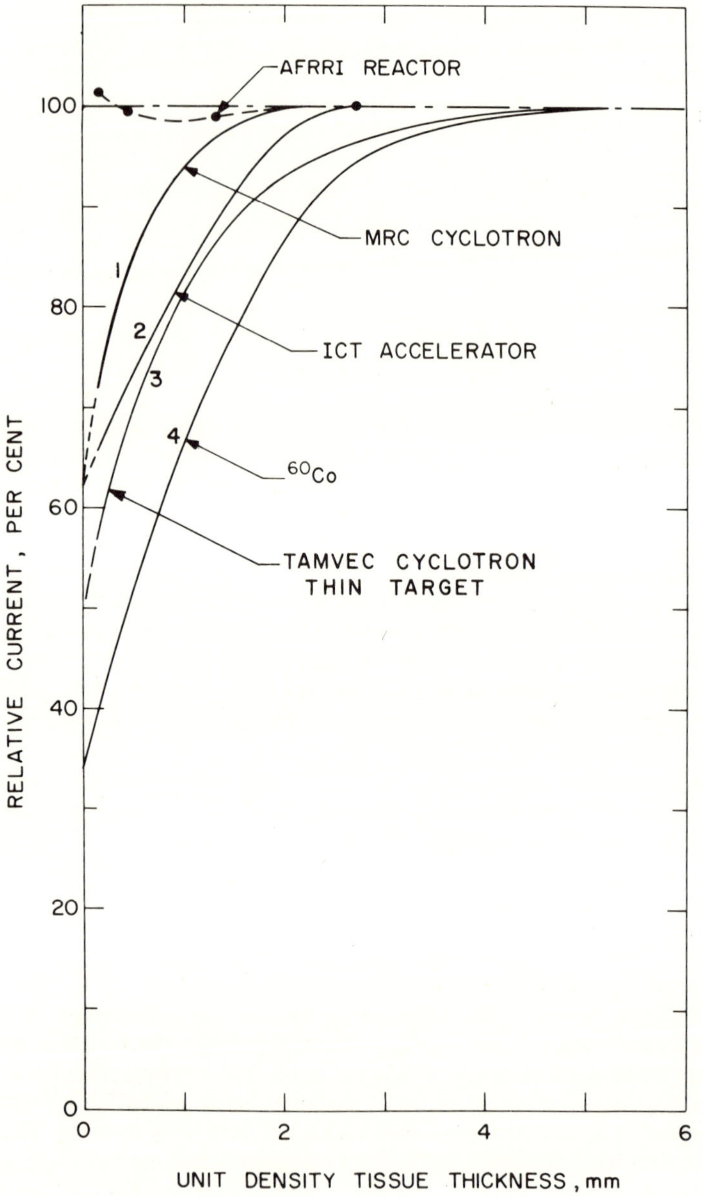

Fig. 8. Comparison of skin sparing curves
Curve 1: MRC Cyclotron; 16 MeV deuterons on 0·8 mm thin Be *target*
Curve 2: ICT Accelerator; 15 MeV neutrons
Curve 3: TAMVEC; 30 MeV deuterons on 0·81 mm thin Be *target*
Curve 4: ^{60}Co

Curve 2 is that obtained at TAMVEC using 30 MeV deuterons to bombard a thick beryllium target. The improvement due to the higher deuteron energy is evident. Curve 3 was obtained using the Insulated Core Transformer Accelerator at the Lawrence Radiation Laboratory, Livermore, California, using the reaction $^{3}H(^{2}H, n)^{4}He$ to produce essentially monoenergetic 15 MeV neutrons. Although this shows an improvement over Curve 2, the increase in penetration is small. Curve 4 was obtained at TAMVEC with 30 MeV deuterons and the 0·81 mm thin beryllium target backed with tantalum. Finally, Curve 5 is for a ^{60}Co γ-ray source[10] at an SSD of 80 cm. This standard SSD for ^{60}Co machines results from the practical limit of specific activity of the ^{60}Co sources utilised, and justifies this comparison of fast neutron sources at an SSD of 125 cm with a ^{60}Co source at an SSD of 80 cm. The comparison shows that for deuteron energies of ~30 MeV and higher, cyclotron-generated neutrons can be obtained which have penetrations equivalent to, or even greater than, those obtained with 15 MeV neutrons or conventional ^{60}Co γ-ray sources.

Skin sparing. Secondary build-up curves are shown in Fig. 8. Each of the Curves 1–4 was obtained from at least eight consistent data points. The dashed portions are extrapolations. The skin sparing curve obtained with 30 MeV deuterons using a 0·95 mm thin beryllium target backed by tantalum is shown by Curve 3. This may be compared to Curve 1, for the MRC Cyclotron using 16 MeV deuterons. The improvement resulting from higher deuteron energy in a thin target may be noted. The skin sparing for 15 MeV neutrons is given by Curve 2 and shows a build-up between that of Curves 1 and 3. It is interesting to note that the initial rise of Curves 1 and 3, obtained for the broad neutron spectra of the reaction $^{9}Be(^{2}H, n)^{10}B$, is quite rounded compared with the much less curved rise of Curve 2, obtained for a narrow neutron spectrum.

Curve 4 shows the skin sparing with ^{60}Co.[11] Although this curve appears to be better from a therapy viewpoint, Curves 1, 2, and 3 are probably satisfactory, especially with the use of multiport therapy techniques.

DISCUSSION

Data have been presented to show the substantial dose rate increases that can be achieved with increased deuteron energy, and also the reductions in dose rate resulting from the use of thin targets. A decrease in target thickness has been shown to yield a more penetrating neutron beam, but it appears that for a given energy there is an optimum target thickness below which the penetration is again reduced by neutrons generated within the beam-stop material. The data also show that all of the fast neutron beams discussed demonstrate satisfactory skin sparing.

The efficacy in radiation therapy of neutron spectra obtained with beryllium targets and cyclotron-generated deuterons depends on both physical and radiobiological parameters. Satisfactory physical parameters, especially depth dose, have been shown to be attainable with accelerators capable of producing deuterons with energies at least as high as 30 MeV. The thin-target investigations have shown that depth dose can be increased somewhat if thin targets are used. This requires machines capable of correspondingly higher beam currents and the use of target assemblies capable of dissipating the additional heat load.

Having determined that the minimum deuteron energy required was in the range 30–35 MeV, informal conferences with three different cyclotron designers were arranged. The consensus among these designers was that the minimum cost of a cyclotron having a 70-in magnet would be ~$1·2M.

These conclusions are based on the reaction $^{9}Be(^{2}H, n)^{10}B$. An investigation[12] is under way concerning the possibility that there may be some other reaction which will produce a neutron spectrum having satisfactory physical characteristics but requiring a cyclotron magnet diameter less than 70 in. However, this investigation has not yet been successful.

ACKNOWLEDGEMENTS

This work was performed as part of a project by a group convened at TAMVEC in order to determine the Oxygen Enhancement Ratio (OER) and the physical characteristics of the neutron spectra produced by deuterons in the energy range 16–35 MeV.

In addition to the authors, the members of the field team were: R. J. Berry (Churchill Hospital, Oxford, England), D. F. Herring (Enviromed, Inc., La Jolla, California), A. C. Lucas and W. Quam (Edgerton, Germeshausen & Grier, Goleta, California).

Our appreciation is tendered to Mr. James Davis, Mr. G. Dehaven, Dr. W. A. McFarlin, and Dr. A. D. Suttle, Jr., of TAMVEC, for their technical and administrative assistance. We are also grateful to Dr. A. D. Kantz of E G and G, Inc., for making TLD measurements to verify beam position.

REFERENCES

1. Physical Aspects of Irradiation, National Bureau of Standards Handbook 85.
2. Bewley, D. K., *Brit. J. Radiol.* **36**, 422, 81, (1963).
3. Bewley, D. K. and Parnell, C. J., *Brit. J. Radiol.* **42**, 281, (1969).
4. Brennan, J. T., *Radiologic Clinics of North America* **VII**, (2) (August, 1969).
5. Tochilin, E. and Kohler, G. D., *Health Phys.* **1**, 332, (1958).
6. Bond, V. P., Carter, R. S., Robertson, J. S., Seymour, P. H., and Hechter, H. H., *Rad. Res.* **4**, 139, (1956).
7. Radiological Research Laboratory, Columbia University, New York, Annual Report on Research Project, NYO-2740-1, Section P-15 (1964).
8. Goodman, L. J., *Health Phys.* **16**, 763, (1969).
9. Janni, J. F., Air Force Weapons Laboratory, Kirtland Air Force Base, New Mexico, Technical Report No. AFWL-TR-65-160, (1966).
10. Depth Dose Tables for Use in Radiotherapy, *Brit. J. Radiol.*, Supplement No. 10 (1961).
11. Watson, T. A., Johns, H. E., and Burkell, C. C., *Radiol.* **62**, 2, 165, (1954).
12. Fleischer, A. A., The Production of Fast Neutrons by Small Cyclotrons. Symposium on Fast Neutron Radiotherapy at the Radiation Research Society Meeting, April 25, 1968, at Houston, Texas.

DISCUSSION

M. A. Chaudhri (M.R.C., Hammersmith): In discussing and comparing various machine sizes for producing neutrons for neutron therapy you have considered only a Be target. The minimum size of cyclotron specified relates to this particular target, without considering the behaviour of such machines with other targets and nuclear reactions. We have done extensive calculations on various neutron-producing targets with different-sized machines. For example, a deuteron target would produce a dose-rate of ~180 rad/min and average energy of ~11·3 MeV at an SSD of 100 cm in the forward direction when bombarded with 16 MeV deuterons and 100 μA beam. Those figures reduce to 30 rad/min/100 μA and 7·5 MeV at E_d = 10 MeV.

L. J. Goodman: It is true that my paper considers only the use of beryllium targets for producing neutrons for neutron therapy, although I did mention the possibility of other target reactions having satisfactory physical characteristics but requiring a cyclotron magnet diameter less than 70 in. The problem with a deuterium target, or perhaps a tritium target, is that it is generally much more difficult to develop such targets which are capable of handling the high beam-power inputs required for this application.

A. A. van Kranenburg (Philips): May I conclude from your talk that as long as no other targets are available than the ones we have now it is rather useless to talk about the use of compact cyclotrons for neutron therapy?

L. J. Goodman: Yes, provided one accepts the criteria we have adopted in this evaluation. It would be necessary to relax these criteria considerably before considering the possibility of using a low-energy, compact cyclotron for neutron therapy.

G. Burton (M.R.C., Hammersmith): There is no indication of the level of γ-contamination in the neutron beams from the beryllium targets of different thicknesses. Does the γ-contamination of the neutron beam increase at higher incident particle energy or with thinner targets? In addition, compensating loss of dose-rate from thinner targets by increasing the beam current does not necessarily lead to increased power dissipation in the beryllium target, since the energy deposited in the beryllium is less. The beryllium is the most critical part of a neutron target assembly, due to its reduced thermal conductivity and the possibility of a poor interface with the backing material, and it is the energy deposited in the beryllium which is important.

L. J. Goodman: The dose-rates and depth doses reported here were totals, i.e. neutron doses plus γ-ray doses. Previous measurements, using 16 MeV deuterons on a thick beryllium target, had indicated that in the presence of the phantom the γ-ray dose at the front was increased to ~11% of the total. Similar measurements at the higher deuteron energies were not made, since it was felt that we should first demonstrate adequate depth dose characteristics for the total beam before investigating the γ-ray contamination of the beam.

Although it is true that increasing the beam current to compensate for loss of dose rate from thinner targets does not necessarily lead to increased power dissipation in the beryllium target, it is nevertheless required that the target assembly be capable of dissipating the increased power load. The target we used lost heat principally via thermal conductivity to the backing material. In my paper I did not wish to give the impression that target heating had been a serious problem, but rather that these heating effects must be considered when the use of higher beam currents is anticipated.

W. Rautenbach (N.P.R.L., Pretoria): What are the possibilities of using a fast, rotating unbacked beryllium target which will be cooled by radiation, removing the transmitted beam with a magnet system and dumping it in a suitable beam catcher? An unbacked target should improve the penetration of the neutrons considerably.

J. T. Brennan (University of Pennsylvania Hospital): In order to get rid of unwanted deuterons which have penetrated the beryllium target one might consider a supplementary magnet downstream from the target. However, the apparatus used in our target was backed by a double-walled copper cooling assembly. The latter would have to be removed from the beam, leaving a beryllium disc target cooled on the edges only. This would be a moderately large target development problem.

W. L. Caldwell (Vanderbilt University): I, and others, certainly appreciate the tremendous efforts which you, Dr. Brennan, and associates have made in evaluating available cyclotrons. My present concern is that perhaps your requirements for a fast neutron beam have been too strict. Life as a radiotherapist would be easier with a dose rate exceeding 10 rad/min and with a 50% depth dose of 10 cm or more. This assumes that neutrons are necessary for the entire course of treatment. Professor Fowler has previously suggested that perhaps neutrons would be most effective if used only at the initiation and conclusion of a treatment course, for example. There is a theoretical basis, of course, for such a technique, which would allow the excellent depth dose characteristics of a megavoltage photon beam to overcome the disadvantages of a fast neutron beam with poor depth dose characteristics. Investigations of such an approach would seem of prime importance. The economic advantages of 'mixed beam' therapy are obvious, should fast neutron beam therapy prove valuable in clinical situations.

L. J. Goodman: It may indeed turn out that a 'mixed beam' approach to radiation therapy would have both economic and therapeutic advantages, This, however, would have to be tested and proven in a clinical trial, and for such a trial it would be definitely preferable to have a beam useful for therapy with neutrons only. Thus, our dose rate and depth dose criteria are those required for such a machine rather than that which may ultimately be required after years of clinical trials and studies.

G. Burton (M.R.C., Hammersmith): Since experience at Hammersmith shows that patient set-up times are of the order of 10 min while treatment times at our dose-rates of 50 rad/min are 2–3 min, one can visualise two or three treatment rooms around one machine, with patient set-up going on in one room while treatment is taking place in another. In this way many more patients per hour may be treated than with low dose-rate machines, and the cost per treatment using a cyclotron made lower.

L. J. Goodman: There may be a possible advantage in having more than one treatment room around one machine, to permit patient set-up in one room while treatment is taking place in another. However, present indications are that there will be problems connected with obtaining sufficient numbers of patients, rather than with having sufficient irradiation facilities.

M. A. Chaudhri (M.R.C., Hammersmith): Could you comment, on the basis of your team's experience, on the minimum required and entirely satisfactory physical characteristics of a beam of neutrons for therapy?

L. J. Goodman: It is difficult to state the physical characteristics of a beam of neutrons for therapy to meet minimum requirements or to be entirely satisfactory, since this depends very much on the point of view of a particular radiotherapist, and on the clinical trial contemplated. As stated previously, for our evaluation we used the criteria of a minimum dose rate of 10 rad/min at a SSD of 125 cm and a minimum depth of 10 cm of tissue for a 50% depth dose.

SESSION 5

Techniques and Instrumentation

Chairman: Dr. C. B. Amphlett

Techniques used in on-line fission experiments with the V.E.C.

I. F. Croall and J. G. Cuninghame

Chemistry Division, A.E.R.E., Harwell, Berks., U.K.

ABSTRACT

Techniques currently in use on the Harwell Variable Energy Cyclotron for on-line fission studies are discussed. Methods of target preparation and thickness measurement are described. Pulse-height measurements and preliminary data processing use a PDP-8 computer programmed to act as an associative storage analyser.

Investigations carried out using these techniques include the effects of increasing incident particle energy on the proton-induced fission of ^{232}Th, and the effects of angular momentum on the fission of the compound nucleus ^{210}Po.

INTRODUCTION

The fission experiments currently being pursued on the V.E.C. are of two main types:

(a) Experiments in which the mass and kinetic energy distributions of the fission fragments and their multifarious correlations are examined. The theoretical interpretation of the results of such experiments casts light on the mechanism of the fission reaction itself.

(b) Experiments on the detection and measurement of the half-lives of fission isomers. These are a comparatively new phenomena whose existence has resulted in a completely new interpretation of the shape of the fission barrier.

As time has proceeded, the precision of the mass and energy distribution measurements has been steadily improved, and it is believed that mass distributions comparable to those achieved by radiochemical measurements could be made, provided sufficient machine running time was taken to get good statistics, and provided a reasonable neutron distribution function was available for the case in point. In fact, measurements currently being made in several countries should greatly increase the number of relevant neutron distribution functions in the course of the next few years. Relative mass yield curves can be obtained by on-line methods with a reduction of the man-hours taken for the corresponding radiochemical curve by a factor which is conservatively estimated to be at least 100.

With these facts in mind, serious thought is now being given to the construction of a neutron source for the V.E.C. which will allow targets to be irradiated in neutron fluxes with spectra resembling those of fast reactors.

BEAM LINE AND SCATTERING CHAMBER

The fission experiments are performed in a 24 in diam. scattering chamber which is permanently located on one of the V.E.C. beam lines. After entering the target room, the beam passes through a measuring box equipped with a Faraday cup, variable horizontal and vertical slits, and a scintillator, and then through a water-cooled graphite collimator system. Six feet further downstream is the second graphite collimator system, which produces an ⅛ in diam. beam which then enters the scattering chamber through a final series of clearing collimators, passes through the target, and is finally stopped in a Faraday cup, 3 ft downstream from the scattering chamber.

The latter is equipped with two arms which can be rotated through 350° about its centre, and a central spindle rotatable through 360° which carries the targets. All these rotations can be performed remotely, and remote indication of the arm and target spindle positions is also provided. These positions can be measured to $\sim \pm 0{\cdot}5^\circ$.

For the mass and energy distribution experiments the gold surface-barrier detectors used to measure the fission fragment kinetic energies are mounted on the rotating arms and their position on them can be preset in both height and distance from the target. The latter is mounted on the target spindle. Collimators to reduce edge effects and ring magnets to deflect electrons are mounted in front of the detectors.

For the fission isomer experiments a special target and detector assembly replaces both the arms and the target spindle.

TARGET ASSEMBLIES

Target spindle for mass and energy distributions. Up to four targets, mounted on $^1/_{32}$ in thick stainless steel holders, are fixed one above the other on a metal holder which slides into the rotating target spindle. Any one of the targets can be brought into the beam position without having to break the vacuum and open the chamber.

Target block for fission isomer work. This consists of an aluminium block, drilled to take two 400 mm^2 area Ortec gold surface-barrier detectors which are mounted on opposite sides of the target or collector position at right angles to the beam direction. Their front faces are 1 in from the target. The position of the target itself depends on whether or not a pulsed beam is used. If it is, the target is mounted on a holder in the position between the detectors. Both prompt and delayed (fission isomer) fission fragments impinge on the detectors, but the counting system is only live during the time when the beam is switched off. Thus the prompt fissions are not recorded.

If, however, a continuous counting mode is required (beam not pulsed), the target is mounted in front of the aluminium block, ¾ in from the collector foil, which is placed in the position between the detectors. Between the target and the collector foil is a $\frac{1}{4}$ in diam. collimator, which does not allow the detectors to see the prompt fissions. Thus only fissions from recoil species which have been stopped by the collector are seen by the detectors. The prompt fissions are then monitored by a third gold surface-barrier detector placed in a forward position relative to the target.

TARGET PREPARATION

Normal targets. Copper-backed nickel foils of thicknesses ranging from a minimum of ~50 $\mu g/cm^2$ are glued to standard stainless steel target holders with epoxy resin glue, with the copper side next to the holder. The holders have a 0·5 in diam. hole in them and the copper is removed from the area of this hole with chromic acid. This method of mounting the foils produces a much smoother, tauter surface than floating the foils on to a bowl of liquid.

Target material is deposited on the nickel to a suitable thickness (usually ~100 $\mu g/cm^2$) by vacuum sublimation before the copper backing is removed. If the target material is attacked by chromic acid it is usually not possible to use the very thinnest (~50 $\mu g/cm^2$) nickel, because it has pinholes through which the acid can reach the target. However, the next thinnest (~100 $\mu g/cm^2$) is normally suitable.

Platinum targets. In a recent experiment it was necessary to use targets of the stable isotope ^{198}Pt. Mass separation of platinum has been very difficult in the past, owing to the fact that in order to obtain a suitable platinum beam from the ion source it was necessary to heat the metal to ~3000°C. However, J. H. Freeman (E.M. Separator Group, A.E.R.E.) has now designed a sputtering ion source for the Harwell small mass separator in which the platinum metal is fixed to a sputtering electrode held at -1 kV. Argon gas is fed into the arc chamber and ionised in the plasma. The positive argon ions then bombard the platinum, causing ejection of neutral platinum atoms which are stripped by the arc and then accelerated into the separator. The separated isotopes arrive at the collector slit with 40 keV of energy.

It is clear that if the nickel foil to be used as the target is bombarded directly with the 40 keV platinum ions, the latter will penetrate too deeply into the foil to make a good target for fission fragment work. Deceleration of the ions is not favoured because of its effect on beam resolution. The solution adopted was to allow the ^{198}Pt ions to bombard a collector plate of some suitable metal. The ions, together with some from the collector plate, will then be sputtered off and can be collected on the nickel target foil placed nearby. Molybdenum was chosen for the target plate, since it gives the most favourable ratio of platinum to unwanted atoms on the target foil. In practice, approximately four molybdenum atoms were collected on the target for every platinum atom.

PARTICLES AND ENERGIES

The variable-energy feature of this cyclotron permits the use of a wide range of particles and energies. In recent studies[1] of the fission of the nucleus ^{210}Po it was necessary to produce the compound nucleus at a given excitation energy using α particles, protons and ^{12}C ions, in order to study the effects of angular momentum on the mass and kinetic energy distributions. The excitation energies selected were ~31, ~44, and ~57 MeV.

Table 1 shows a typical combination of target and bombarding particle used in this work. The mean square angular momentum is shown to indicate the range obtained.

For the ^{12}C beam, due to cyclotron difficulties, the required energy of 77·2 MeV could not be directly obtained. However, an adequate 88 MeV ^{12}C beam

Table 1

Reaction	*Particle Energy* (MeV)	*Excitation Energy* (MeV)	$\langle I^2 \rangle$
$^{209}Bi + p \longrightarrow {}^{210}Po$	52·7	57·6	94
$^{206}Pb + \alpha \longrightarrow {}^{210}Po$	63·8	57·2	435
$^{198}Pt + {}^{12}C \longrightarrow {}^{210}Po$	77·2	58·5	1240

was available, and a suitable aluminium foil degrader (6·72 mg/cm^{-2} thick) was placed in the first water-cooled collimator in the beam line.

The beam currents used in this work were limited by scattering in our targets; for protons and α-particles, currents of the order of 100 nA on target were commonly used. This corresponds to ~1 μA before the collimating system. For ^{12}C ions, scattering in the targets again limited useful beams to about 15 nA on target.

PULSED BEAMS

For the fission isomer experiments two techniques are adopted, as discussed earlier. Two systems of beam pulsing are available, one pulsing the radio-frequency supply, the other the ion-source arc supply. The RF pulser enables beam-on times of the order of 200 μs upwards to be obtained, with off times > 200 μs. However, some beam is found after the main pulse, and in general useful on/off times are of the order of 400 μs/400 μs and longer. Ion source pulsing gives on/off times of the order of 100 μs/100 μs, but the present system suffers from the limitation that off times should be as short as possible, a duty cycle of ~50% being required.

A third method of effectively having a very short pulse is to work between beam bursts at the radio-frequency rate. The range of machine frequencies is from 7 to 22 Mc/s; effective acceleration takes place over ~40° of RF phase, so that beam pulses of the order of 10 ns every 100 ns can be obtained. The necessary fast-timing electronics is at present under development. This will enable us to examine very short half-life isomers.

DETECTORS AND ELECTRONICS

Gold surface-barrier detectors placed at calculated laboratory angles corresponding to 90° and 270° in the centre-of-mass system relative to the projectile beam are used to measure the pulse-heights of the two coincident fission fragments. One detector of 100 mm^2 active area placed 5 cm from the target defines the laboratory angle at which the fission events are measured. The second detector, also at 5 cm from the target, is much larger, 400 mm^2, to ensure that greater than 90% of the coincident complementary fragments would be detected.

Hybrid charge-sensitive preamplifiers coupled to a conventional fast slow coincidence system with a resolving time of 50 ns form the electronic system. The gain stability of the entire electronic system and the energy resolution broadening due to the high count-rates of inelastically-scattered beam particles is measured directly by observing the pulse-height distributions from a high-stability pulse generator. These pulses are fed through the preamplifiers and are finally recorded simultaneously with the fission data. Beam currents are controlled so as not to exceed an acceptable broadening of the generator pulse-height destributions.

PULSE HEIGHT ANALYSIS

The output fission pulses from the electronic system are fed to two Harwell-type 2110 amplitude-to-digital converters, the outputs of which are fed in turn to a pair of binary scalars in an interface to a PDP-8 data processor. The PDP-8, with 4096 12-bit words of fast core store and magnetic tape back-up store, is programmed as an associative storage analyser. The effective resolution is 128 X 128 channels, but there are available only 2048 storage locations for data. Using an associative storage program[2] these locations are allocated to events as they occur. Data from fission experiments are clustered and it is found that adequate resolution can be obtained using this technique. Plan and isometric oscilloscope displays of the data are used to assess the progress of the experiments.

DETECTOR CALIBRATION AND TARGET THICKNESS MEASUREMENTS

The detectors are calibrated using the mass-dependent energy calibration procedure described by Schmitt *et al*[3] using standard ^{252}Cf fission-fragment spectra.

In most previous work, target thicknesses have been calculated from fission fragment range-energy data. In the ^{210}Po work referred to earlier we adopted a direct measurement technique. For a given target and incident beam the average single kinetic energies of the fragments, $<E_f>$, were determined with the target at different angles, θ, between the target plane and the detector. This was achieved by rotating the target over the range $\theta = 20°$ to $60°$ with the fission detector kept at a constant angle relative to the particle beam.

Using the relation $<E_f> = <E_{f90}> - <\Delta E_{90}> (\sin\theta)^{-1}$, where $<E_{f90}>$ refers to the average kinetic energy for fragments emitted at 90° to the target and $<\Delta E_{90}>$ to their average energy loss, $<E_f>$ is seen to have a linear dependence on $(\sin\theta)^{-1}$. From graphs of $<E_f>$ against $(\sin\theta)^{-1}$, the slope of the line gives the energy loss $<\Delta E_{90}>$. Examples of these plots are given in Fig. 1 for two different targets. Good linearity was found over the range of angles used.

Energy losses in the nickel backing were found directly by observing the change in energy for fragments emitted from the target through the backing and directly from the front of the target, i.e. by rotating the target through 180°.

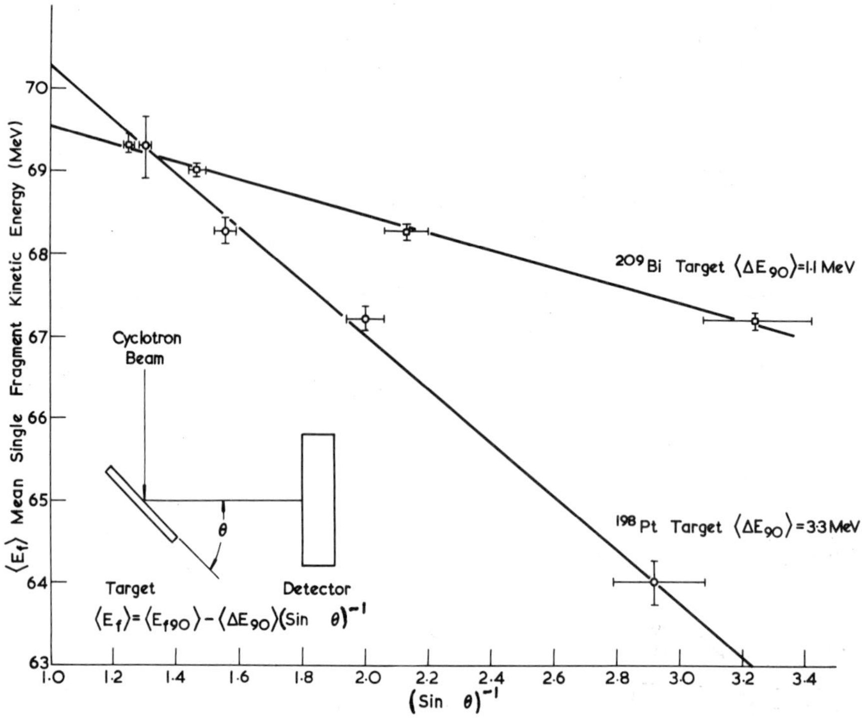

Fig. 1. Examples of target thickness plots

RESULTS AND DISCUSSION

Using the techniques described earlier, two experimental programmes have been completed. The first was an investigation[4] of the effects of increasing proton energy on the proton-induced fission of ^{232}Th. The mass yield, kinetic energy and variance of mean total kinetic energies were studied as a function of incident proton energy over the range 13 to 53 MeV. The distributions were found generally to be monotonic functions of excitation energy. The change from asymmetric fission to symmetric fission at the highest excitation energies was observed, and it was found that some asymmetric fission was still taking place at the maximum proton energy used. Although they do not finally settle the question of whether there are two modes of fission, or whether the fragment shell hypothesis is a better approach to the mechanism, the results seem to favour the former.

The second series of experiments was a study[1] of the effects of angular momentum on the fission of the compound nucleus ^{210}Po. As mentioned earlier the same compound nucleus can be formed by three different reactions, with the same excitation energy if the energies of the incident particles are correctly chosen, but with very different angular momentum distributions. The mass distributions were found to be independent of the angular momentum of the compound nucleus. The mean total kinetic energies were observed to increase with increasing angular momentum and with increasing excitation energy. A full discussion of these results will be found in ref. 1.

The addition of a disc store to the PDP-8 computer and the recent completion of a 4-parameter interface has widened the scope of these techniques. A new modular data collection and display programme is under preparation and will then enable more complex experiments to be carried out, for example the measurement of two energies and one time-of-flight of the fission fragments.

REFERENCES

1. Unik, J. P., Cuninghame, J. G., and Croall, I. F., Proceedings of I.A.E.A. Conference on Physics and Chemistry of Fission, Vienna, to be published. Paper SM-122/58 (1969).
2. Croall, I. F. and Best, G. C., to be published (1969).
3. Schmitt, H. W., Kiker, W. E., and Williams, C. W., *Phys. Rev.* **137**, B837, (1965).
4. Croall, I. F. and Cuninghame, J. G., *Nucl. Phys.* **A125**, 402, (1969).

DISCUSSION

K. V. Ettinger (Birmingham University): Will you please explain how you use the PDP-8 computer as an associative memory pulse-height analyser? Is this equivalent to the active zone memory system?

I. F. Croall: The principle of the associative storage analyser is that channel storage is only created for events as they occur. For example, we have only 2048 channels available for data in our PDP-8 fast core, but we work with a 128 × 128 array, only allocating storage for events as they occur. Fission data are clustered, and we find that we have sufficient data space for our experiments. In a conventional analyser we would require 16 384 core locations to store a 128 × 128 array. The basic principles are similar to the technique called active zone encoding.

A. B. M. Golam Mostafa (Birmingham University): Is there any significant effect due to background neutron-induced fission; if so, how is this distinguished from charged-particle induced fission?

I. F. Croall: There are no significant effects of this kind.

G. N. Walton (A.E.R.E.): How is the centre-of-mass effect treated in practice?

I. F. Croall: The small (100 mm^2) detector was fixed relative to the beam at the calculated laboratory angle corresponding to the centre-of-mass angle for the system under investigation. The large (400 mm^2) detector was then moved in angle $\pm 10^\circ$ relative to its calculated angle, and the angle giving the best coincidence efficiency (>90%) was found for use in the main data collection run.

On-line nuclear spectroscopy of short-lived isotopes*

Ronald D. Macfarlane†
Texas A & M University, College Station, Texas, U.S.A. and Niels Bohr Institute, Copenhagen, Denmark

Neil S. Oakey and Richard A. Gough
Texas A & M University, College Station, Texas, U.S.A.

ABSTRACT

Some details of the helium-jet recoil transport method are discussed, particularly with respect to the speed of the system for transporting short-lived radioactive species. Experiments are described where specific hot atom reactions are used to isolate nuclear reaction products. The possible adaptation of the helium-jet recoil transport method for on-line mass separation is also discussed and some results are presented on helium-product separation and product ionisation.

INTRODUCTION

Modern cyclotrons with high particle energies and intense currents are ideally suited for studying nuclei close to the limit of nuclear stability. Problems encountered with rapid transport of nuclear reaction products to low background areas, fast chemical separations on millisecond time scales and 'on-line' mass separation present a challenge to the nuclear chemist.

We have worked for several years on some of these problems and have obtained information on many new radioactive species. The helium-jet recoil transport technique which is described below has been most useful in our programme.[1] We report here some recent developments in the method and also describe some experiments in progress on the use of this method for 'on-line' chemical and mass separation.

EXPERIMENTAL

Description of the system. The function of the helium-jet recoil transport method is to provide a means for the continuous rapid transport of nuclear

*This work was supported by the U.S. Atomic Energy Commission and the Robert A. Welch Foundation.

†J. Simon Guggenheim Memorial Fellow, 1969–70.

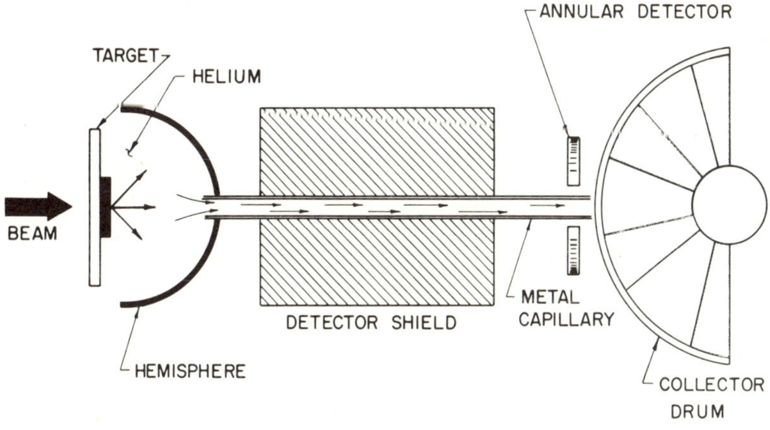

Fig. 1. Schematic diagram of the main features of the recoil transport system

reaction products from a region of high radiation field to a relatively low background region over a distance ranging from 5 to 100 cm. Transport times in the millisecond range have been achieved. Fig. 1 is a schematic diagram of our system in its present form which is particularly designed for optimum speed. The target is situated in front of a water-cooled copper hemisphere (0·4 cc volume) located in a reservoir of helium. During bombardment, recoils ejected from the target fill the hemisphere and are thermalised by the helium. They are then pumped, along with the helium, through a capillary tube into an evacuated chamber. The helium jet emerging from the capillary tip impinges on a collector drum and the recoils are adsorbed on the surface. The helium is pumped off by a high-capacity diffusion pump. An annular detector through which the capillary passes is used to measure radiations emitted by the adsorbed recoils. The beam terminates in the hemisphere and the detector shield adsorbs and scatters radiation from the beam deposition.

RESULTS AND DISCUSSION

Speed and efficiency of the system. The performance of the system in different experiments has been reported by us in a paper soon to be published.[2] We shall summarise some of the results here. Fig. 2 shows some time distribution curves of the α-counting rate of a 3-ms. activity, ^{214}Fr, produced in a heavy-ion reaction using the pulsed beam of the Yale heavy-ion accelerator. As shown in Fig. 2, the hemisphere size determines the evacuation time; the smaller the volume the greater the speed. An optimum volume was found below which effects of gas turbulence began to play a role. The time distribution curve for recoil collection (corrected for decay) is shown in Fig. 3. It has many components and reflects the complicated flow pattern in and around the region of the hemisphere. The lower limit on the half-life of a nuclide which can be studied is determined by how steeply the curve falls off at the very short times. A 'time-threshold' exists which corresponds to the minimum flight time of the recoil moving in the capillary at the velocity of sound in helium (965 m/s). For this particular system the flight time was 0·2 ms. Fig. 4 shows a curve for our system of the fraction of recoils which survive the transport before decay, as a function

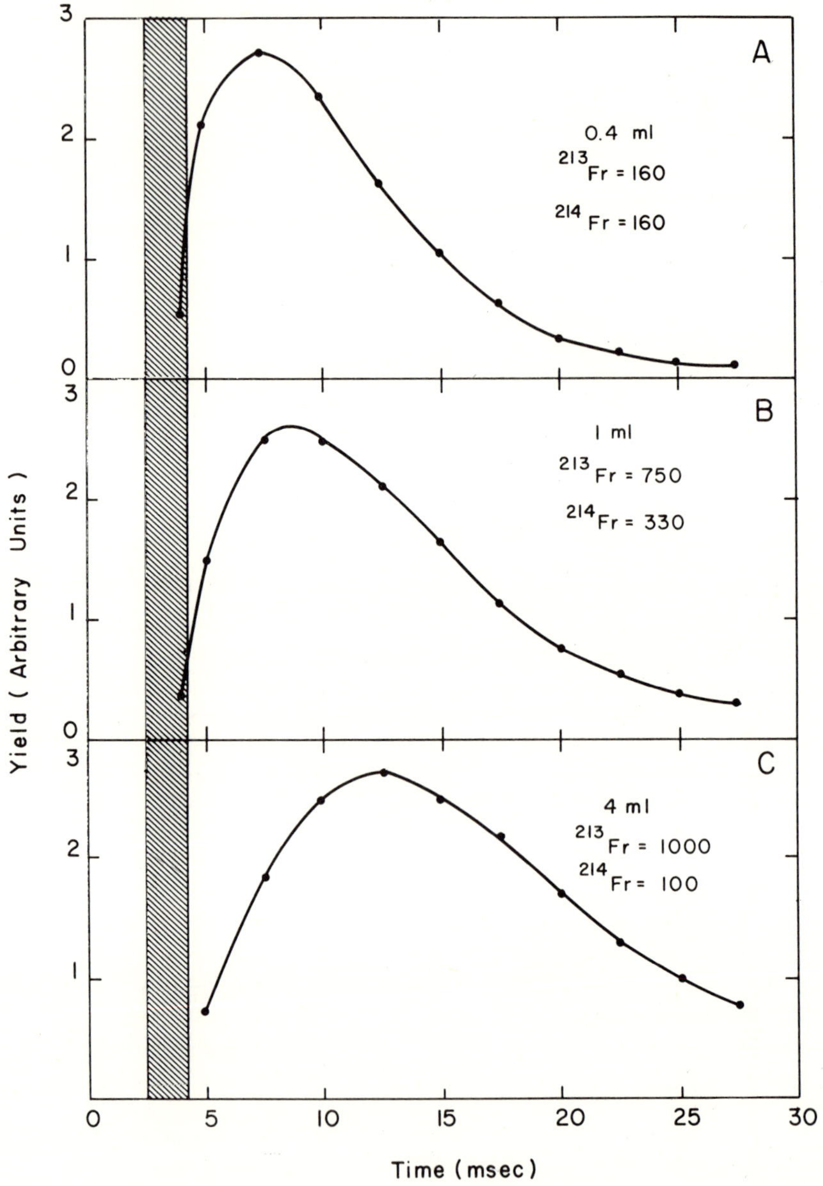

Fig. 2. Time distribution curves of the ^{214}Fr *α-decay rate for different hemisphere volumes: A–0·4 cc, B–1 cc, and C–4 cc. Relative yields of* ^{214}Fr and ^{213}Fr *are indicated on each figure. The shaded area denotes the duration of the beam*

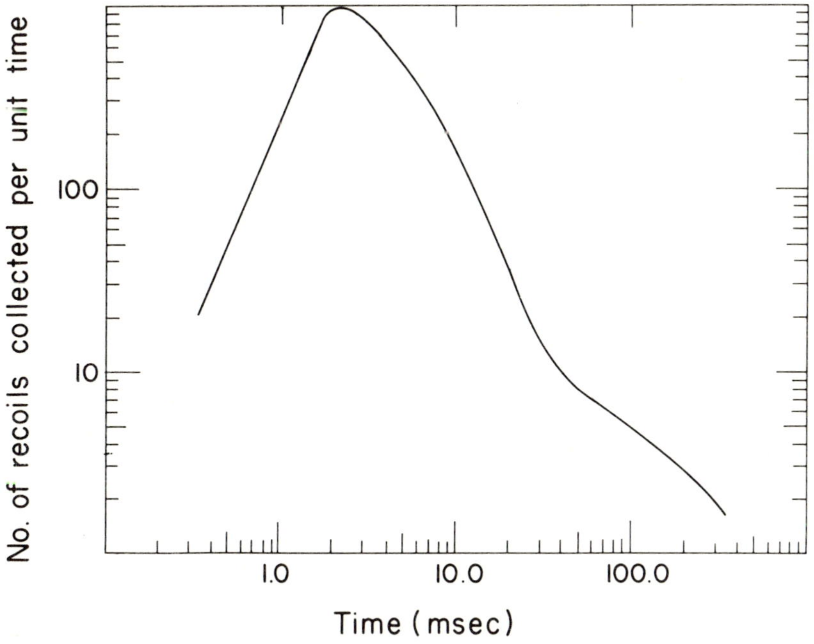

Fig. 3. Time distribution curve for the collection of nuclear reaction recoils by gas sweeping A 0·4 cc hemisphere was used with 2 atm of He *as the carrier gas*

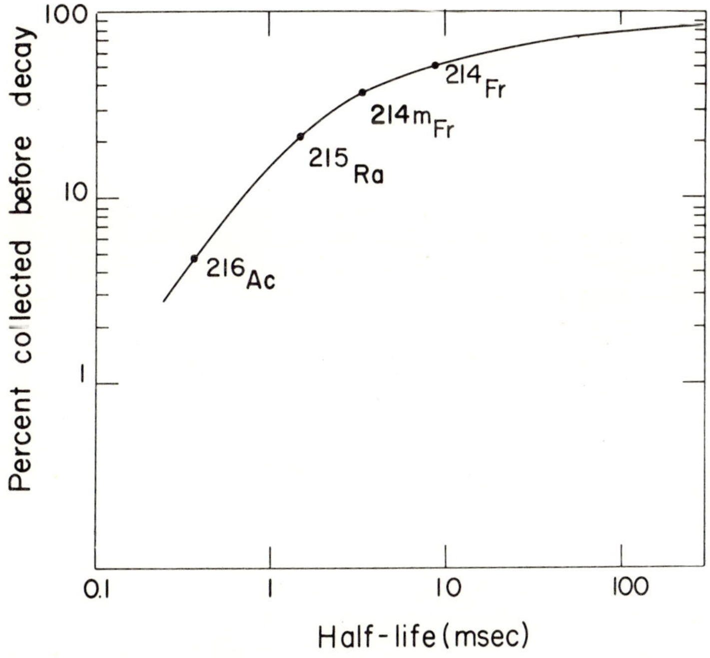

Fig. 4. Fraction of recoils which are collected before decay as a function of half-life for the distribution curve shown in Fig. 3

of half-life. If the capillary tube is reduced in length it should be possible to achieve a lower limit of 10 μ s for the half-life of a nuclide which we can study if it is produced in good yield (> 50 mb).

On-line chemical separation. The helium-jet method described above is generally not dependent on the chemical nature of the species. There are some isolated exceptions, such as the rare gases which are not adsorbed at room temperature, and the enhancement of the yield for At isotopes when traces of O_2 are added to the He.[3] We have made no major effort to perform chemical separations, because the radiations emitted by the species (usually α-emitters) have been highly specific and readily identified by nuclear means. For example, in a study in progress, we have been able to simultaneously identify 30 nuclear reaction products produced in the reaction ^{208}Pb + ^{12}C covering a span of 5 elements using known α-decay properties. The analysis is extremely complex, and any further detailed work in this region will require 'on-line' chemical and/or mass separation. We have encountered the same difficulties in the light elements in studies of β-delayed particle emitters. It is on these nuclides that we have made a small amount of progress. In studying the decay of ^{12}N, which can be produced cleanly by the reaction ^{12}C (p, n) ^{12}N at low proton energies, the ^{12}N recoil does not adhere to the collector when it is entrained in the helium stream. If a trace of CH_4 is added to the He, the ^{12}N recoil reacts with CH_4 during the thermalisation stage, producing HCN.[4] This is adsorbed very efficiently on the collector when it is cooled to liquid nitrogen temperature.
Good β-delayed α-spectra for ^{12}N have been obtained using this method. The addition of impurities to helium carrier gas, to control the formation of nuclear species in a specific chemical form, has been used with some success in fission-product studies but it has only been applied to a few elements, mainly the halogens.[5]

On-line mass separation. We are attempting to adapt the helium-jet method for on-line mass separation which, we hope, will have some capabilities for chemical separation as well. The results reported here are fragmentary and no workable system has yet been achieved. Two interesting observations have been made on this problem in the past few months which relate to the separation of the recoils from the helium and the development of a different type ion source. Fig. 5 shows schematically the system used to separate the recoils from the helium. It is based on the principle that in the expansion of a helium matrix, containing a heavy component, from a supersonic nozzle into vacuum (as we have in our experiments), the heavy component expands in a cone which is peaked more in the forward direction than for helium. The angle of the cone is directly related to the mass of the component. Furthermore, the heavy component is accelerated to the same velocity as the helium, which means that it gains energy by the ratio of its mass to the mass of He. These phenomena are discussed in detail in a review article by Anderson *et al.* on the properties of 'seeded' nozzle beams.[6] Because of the difference in angular divergence, the helium can be skimmed from the heavy component using a cylindrical cone with a hole in the tip, as shown in Fig. 5. This system has been tested with rare earth and ^{20}Na recoils, and it has been possible to transmit 90% of the recoils through the skimmer with a 1000-fold reduction in He pressure between the two vacuum chambers (50μ to 5×10^{-5} torr). The chemical nature of the recoils in the helium is not known. It is possible that they may be associated with aerosols (microscopic dust particles), and could have very large effective molecular weights. The fact that the ^{20}Na component was so well collimated after helium expansion strongly suggests that the mass of that component is much greater than 20.

The other observation recently made relates to ion source development. Fig. 6 shows a schematic diagram of the helium-jet system incorporating a flowing gas discharge. In this system, the transport capillary is made of glass and terminates in a series of alternating metal and glass capillary sections. With the helium moving through the capillary system at sonic velocity, an electric field is applied between the two metal capillary sections. A well-defined gas discharge is

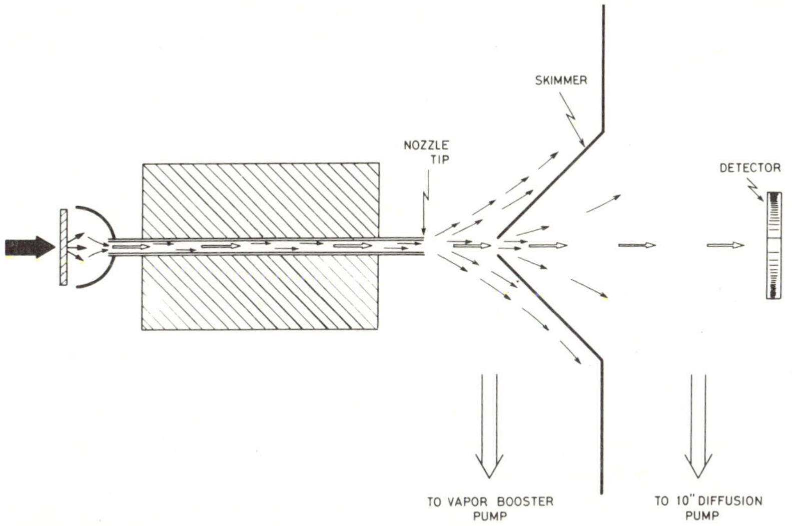

Fig. 5. Schematic diagram of the system used to separate He *carrier gas from the recoil beam. The nozzle diameter was 0·6 mm, the skimmer orifice diameter 1·2 mm, and the skimmer nozzle distance 1 cm*

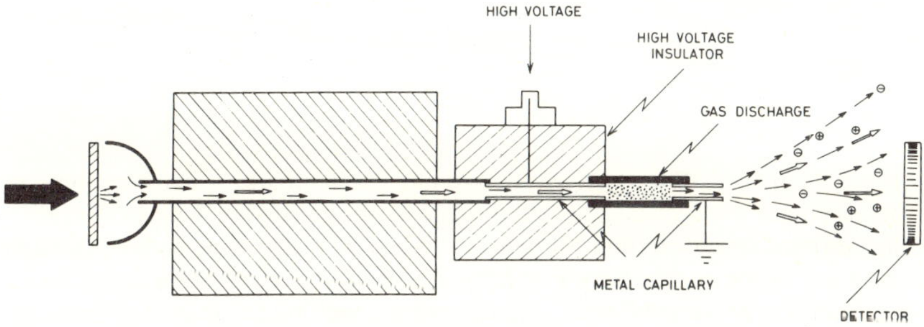

Fig. 6. Schematic diagram of a flowing gas discharge ion source

established in the intermediate glass capillary section at a gas pressure close to 2 atm over a distance ranging from 1 to 2 cm. It is probable that the rapid flow of the gas through the capillary may be necessary in establishing the discharge at these high pressures. Radioactive recoils of ^{150}Dy and ^{20}Na are then ejected into the gas stream during a cyclotron bombardment, and the fraction ionised is determined by loss due to space-charge blow-up after expansion of the gas into vacuum. A radiation detector located on the axis of the capillary channel detects the neutral component of the radioactive beam that emerges from the tip. The neutral component is highly forward directed and the detector can be located at a considerable distance from the nozzle (20 cm). The power into the gas discharge was varied and a measurement made of the loss in intensity of the neutral beam. The results are summarised in Fig. 7. Between 0 and 10 watts, little change was observed in the neutral beam intensity. Above the latter power level, a sudden drop in intensity was observed for both ^{20}Na and ^{150}Dy, and above 40 watts total power the neutral beam was reduced to one tenth of its original value. To ensure that the recoils were not being adsorbed on the walls of the ioniser, a foil was placed within 1 mm of the exit nozzle and the ^{150}Dy throughput was monitored at the 40 watt power level. Essentially all of the recoils entering the

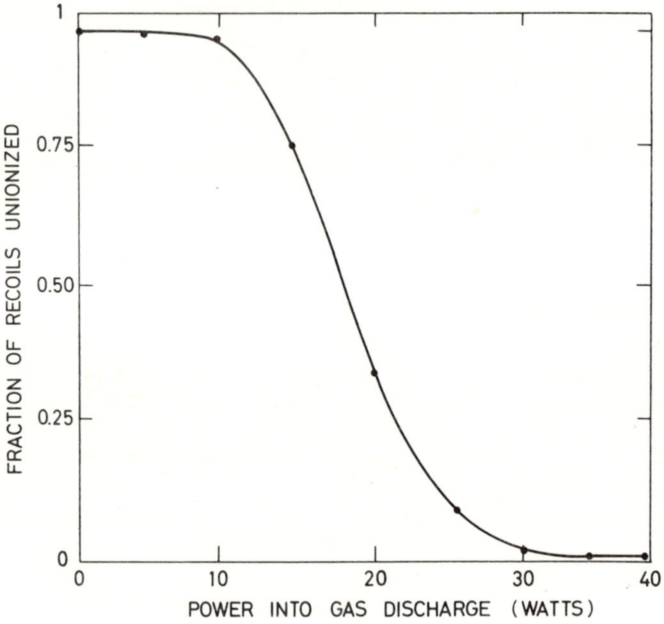

Fig. 7. Effect of discharge power on fraction of recoils ionised

ioniser were ejected from it after traversing the gas discharge. The identification of the nature of the ionic species produced has not yet been established.

In another experiment employing both ioniser and skimmer, a small quantity of neon was added to the helium and the gas, after passing through the ioniser and skimmer, was injected into a quadrupole mass filter. Singly-charged ions of ^{20}Ne were detected, but no ^{4}He ions from the carrier gas were observed. From other work on ionisation with a He gas discharge, it is felt that the ionising species is a metastable state of He at ~20 eV. If this is so, the equilibrium charge state for a particular element may be uniquely determined in the gas discharge by its ionisation potential relative to the energy of the metastable state of He. Thus, if a separation of ^{20}Na from ^{20}Mg is desired, it might be performed using the He discharge, where Na^{+} and Mg^{2+} would be produced. Separation would take place at the quadrupole mass filter, where ^{20}Mg would appear at the mass 10 position. At this point one could speculate further on how the addition of reactive impurity gases to the helium would effect the mass spectrum of the products. The possibility of greatly extending our knowledge of hot atom reactions throughout the periodic table is evident.

ACKNOWLEDGEMENT

We are grateful to the staffs of the Texas A & M Cyclotron Institute and Yale heavy ion accelerator for their interest and support. One of us (RDM) thanks the staff of the Niels Bohr Institute for their kind hospitality.

REFERENCES

1. Macfarlane, R. D. and Griffioen, R. D., *Nucl. Instr. and Methods* 24, 461, (1963).
2. Macfarlane, R. D., Gough, R. A., Oakey, N. S., and Torgerson, D. F., The Helium-Jet Recoil Method, *Nucl. Instr. and Methods* (in press).
3. Treytl, W. and Valli, K., *Nucl. Phys.* A97, 405, (1967).
4. Dubrin, J., MacKay, C., and Wolfgang, R., *J. Chem. Phys.* **44**, 2208, (1966).
5. Herrmann, G. and Denschlag, H. O. Rapid Chemical Separations, *Ann. Rev. Nucl. Sci.* **19**, 1969, (in press).
6. Anderson, J. B., Andres, R. P., and Fenn, J. B., *Advan. Chem. Phys.* **10**, 275, (1966).

DISCUSSION

J. G. Cuninghame (A.E.R.E.): I note that the tube down which the recoils pass is head-on to the beam. This means that the beam can also pass down it into the detector area, which seems an undesirable feature? Can you explain why the recoil tube is not turned through, say 90°, to avoid this? I believe I have seen this in some similar pieces of equipment.

R. D. Macfarlane (communicated): The fraction of the beam which passes down the capillary tube is extremely small and has given no problem. We have found that collection at 0° gives the fastest collection time, and for experiments where this is important this is the configuration used. 0° collection gives some serious problems at higher beam currents ($>0{\cdot}5\ \mu A$) due to local heating effects, and for this reason 90° collection is preferred where possible.

H. G. Blosser (Michigan State U.): The performance of the skimmer implies that the recoil ions pick up the same longitudinal velocity as the helium atoms but a much smaller transverse velocity. Do you understand the cause of this asymmetry?

R. D. Macfarlane (communicated): The reason for the asymmetry is a well-understood phenomenon associated with the flow of 'seeded' beams from a supersonic nozzle. For a detailed account of this effect, I would like to refer you to reference 6 of this paper.

Cyclotron irradiation of extended objects

J. F. B. Dealler and K. V. Ettinger

Department of Physics, University of Birmingham,
P.O. Box 363, Birmingham 15, England

ABSTRACT

The problems of achieving uniform irradiation of objects which are large in comparison with the size of the cyclotron beam are discussed, and the method of assessing uniformity is described. Techniques of achieving a uniform distribution of radiation over a large specimen placed outside the vacuum system are proposed, based on a method of applying random movements to the target.

INTRODUCTION

Many methods used in activation analysis and radiation damage studies presuppose either that the radiation field is uniform over the region of interest or that the variations are easily calculable. In addition, in problems connected with the wear of materials it is often necessary to irradiate large and convoluted surfaces with a prescribed dose distribution $D(x, y)$.

For cyclotron irradiations, one source of non-uniformity, viz. the range of the particles in the absorbing medium, is well known. The present contribution is concerned with some methods of irradiating extended objects from the viewpoint of delivering a uniform surface dose to the specimen.

Irradiations in vacuo and in atmosphere. Large objects can seldom be placed within the vacuum system, and a more practical approach is to extract the beam to atmosphere through a thin window. If the pathlength in air is short, the energy losses need not exceed 500 keV or so, although energy uniformity will not be so good and some scattering of the beam will take place.

Using vacuum windows of nickel, copper, and Havar foils beam currents up to 10 μA of protons can be extracted. The foil thickness is limited by energy losses on the one hand and by tensile strength on the other; in practice a foil 3 μm thick and 10 mm diam. will last many hours at 10 μA provided it is properly cooled.

Scanning. Once in air, the beam can strike the target. It is evident that it is impossible to scan the surface of a large object by deflecting the beam, and therefore a method must be devised of moving the target instead. Devices for moving the irradiated object already exist in the form of coordinate tables used in modern machine tools. Alternatively it may be practical to place the object in a simple jig and rotate it. One solution used to irradiate balls from ball-bearings is to balance the irradiated ball on three jets of air (J. H. Fremlin and G. Mostafa).

Curve following and contour filling. Fig. 1 illustrates the two types of irradiation that can be required when the object for irradiation is larger than the beam spot.

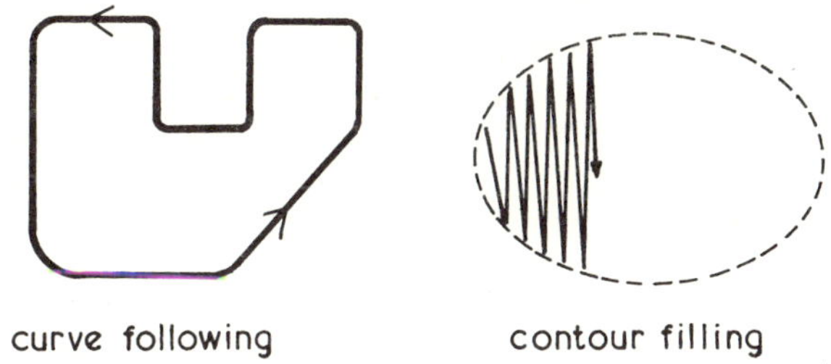

Fig. 1

Modern coordinate tables which are designed to be driven under tape control can be used in both modes; programming details for such tables will be treated elsewhere. In principle such an arrangement can give not only a uniformly-distributed dose but also a specified dose distribution. Ideally the table and its servosystem should have a zero static positioning error and zero velocity error for curve-following and ramp scanning, but if random scanning is envisaged, as we discuss later, acceleration errors should also be zero.

The speed of response of commercial tables is usually sufficiently rapid and the available translational speeds (up to 125 $cm.s^{-1}$) should be adequate.

Targets in vacuo. Uniform irradiation in vacuo presents different problems. Because of the difficulty in maintaining vacuum, only movement by rotation or vibration is practical. The beam may be scanned, but the extent of the scan is limited to a few degrees of arc both by practical problems of beam bending and by the need to minimise beam distortion.

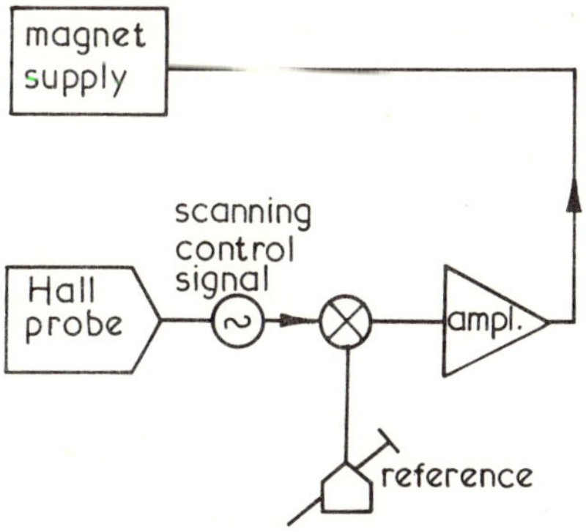

Fig. 2. Adaptation of the magnetic field stabiliser for scanning with the beam bending magnet

EXPERIMENTAL

Scanning in vacuo. We have been able to obtain a satisfactory deflection in one plane by varying the beam in the bending magnet, which is corrected for first-order aberrations. In order to achieve this a control voltage was introduced into the feedback loop stabilising the field of the magnet (Fig. 2). When the rise-time of the applied waveform is much longer than the time constant of the magnet, L/R, then the magnetic field follows the control voltage.

The polarity of the feedback loop can be reversed to produce a self-oscillating magnet, but the results are difficult to forecast as the oscillations are very non-linear. Our frequency was ~0·1 Hz.

Distribution measurements. A simple and well-known method of obtaining uniform irradiation is to defocus the beam. We have attempted to check the dose distribution obtained in this way by using a dosimetric method. Some time ago Artandi[1] introduced the use of polyvinylchloride (PVC) film for radiation dosimetry, and Maul, Holm and Draganic[2] developed a technique which gave reproducible results. The particular type of film used by Holm and others at Risö shows its main response to radiation by producing optical blackening in the ultraviolet region, while we have used an unplasticised PVC film, 'Craytherm' (Greenwich Plastics Ltd.), which turns brown when irradiated. The density of coloration, as measured with while light, is monotonically related to dose over a wide range. The film is developed for 30 min at 80°C. With cyclotron beams a beam density of ~1 μA cm^{-2} for a second or so gives good results.

After development a profile of the beam is seen. For preliminary work it is sufficient to heat the material on a hot plate till development occurs. For quantitative work the film is scanned with a microdensitometer which has been previously calibrated by measuring PVC samples which have been irradiated with known doses and heat-treated at the same time as the experimental material. Alternatively, the same calibration can be used for samples developed at different times, provided that one dose point on the calibration curve is repeated.

Since the calibration curve has the form

$$\Delta = f(kD) \tag{1}$$

where Δ is optical density, D the radiation dose, and k a constant which depends upon the conditions of heat treatment, a simple transformation can be made. In estimating beam uniformity we are interested primarily in relative measurements, but because of the non-linear response of the material it is necessary to work in relative doses rather than optical densities. The significance of the non-uniformities has been estimated in dose terms and appears as the right-hand plot in Fig. 3.

Criteria of non-uniformity. For some purposes any region which falls outside a given dose range may be unacceptable, i.e.

$$(D_{mean} - D_{min}) \text{ or } (D_{max} - D_{mean}) > C \tag{2}$$

where C is the maximum permissible dose departure. In this case the optical

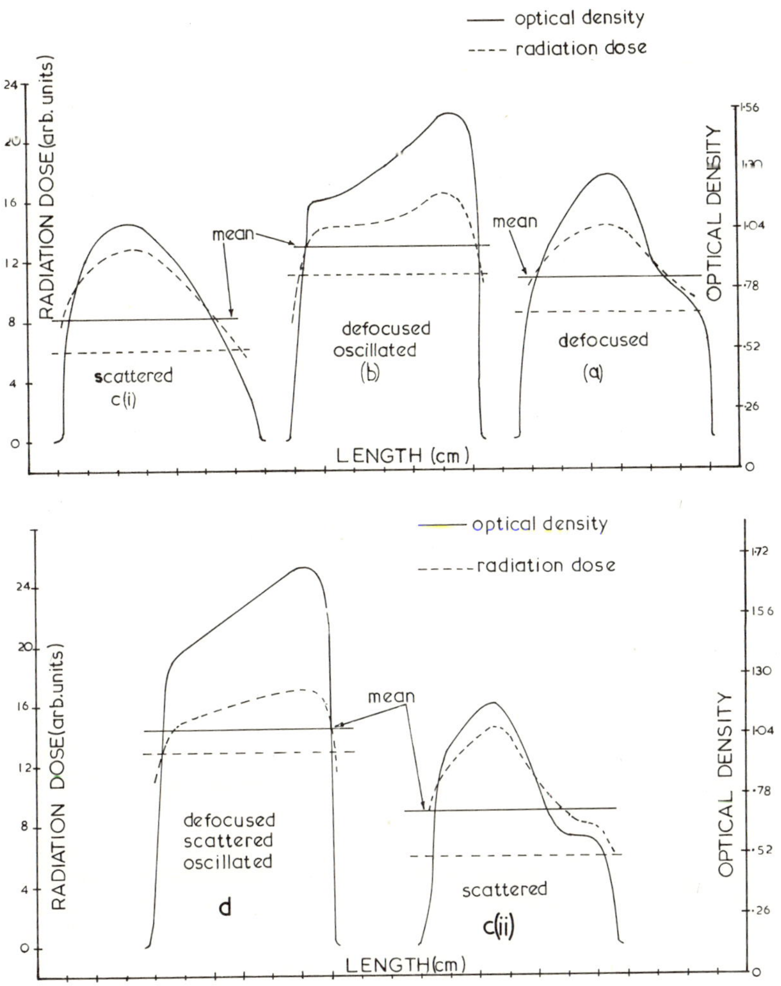

Fig. 3(a, b). Microdensitograms of irradiated foils and dose profiles

density corresponding to the limiting dose can be calculated and sought. In other cases the non-uniformity may be unacceptable in average terms, i.e.

$$\frac{\int |D_{\text{mean}} - D_{\text{observed}}| \, dA}{D_{\text{mean}} \cdot A} > C \qquad (3a)$$

or in r.m.s. terms

$$\frac{\sqrt{\int (D_{\text{mean}} - D_{\text{observed}})^2 \, dA}}{D_{\text{mean}} \, A} > C \qquad (3b)$$

In practice (3a) gives results close to those of (3b) but is much simpler in application. By taking a series of scans we may write

$$\frac{\int |D_{mean} - D_{observed}|\,dP}{D_{mean} \cdot P} \approx \frac{\int |D_{mean} - D_{observed}|\,dA}{D_{mean} \cdot A} \qquad (4)$$

where P is the total length of the photometric scan.

Degree of non-uniformity in some practical cases. Some commonly-adopted methods of improving uniformity have been examined using the foil technique.

(a) Beam defocused.
(b) Beam defocused and oscillated in the vertical direction at a frequency of about 0·1 Hz.
(c) Scatterer (10 μm Al) placed in the beam
  (i) 3 m from point of measurement
  (ii) 20 cm from point of measurement.
(d) As (b), but with 10 μ Al added at 20 cm.

The subjective impressions of these foils are supported by the photometric measurements, which gave the following results:

Method / *Foil*	*Absolute deviation, % (Eqn (2))*	*Average, % (Eqn (3a))*
a	32	18
b	27	13
c(i)	55	29
c(ii)	31	27
d	18	11

DISCUSSION

As might be expected, these results show that non-uniformities can be severe, but that even a crude level of scanning makes a considerable improvement. It is also clear that any regular scanning of the beam may well introduce a pattern into the radiation distribution. In order to overcome this problem we propose the use of random scanning of the beam or target as appropriate.

The scanning process is described in general terms by the parametric equations

$$x = x(t) \qquad y = y(t) \qquad (5)$$

e.g. for bidimensional ramp scanning with periods τ_x and τ_y the parametric equations will be

$$x = x_o + a_x(t/\tau_x - E(t/\tau_x)) \qquad y = y_o + a_y(t/\tau_y - E(t/\tau_y)) \qquad (6)$$

where a_x and a_y are the full magnitudes of scan in the directions X and Y respectively and $E(t/\tau)$ denotes the integer part of t/τ.

Considering random scanning as a Poisson process, requiring the probability density of position of the beam spot to be uniform over the whole of the irradiated area S, equations (5) become stochastic functions

$$x = X(t) \text{ and } y = Y(t)$$

with uniform probability densities within *S*. For a rectangular area

$$p_x = \begin{cases} 1/(b-a) & \text{if } a \leqslant x(t) \leqslant b \\ 0 & \text{if } x < a \text{ or } x > b \end{cases}$$
$$p_y = \begin{cases} 1/(d-c) & \text{if } c \leqslant y(t) \leqslant d \\ 0 & \text{if } y < c \text{ or } y > d \end{cases} \quad (7)$$

where *a, b, c and d* limit the rectangle in question.

If the object is not rectangular then a mask should be used, either materially or as a set of mathematical limitations on the scanning wave-form (Fig. 4).

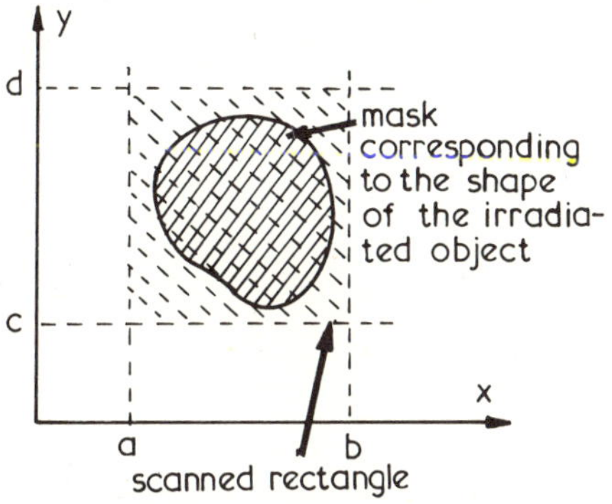

Fig. 4

A suitable system for producing the desired coverage is shown in Fig. 5. The 'limit check' restricts the deflection of the beam to the interior of the rectangle. The random signals are provided by a noise source (which could be made to provide output signals within specified limits only).

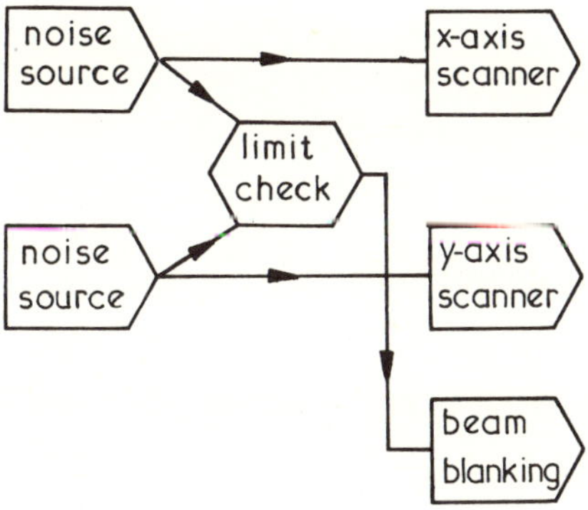

Fig. 5. Block diagram of a random scanning system with continuous noise sources

Most analogue random signal generators provide a Gaussian probability distribution which must be transformed into the uniform distribution needed for our purpose; methods of transformation are described in the literature.[3, 4]

If the boundaries of the object can be described by

$$y = f_1(x)$$

with

$$y = f_2(x) \tag{8}$$

and

$$f_1(a) = f_2(a); f_1(b) = f_2(b) \tag{9}$$

and if $f_1(x)$ and $f_2(x)$ are single-valued, then an 'electrical mask' can be set by blanking the beam when the coordinates of the random point fall outside the limits specified by eqns (8) and (9). The important question is the uniformity of the beam delivered to the object. The variance of the dose is given by

$$\sigma_D^2 = \frac{Ts}{S\theta}\left[1 + \frac{2}{T}\int_0^T \left(1 - \frac{\tau}{T}\right) R_s(\tau) d\tau\right] \tag{10}$$

where s = cross-section of the beam

S = area of the irradiated surface

T = exposure time

$R_s(\tau)$ = autocorrelation function of the random process used to generate the position of the spot

θ = average time for the beam to be displaced by one beam diameter.

It is assumed that the beam intensity is not dependent upon the coordinates of the spot and that it is constant in time. Otherwise, the total variance of the delivered dose is approximately equal to the sum of (10) and the variance of the intensity fluctuations of the beam.

Most noise generators used as sources of random waveform are characterised by an autocorrelation function of the type shown in Fig. 6. It can be shown

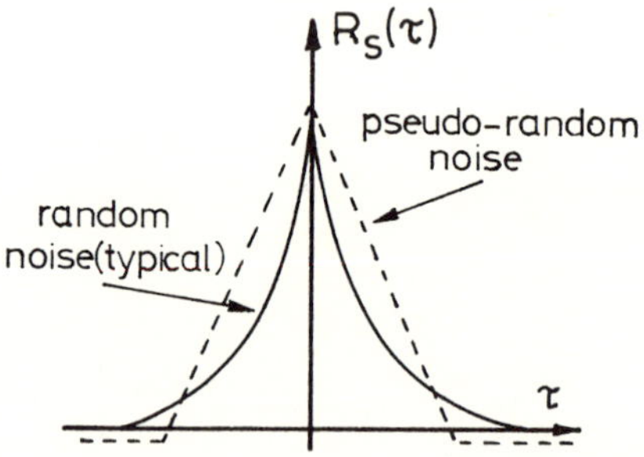

Fig. 6. Typical autocorrelation functions of noise sources

from Eqn (10) that the variance of the dose distribution is given by

$$\sigma_D^2 = \frac{Ts}{S\theta} \tag{11}$$

provided that T is sufficiently great.

The relevant standard deviation (in %) is

$$\delta = \sqrt{\frac{\theta S}{sT}} \times 100 = \text{const}\sqrt{\frac{\theta}{sT}} \qquad (12)$$

Since the time θ is a linear function of $\sqrt{s}$ the expression for the uniformity of the irradiation takes the form

$$\delta = \frac{\text{const}}{\sqrt{(dT)}} \qquad (13)$$

where d = diameter of the beam spot, thus confirming the general conclusion that defocusing the beam improves the uniformity of irradiation.

It is not necessary to have two separate noise sources to provide waveforms for the directions of beam displacement. A single source can be used provided a delay is inserted in one channel to reduce the correlation function to practically zero. If, instead of a noise source with continuous output, random signals are obtained from a digital noise source, it should be noted that the probability distribution of the generated numbers may be uniform but the actual irradiation of the target deviates from ideality. The finite time taken by the beam to be displaced from one position on the target to the next may increase the dose to the central part of the scanned area. Use of an on-line computer, even one of very modest size (4K memory), permits the generation of random numbers with a desired distribution law, and at the same time electronic masking during beam displacement can be introduced (Fig. 7). Pseudo-random noise generators[5, 6, 7] are particularly suitable for random scanning, due to the exceptional stability of their distribution function.

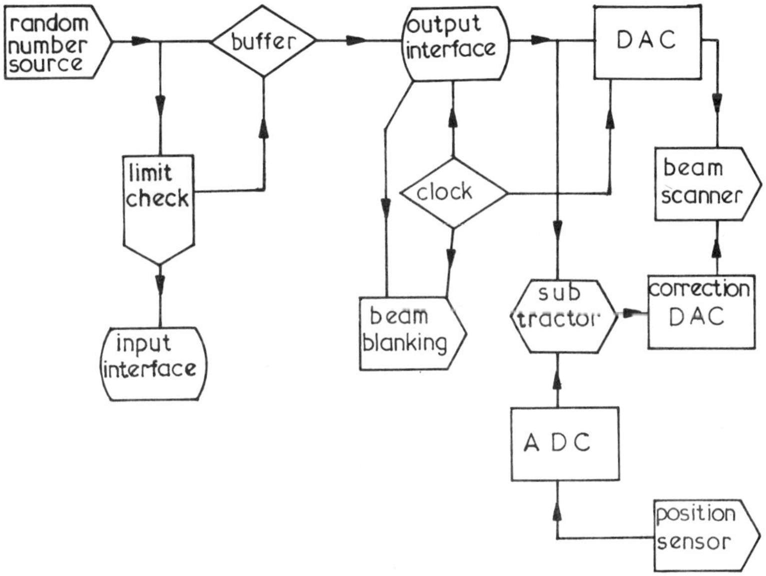

Fig. 7. Block diagram of a digital, discrete random scanning system (only one direction of scanning shown for simplicity)

Eqn (12), derived for the analogue system, applies also to digital systems, θ simply being the clock period, while for the continuous system θ usually has to be estimated. The servo-system controlling the movement of the coordinate table may be characterised by a response time τ_r. The latter is usually defined as a time taken by the response to the unit step, measured between 0·1 and 0·9 limits. In that case

$$\theta \approx 2\frac{d}{\sqrt{S}}\tau_r \tag{14}$$

The rise-time for a simple system is related to the amplitude-frequency characteristics by the equation

$$\tau_r = \frac{0{\cdot}45}{B} \tag{15}$$

where B is an effective bandwidth; the lower frequency limit must be 0. The coordinate table may be equally well characterised by the highest drive-speed indicated by the manufacturer, and the time needed to traverse the diameter of the beam is then

$$\theta \approx \frac{d}{v_{max}} \tag{16}$$

Example. Find the time needed to irradiate an area of 250 cm^2 with a beam of cross-section 1 cm^2 with non-uniformity <5%, when the response time of the table is 0·1 s.

$$\theta = 2 \times \frac{1{\cdot}12}{\sqrt{250}} \times 0{\cdot}1 \approx 14 \text{ ms}; \quad T = \frac{1{\cdot}4 \times 10^{-2} \times 2{\cdot}5 \times 10^{2}}{2{\cdot}5 \times 10^{-3} \times 1} = 1400 \text{ s}$$

This example shows the principle of random scanning to be wasteful in respect of accelerator time. However, checking the uniformity of irradiation by adjustment of periodic scanning consumes as much time and cannot always be done in advance.

So far we have assumed a uniform density of the beam within the beam spot. This is rarely the case, but it is in any case smoothed out by the random scanning. For the calculations an effective beam spot area should be used throughout, defined as

$$s = \frac{\int I \, ds}{I_{mean}} \tag{17}$$

where I and I_{mean} are local and mean beam current densities. The beam spot size can be determined easily with the help of PVC foil.

We have tested the principle and validity of the method using a cathode-ray tube and a photographic record to simulate the actual irradiation. With a commercial random-noise generator based on a Burr-Brown noise module the results indicated the expected degree of uniformity, which was as good as the probability distribution of the noise source, i.e. better than 3%. Photographs of the screen indicate no local brightening, apart from obvious streaks due to the Polaroid developing process.

In our opinion the random scanning method provides a useful complement to the periodic and quasiperiodic (Lissajous) methods of scanning large areas of irradiated targets.

REFERENCES

1. Artandi, C. and Stonehill, A. A., *Nucleonics* **16,** (5), 118, (1958).
2. Maul, J. E., Holm, N. W., and Draganic, L., Risö Report No. 31, Danish Atomic Energy Authority.
3. Rotkop, L. L., Statisticheskiye metody issledovaniya na elektronnykh modelakh, 'Energiya' Publ. House, Moscow (1967).
4. Korn, G. A., Random process simulation and measurements, McGraw-Hill Book Co., New York (1966).
5. Kramer, C., *Electronic Engng.* **37,** 465, (1965).
6. Anderson, G. C., Finnie, B. W., and Roberts, G. T., *Hewlett-Packard Journal,* September, 1967, (2).
7. Abbatista, N., Marino, D., and Plantamura, V. L., *Nucl. Inst. Meth.* **60,** 317, (1968).

DISCUSSION

J. G. Cuninghame (A.E.R.E.): You have clearly described how to irradiate irregular two-dimensionsal objects, and also regular three-dimensional objects such as a ball bearing. I am not clear how you would irradiate an irregular three-dimensional object.

K. V. Ettinger: So far we have been mainly concerned with two-dimensional scanning. It may be possible to introduce a relief of a three-dimensional surface into the computer memory and to generate suitable three-dimensional coordinates by the Monte Carlo method. There may of course be holes, cavities, etc., which are inaccessible to the beam and which cannot receive any direct radiation. Nevertheless, we think that for a surface which is accessible to the beam random scanning is the only feasible method of irradiation.

A simple method for measuring fast neutron fluxes

E. Bruninx

Philips Research Laboratories, N. V. Philips' Gloeilampenfabrieken, Eindhoven, The Netherlands

ABSTRACT

The proposed method uses a paraffin wax sphere with an indium foil at its centre. The whole assembly is shielded by cadmium foil to cut off thermal neutrons.

Depending on their initial average energy, the fast neutrons incident on the box are more or less thermalised and activate the indium foil via the (n, γ) reaction to ^{116m}In. The induced activity is directly proportional to the neutron flux. The response of the detector is given as a function of the average neutron energy.

Calibration of the induced ^{116m}In activity (in dis s^{-1}) with absolute neutron fluxes (in $ncm^{-2}s^{-1}$) allows the direct measurement of an unknown neutron flux. A simple method is used to correct for the presence of low-energy back-scattered neutrons.

The method is finally used to measure the neutron flux produced by 18 MeV deuterons on a thick beryllium target, in conditions where severe backgrounds of back-scattered neutrons prevail.

INTRODUCTION

The fast neutron flux produced by an accelerator (e.g. cyclotron) can usually be measured by a number of techniques.[1]

If the fast neutron spectrum is known, nuclear reactions with known excitation functions will yield the value of the fast neutron flux by measuring the absolute disintegration rate of the nuclide formed. If the neutron spectrum is not known, counters with a nearly flat response over a large energy range can be used. The so-called 'long counter' is by far the most extensively studied instrument for this type of work. This consists of a BF_3 counter surrounded by a mass of hydrogenous material, in which the fast neutrons are thermalised and then counted by the BF_3 tube.

This paper describes a highly-simplified version of this instrument in which the BF_3 tube is replaced by an indium foil contained in a paraffin wax sphere.

The absolute calibration of the instrument allows the direct determination of fast neutron fluxes.

The original version of this detector was described by Stephens and Smith,[2] who used it principally for surveys of stray neutrons around accelerators. Only a very crude calibration of the detector was undertaken. A better calibration was obtained by Simpson,[9] but for a sphere of larger diameter. In both cases only relative efficiencies were measured.

EXPERIMENTAL

The detector used in our experiments consists of a cadmium box with a removable lid (thickness of cadmium foil 0·8 mm). The external dimensions of the box are 15·7 × 15·7 × 15·7 cm. Inside this box two paraffin wax

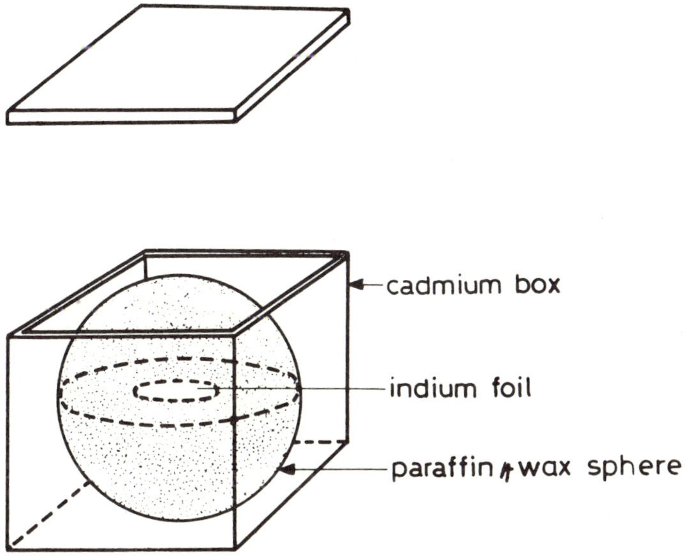

Fig. 1. Schematic drawing of the cadmium box and paraffin wax sphere

half-spheres can be fitted (diam. 15 cm). In the centre of the flat face of one of these halves is a shallow depression in which is placed an indium foil (diam. 18 mm, thickness 0·12 mm, weight ~200 mg).

Principle of operation. The closed box containing the indium foil sandwiched between the two paraffin wax half-spheres is irradiated in the neutron field at some distance R from the source. The position of the indium foil with respect to the incident neutron beam is irrelevant. Fast neutrons falling on the detector will be thermalised in the paraffin wax and will activate the indium foil to ^{116m}In by an (n, γ) reaction. Thermal neutrons incident upon the box will be absorbed by the Cd shield.

The necessary equations relating the activity of the indium foil to a given neutron flux at some distance from the neutron source are as follows:

$A = K(E)\phi$ (1)

A = activity of the indium foil in $cs^{-1}(mg\ In)^{-1}$, corrected for finite length of bombardment.

$K(E)$ = constant depending upon the average energy of the incident fast neutrons. Dimensions: $cn^{-1}\ cm^2(mg\ In)^{-1}$.

ϕ = neutron flux in $ns^{-1}\ cm^{-2}$.

Due to the presence of shielding, neutrons will be scattered all round and the detector will obviously be activated by these scattered neutrons. The measured activity is therefore due to both direct neutrons (from the fast neutron source) and scattered neutrons. The equation expressing these two effects is as follows:

$A' = Gr^{-2} + a$

A' = measured activity of the indium foil in $cs^{-1}(mg\ In)^{-1}$, corrected for finite length of bombardment.

G = constant, with dimensions $cs^{-1}\ cm^2(mg\ In)^{-1}$.

r = $R - \Delta$ = distance in cm between the neutron source and the effective centre of the paraffin wax sphere.[3] R is the distance between the neutron source and the geometrical centre of the sphere; $\Delta = 0{\cdot}7$ cm, is assumed to be independent of the neutron energy.

a = activity in the indium foil due to back-scattered neutrons (in the same units as A').

Similarly, the flux is given by

$\phi = Qr^{-2}$ (3)

ϕ = neutron flux in $ncm^{-2}s^{-1}$.

Q = constant, with dimensions ns^{-1}.

r = distance in cm.

The flux calibration constant $K(E)$ in Eqn (1) is easily found by combining Eqns (2) and (3).

$$\frac{A' - a}{\phi} = \frac{A}{\phi} = \frac{G}{Q} = K(E) \qquad (4)$$

Two steps are thus necessary in determining the constant $K(E)$.

(a) Measurement of the activity of the indium foil at different distances from the neutron source in order to evaluate the term 'a' and the slope G in Eqn (2);

(b) $K(E)$ can be obtained either from the corrected activity A and the flux at a distance r, or, more simply from G and Q. The flux in these measurements must be known.

These measurements, with a set of neutron sources having different average energies, result in a suitable curve for $K(E)$ as a function of neutron energy. We used an Am-Be source and neutrons from a 14 MeV generator. Further work is in progress using neutron sources of other energies.

An unknown fast neutron flux can now easily be determined by a similar set of measurements as described under (a).

Step (b) must be slightly modified as follows: the activity A (corrected for background neutrons!) at a distance r from the neutron source (or target) is simply divided by $K(E)$ to yield directly the value of the flux in $ncm^{-2}s^{-1}$ (see Eqn 4). $K(E)$ is taken at the appropriate neutron energy.

Counting. The induced ^{116m}In activity can be measured either by β- or γ-counting. For very low neutron fluxes a G.M. tube or, even better, a low background β-counter should be used.

All the indium foils were counted on an aluminium planchet (thickness 0·5 mm at 36·5 mm from a Philips G.M. tube No. 18.536 (window diam. 27·8 mm, window thickness 1·5–2·0 mg cm^{-2})). Occasionally, when the activities were very low, a low-background counter was used. In all cases the measured activities were normalised to those measured on the G.M. tube. As mentioned above, all the indium activities are expressed as counts/mg indium. The G.M. tube was calibrated for these particular dimensions of the indium foils by means of a calibrated NaI (T1) crystal. The branching ratios for the γ-rays were obtained from ref. 4. The half-life was taken to be 54 min.

The efficiency of the G.M. tube for converting the β-counting rate into the disintegration rate of ^{116m}In was $3{\cdot}44 \times 10^{-2}$. All counting rates in Eqns (1), (2) and (4) can thus be converted to disintegration rates using the above-mentioned β-efficiency. The constant $K(E)$ can also be expressed in dis $n^{-1} cm^2 (mg\ In)^{-1}$.

Neutron sources. The Am-Be source had a neutron emission rate of $2{\cdot}5 \times 10^6$ ns^{-1}. The 14 MeV neutron generator was calibrated with aluminium foils using the reactions $^{27}Al(n, \alpha)^{24}Na$ and $^{27}Al(n, p)^{27}Mg$. The activities formed in these nuclear reactions were likewise measured on this G.M. counter calibrated for these particular dimensions of the foils.[5] The cross-sections of these nuclear reactions were taken from refs 6–8.

RESULTS AND DISCUSSION

Determination of G and K(E). Since the detector is used in surroundings that are not always ideal (i.e. far from walls and other neutron-scattering objects) we chose three different conditions in which the measurements with the detector were made (Fig. 2).

In position S the closest wall was at a distance of ~1·5 m. Position T was chosen to maximise the contribution of back-scattered neutrons. An additional measurement was made in position C, in which the Am-Be neutron was placed in the target head used in the cyclotron for the production of fast neutrons.[5] The 14 MeV neutron generator was housed in a water shield, and all measurements were done inside this shield (inside dimensions: 1·80 × 2·40 m). Virtually all measurements were taken in triplicate at different distances from the neutron source and the results were fitted by least squares to Eqn (2).

Fig. 3 shows the results obtained with the Am-Be source for the three different positions.

We can draw the following conclusions. The indium activity due to the back-scattered neutrons is not negligible, especially at the larger source-detector distances. This background can therefore not be neglected in any kind of flux measurement. The slope G for the measurements (S and T) remains virtually the same. The situation is much more complicated when the neutron source is surrounded by a large mass of metal (position C), as the value of the slope G changes markedly compared to position S. It is not yet clear what is the relative importance of the degradation of the neutron energy on absorption and possible scattering effects.

This change in G in different surroundings will finally be one of the limitations to the accuracy of a fast neutron flux determination using the absolute values of $K(E)$ and Eqn (1). The effect can be circumvented to some extent by simulating as accurately as possible the surroundings in which the neutrons are generated.

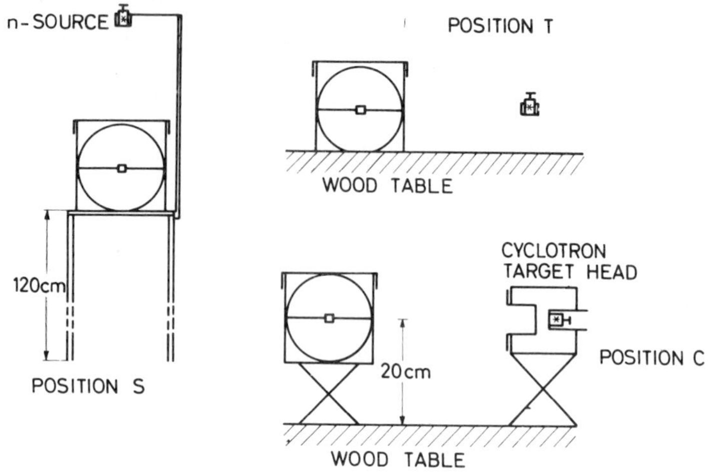

Fig. 2. Different surroundings used in the measurements with the calibrated Am-Be *neutron source*

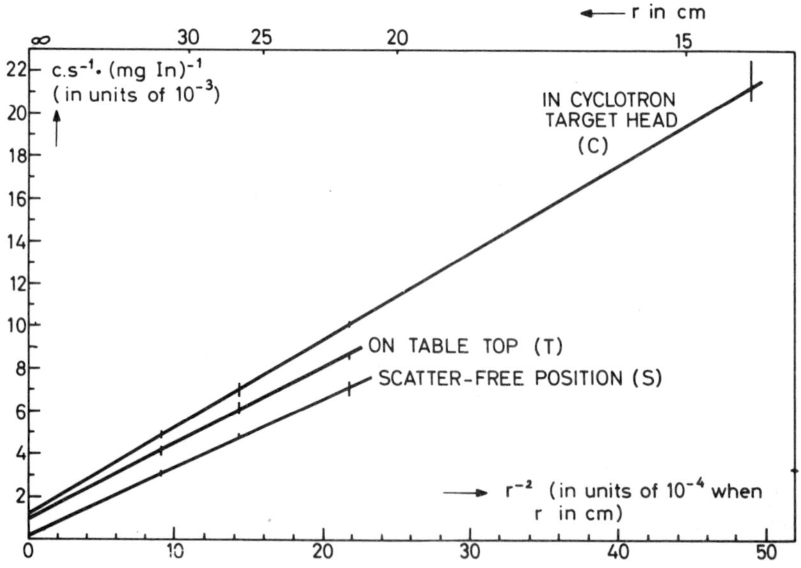

Fig. 3. Activity A' of the indium foil measured in different surroundings at various distances

The values of G for the different measurements are given in Table 1.

Once we know A or G we can easily compute $K(E)$ from the known neutron flux or from Q. The values are given in Table 1 and in Fig. 4. For the Am-Be source the value obtained in position S is plotted in Fig. 4. The values of $K(E)$ were normalised to the $K(E)$ value of the Am-Be source (E_{av} = 4·3 MeV and compared with the measurements of Simpson[9] (normalised to his $K(E)$ value at 4·5 MeV from Po-Be neutrons). The latter were relative measurements and for a

Table 1. VALUES OF *G* AND *K(E)* FOR DIFFERENT NEUTRON ENERGIES

Neutron source		*Type of surrounding*	*G*	*K(E)*
Type	E_{av}			
Am-Be	4·3	*S*	3·2	$1{\cdot}59\ 10^{-5}$
		T	3·5	$1{\cdot}78\ 10^{-5}$
		C	4·0	$2{\cdot}02\ 10^{-5}$
(D, T)	14·0	Water Shield	$17{\cdot}8 \times 10^3$	$8{\cdot}5\ \ 10^{-6}$

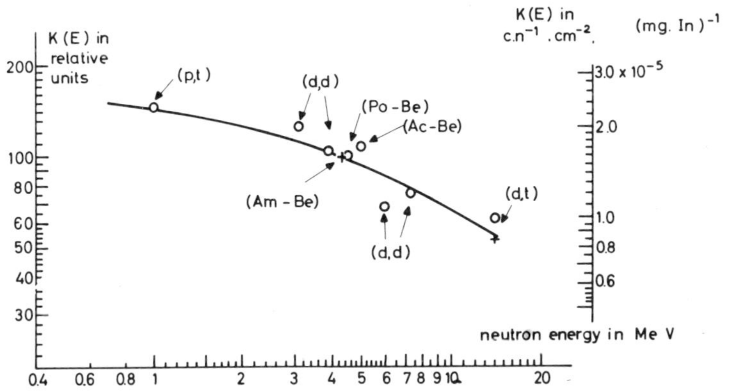

Fig. 4. Value of K(E) as a function of the neutron energy
o: values for ref. 9; + this work

larger paraffin wax sphere. This difference in size will only show up for high and for very low energy neutrons.[10]

The right-hand ordinate gives the value of *K(E)* in $cn^{-1}\,cm^{2}(mg\ In)^{-1}$ for the G.M. counter used in our experiments. Using the efficiency value of this counter, the values of *K(E)* can be computed for any other type of β- (or γ-) counter.

Due to the limited size of the detector the response [or *K(E)*] becomes more energy-dependent as compared with, e.g. a long counter. The detection of high-energy neutrons is therefore less efficient than that of low-energy neutrons.

Application to fast neutrons produced by 18 MeV deuterons on a thick beryllium target. Irradiations were performed with a fixed-energy cyclotron, using the internal probe target. The neutron flux was measured at various distances outside the vacuum chamber but inside the magnet yoke by means of the paraffin wax detector (Fig. 5.). And irradiation lasted usually for a few minutes at beam intensities between 1 and 10 μA.

As an independent check on the flux measurement we used the $^{27}Al(n, \alpha)^{24}Na$

reaction. Since the fast neutron spectrum for this reaction is known,[11] we calculated the average cross-section to be 42 mb.

The results obtained with the paraffin wax detector (Fig. 6) were fitted to Eqn (2) in order to obtain the value of Q.

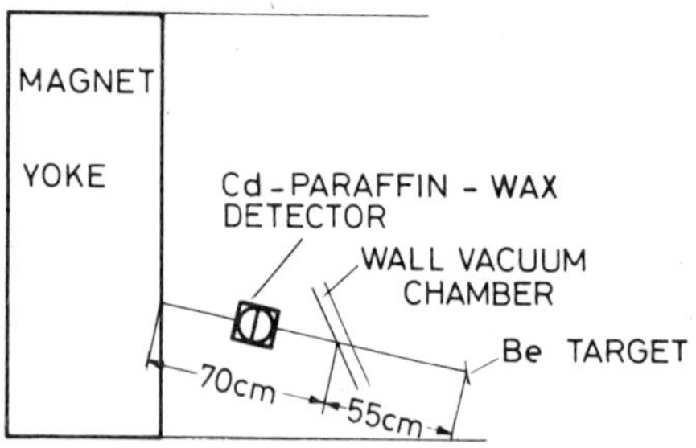

Fig. 5. Set-up for flux measurements at the cyclotron

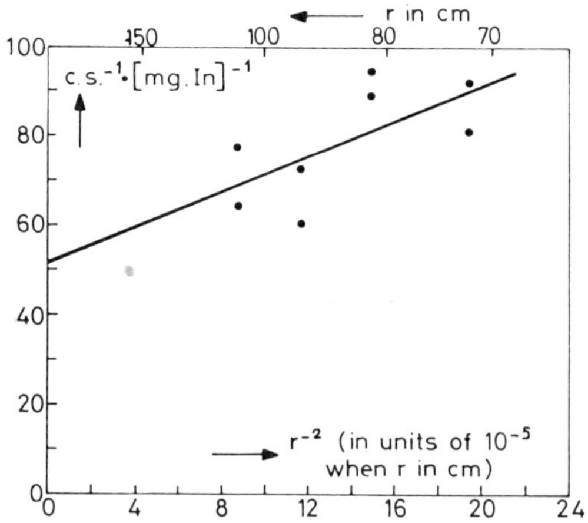

Fig. 6. Activity A′ of the indium foil at various distances in the cyclotron irradiations

Thus

$$A' = 19{\cdot}8 \times 10^4 \times r^{-2} + 51{\cdot}8$$

All activities A' were normalised to a beam intensity of 1 μA. From Eqns (4) and (3) we can deduce

$$\phi r^2 = Q = \frac{G}{K(E)} = \frac{19{\cdot}8 \times 10^4}{1{\cdot}21 \times 10^{-5}} = 1{\cdot}64 \times 10^{10}\ \mathrm{ns^{-1}(\mu A)^{-1}sr^{-1}}$$

The value of $K(E)$ is read off from Fig. 5 using an average neutron energy of 7·7 MeV.[11]

The flux values based upon the measurement of the ^{24}Na activity at three different distances (55, 75, and 90·8 cm) yielded respectively.

$$(1{\cdot}32,\ 1{\cdot}24,\ 1{\cdot}27) \times 10^{10}\ \mathrm{ns^{-1}\ (\mu A)^{-1}\ sr^{-1}}$$

The flux measured by Allen *et al*[12] for 15 MeV deuterons on a thick beryllium target is

$$1{\cdot}06 \times 10^{10}\, \mathrm{n\,s^{-1}(\mu A)^{-1}sr^{-1}}$$

The value is slightly higher than the one given in the literature owing to the adoption of a better cross-section value. When we take into account the increase in neutron flux in going from 15–18 MeV we may conclude that the agreement is satisfactory despite the severe background due to stray neutrons.

REFERENCES

1. Marion, J. and Fowler, J., Fast Neutron Physics, p.623, Interscience Publishers, N.Y. (1960).
2. Stephens, L. and Smith, A., U.S. Atomic Energy Commission Report U.C.R.L. 8418 (1958).
3. Aten, A. H. W. Jr., Heertje, I., Bonn, R., Mosk, J., van Westen, N., and Kapteyn, J., *Health Phys.* **11**, 1094, (1965).
4. Hansen, P. G., Nielsen, H., and Wilsky, K., *Nucl. Phys.* **30**, 140, (1962).
5. Bruninx, E. and Crombeen, J., *Int. J. Appl. Rad. and Isotopes,* **20**, 255, (1969).
6. Hughes, D. and Schwarz, R., BNL 325, second edition supplement No. 2, May 1964.
7. Mani, G., *Nucl. Phys.* **19**, 535, (1960).
8. Butler, J. and Santry, D., *Can. J. Phys.* **42**, 372, (1963).
9. Simpson, P., NIRL/M/70, 1964.
10. Mijnheer, M., private communication.
11. Bruninx, E., to be published.
12. Allen, A. J., Nechay, S., Sun, J., and Jennings, B., *Phys. Rev.* **81**, 536, (1951).

Target systems for radioisotope production on the Medical Research Council cyclotron

D. D. Vonberg, L. C. Baker, P. D. Buckingham, J. C. Clark, K. Finding, J. Sharp, and D. J. Silvester

Medical Research Council, Cyclotron Unit, Hammersmith Hospital, Ducane Road, London, W.12, England

ABSTRACT

Target systems used for short-lived radioisotope production on the M.R.C. Cyclotron are described in detail and the factors affecting efficient production discussed. Choice of target material, target cooling, radioisotope recovery, suitable beam energy and target thickness for optimum yield of the required isotope, are all considered. The problems of power dissipation in the internal target using beam currents up to 300 μA are fully discussed. Reasons for using the static 'grazing incidence' target in preference to rotary or vibratory target systems are stated and methods of internal target preparation described.

Beam extraction through a thin vacuum-retaining foil permits a range of external targets to be bombarded using beams of low power density and also allows the use of energy-degrading foils. A method is described of indicating the external beam distribution during bombardment and details are given of external targets using solid, liquid and gaseous target materials. Recently-developed targets for radioactive gas production are described and typical performance figures quoted.

INTRODUCTION

The Medical Research Council cyclotron,[1,2] which was designed and built between 1950 and 1955, operates for 14 h daily on five days a week. About one third of each day is used for experimental radiobiology or fast neutron radiotherapy; the remainder is devoted to producing radioisotopes for medical use, which often but not invariably because of their short half-life cannot be obtained commercially. Much has been published on the uses of these isotopes,[3,4] and sometimes the methods of producing them have been included, but with very few exceptions the cyclotron target systems have not been described in detail. This paper discusses the factors that have to be considered in designing suitable targets for efficient isotope production, and describes some of those currently in use on the M.R.C. machine.

In general, four main factors must be considered:

(1) The physical properties and chemical purity of the target material, which may be solid, liquid or gas.

(2) The provision of adequate cooling to control the temperature rise of the target surface.

(3) The method for recovery of the required radioisotope from the target during or after bombardment.

(4) The selection of a beam energy and target thickness that will maximise the yield of the required radioisotope and minimise the yield of possible impurities.

The M.R.C. machine is a fixed-energy, classical cyclotron capable of accelerating deuterons to 16 MeV and α-particles to 32 MeV. Internal beam currents up to 500 μA and extracted beams up to 100 μA can be obtained without special tuning. In practice however, the internal and external beam currents normally used are rarely greater than 300 μA and 50 μA respectively, because of the target limitations to be discussed.

The four external target positions are shown in Fig. 1. A beryllium target on beam 1 provides the fast-neutron beam used in an adjacent room for radiotherapy and activation analysis, and beam 2 is used for radiobiology experiments in the same room. Beams 3 and 4 are used in the cyclotron chamber for isotope production, as is the internal target position shown in Fig. 2. An automatic changing device

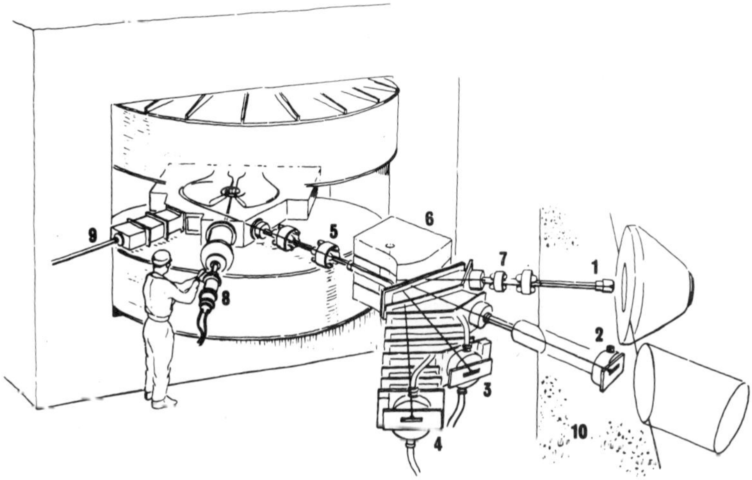

Fig. 1. M.R.C. Cyclotron–External Target Arrangements:
1. Be *target producing fast neutrons beam for radiotherapy, radiobiology and activation analysis)*
2. Beam 2: (Charged particles for radiobiology and activation analysis)
3. Beam 3: (for isotope production)
4. Beam 4: (for isotope production)
5. Beam shaping magnets
6. Beam bending magnet
7. Beam 1 focusing magnets
8. Ion source equipment
9. Internal target equipment
10. Wall dividing cyclotron chamber and 'radiation room'

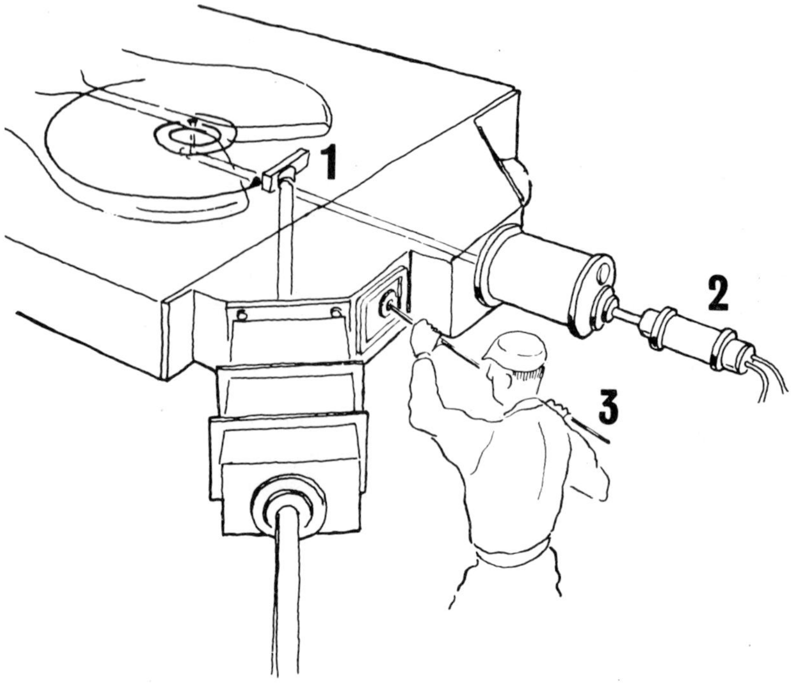

Fig. 2. M.R.C. Cyclotron–Internal Target Arrangement:
1. Grazing incidence target assembly on support tube
2. Ion source equipment
3. Manual probe for maintenance of dee-chamber

enables each of two alternative targets to be selected remotely for bombardment on beam 3, while all internal targets may be loaded and unloaded automatically without the need for personnel to enter the cyclotron chamber. The latter facility is to be extended to beam 3 in the near future.

INTERNAL TARGETS

The internal beam is normally intercepted by the target at approximately 60 cm radius, beyond which the beam current falls rapidly. At this point the beam is in the form of a ribbon 1 cm high × 3 mm wide. In order to use the high beam currents up to 500 μA regularly available in the internal beam, the target must be designed to dissipate the high power density of the beam. At 15 MeV a 500 μA beam of dimensions 1 cm × 3 mm has a power density of 20 kW cm^{-2}. Almost all the power reaching the target appears as a rise in temperature of the target coolant. The temperature differential between the front surface struck by the beam and the cooled back surface depends on the thermal conductance of the target and the power density in the beam.

In a copper target of thermal conductivity 0·92 cal cm^{-1} s^{-1} deg C^{-1}, a surface power density of 20 kW cm^{-2} results in a temperature gradient of 5×10^{3} °C cm^{-1}. In a poorly-conducting metal the gradient would be much higher, e.g. over 20 times higher for antimony. Non-metals generally have thermal conductivities much lower still, resulting in temperature gradients of the order of 1000 times higher than that in copper.

The hottest surface of the target must be kept below the temperature at which the vapour pressure of the target material rises sufficiently to allow appreciable evaporation either of the target itself, or of the products of the nuclear reaction. If it is assumed that the back of the target is kept between 100 and 150°C by cooling water at a pressure of ~7 kg cm^{-2}, then at 20 kW cm^{-2} melting would occur if the thickness exceeded 2 mm for copper or 0·1 mm for antimony. Target construction therefore would present serious mechanical problems except for materials of high conductivity like copper. Moreover, the thickness of 0·1 mm is less than the particle range at 15 MeV, and this would affect the production efficiency in some cases.

Perhaps a more serious problem would be that of water cooling. At thicknesses of this order the power density at the cooled surface will be similar to that in the beam itself. However, the maximum power which could be removed by water cooling under conditions of surface boiling was found in experimental tests to be only about 4·5 kW cm^{-2}. On several counts, therefore, targets perpendicular to the internal beam of the M.R.C. cyclotron are not practicable.

The two methods commonly employed to reduce the power density at the target surface and at the cooled surface are either to present the target to the beam at grazing incidence, as depicted in Fig. 2, or to move the target or the beam continuously.

Rotating targets have been used at Hammersmith, but have been abandoned because of the complicated mechanism required to provide:

(a) the high rotational speed necessary to avoid mechanical stresses from thermal cycling, and

(b) sufficient mechanical power to overcome the eddy current drag on the target assembly in the magnetic field.

Since these two requirements are in direct opposition, and since such targets are inherently more expensive to produce than their static counterparts, rotating or vibrating target systems have not been developed. Thus the M.R.C. machine uses a grazing or oblique incidence target in the form of a flat plate arranged tangentially to the beam. From purely thermal considerations, since the power density is reduced by a factor of $1/\cos\theta$ (Fig. 3), such oblique incidence targets can be thinner than perpendicular incidence targets by a factor of $\cos\theta$, where θ

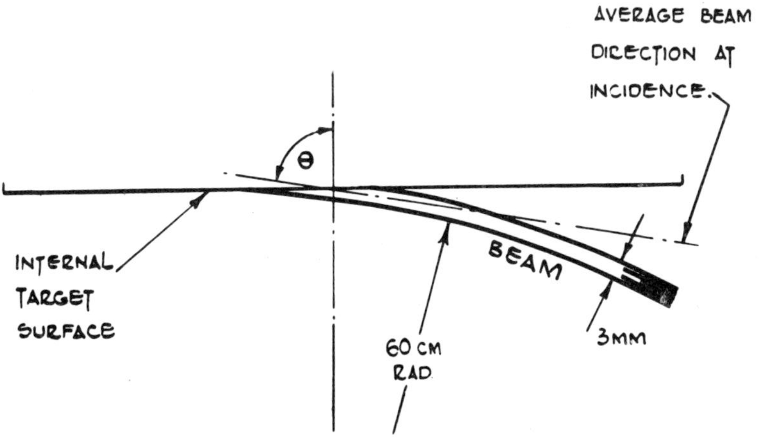

Fig. 3. Internal target geometry

is the angle between beam direction and the normal to the target face. The effective thickness of the target presented to the beam is increased by $1/\cos\theta$ and the former becomes a thick target in the nuclear sense in practical cases.

Since the beam has a finite radial width, θ is not the same for all particles. If the inner edge of the beam is tangential to the target the average value of $1/\cos\theta$ is about 10 in the M.R.C. machine.

For chemical processing the smaller the amount of target material present the better; in practice target thicknesses between 0·025 and 0·050 mm are found to give good yields on the grazing incidence target.

Clearly a target 0·050 mm in thickness needs mechanical support to retain the coolant; this is provided by a target backing plate which is in intimate thermal contact with both the target material and the coolant. The plate is of high thermal conductivity, copper or aluminium being chosen according to the chemical process to be employed in separation, for some of the backing plate material usually appears as a contaminant.

The target back is smooth for optimum cooling, and a water velocity of 4 m s^{-1} is obtained by narrowing the waterway at the back of the target plate to 0·25 mm. The direction of flow is perpendicular to the long dimension of the beam strike area. Because the effectiveness of the cooling depends on a high water velocity, it is essential to avoid distortion of the target plate, which would widen the gap and reduce the velocity. In practice copper or aluminium target plates need to be at least 2 mm thick to avoid distortion under the combined effect of heat and water pressure.

A thin target plate (1·5 mm) permits a high heat transfer rate through the plate, but affords little cooling due to conduction parallel to its surface. Conversely, a thick plate (6·5 mm) has a low heat transfer rate through the plate but a high cooling capacity due to conduction parallel to its surface. Maximum heat dissipation is therefore achieved at an optimum target plate thickness. Experimental measurements on copper target plates show this to be about 5 mm for applied power densities between 0·3 and 5 kW cm^{-2}. However, 3 mm was

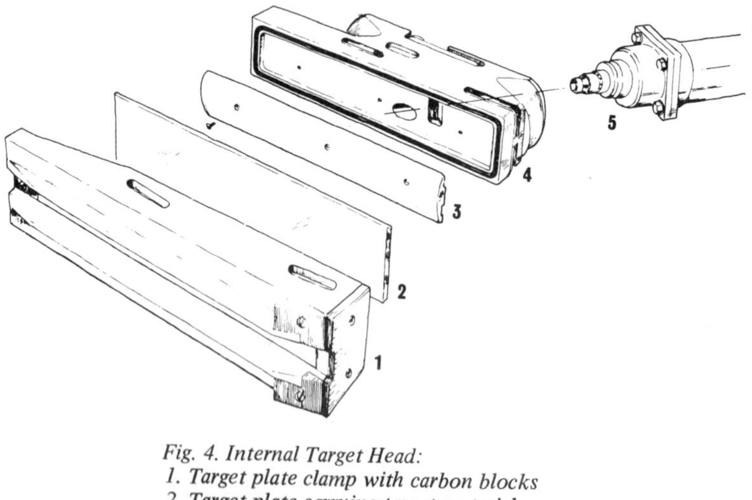

Fig. 4. Internal Target Head:
1. Target plate clamp with carbon blocks
2. Target plate carrying target material
3. Water flow spoiler
4. Target head block with 'O' ring and water ways
5. Support tube with water connections

finally chosen to simplify handling and reduce cost, without serious reduction in effectiveness.

Fig. 4 shows in exploded view the final form of the target assembly incorporating the features described above. Several other features have been added as a result of practical operating experience over a period of seven years. For example, as the machine tuning is adjusted the centre of curvature of the beam moves and the strike area moves along the plate. If the edge of the beam should approach either end of the target it strikes a carbon block, causing it to glow, giving visible warning on a CCTV monitor. As mentioned above, a remote automatic handling machine removes the target assembly from the vacuum system, detaches it from its water supplies, and transfers it by railway to a distant processing cell where the assembly is dismantled. The target plate is then removed and machined or chemically treated to remove the target layer, from which the required isotope is then extracted. The total bombardment on each target in μA h is recorded by a device which integrates the rise in water temperature against time.

In preparing targets for bombardment, a layer of target material 0·025–0·050 mm thick is applied to the target plate by a method selected to provide the optimum bonding. For example chromium (for ^{52}Fe production) is electroplated on aluminium, nickel (for ^{62}Zn production) is deposited by metal spraying, whilst lead (for ^{206}Bi production) is fused onto copper in a vacuum furnace.

EXTERNAL TARGETS

(a) General. Clearly, internal targets for high beam currents can be prepared only from a limited range of materials with suitable physical properties. The M.R.C. cyclotron has the design restriction that such targets can only be bombarded at maximum beam energy and, as has been shown, are necessarily thick in the nuclear sense. Fortunately these limitations can be overcome by using the extracted beam, albeit at the expense of lower maximum currents.

Once extracted from the dee-chamber, the particle beam is defocused by quadrupole magnets and can be made to cover a relatively large area (5–10 cm wide and 1 cm high) at the external target positions on beams 3 and 4. Having reduced the power density in this way, targets can be presented normal to the beam instead of at grazing incidence, and it is then possible to vary the thickness of target material which is bombarded. The energy of the beam incident on the target surface may also be varied by interposing energy-degrading foils of the appropriate thickness. These two variables can be very important; for example, in ^{123}I production from natural antimony the incident α-particle energy is reduced to 25 MeV, and the energy lost in the antimony is restricted to 2 MeV by limiting its thickness to 0·025 mm. This minimises contamination of the ^{123}I produced by an (α, 2n) reaction by ^{124}I from (α, n) and (α, 3n) reactions.[5]

Although it is not always desirable to degrade the beam energy, it is usually necessary to interpose a thin foil between the incoming beam and the target material in order that the latter, if it be a powder, liquid or gas, should be contained. Similarly, a foil is needed at the end of the external beam tube to maintain the vacuum. In principle a single foil could combine both these functions, but in practice target foils are often found to be very vulnerable to corrosion by hot target materials, and so separate foils are used. On the M.R.C. machine, 0·025 mm titanium is used on the beam tube, as this has high mechanical strength and thermal conductivity. Target foils are usually of aluminium or copper, unless there are special reasons for avoiding these materials. Both foils must be efficiently cooled, or they may be perforated by the beam; this is achieved by circulating cooled dry air between them and by water-cooling their supporting frames. A grid of 25 0·2 mm diam. tungsten wires having a 5 mm spacing is supported 2·6 cm on the vacuum side of the titanium foil, and a CCTV system is used to show the beam distribution indicated by incandescence of the wires intercepting the beam. Since all external targets are mounted in such a way that they are electrically insulated from the cyclotron and its beam tubes, beam currents can be measured using a microammeter and simultaneously integrated.

(b) Solids. The standard target plate used to support solid target materials during bombardment in the external beam is similar to that used for internal bombardment, and they may be prepared in the same way, e.g. antimony (for ^{123}I production) is electroplated on copper. However, many alternative methods of preparation are

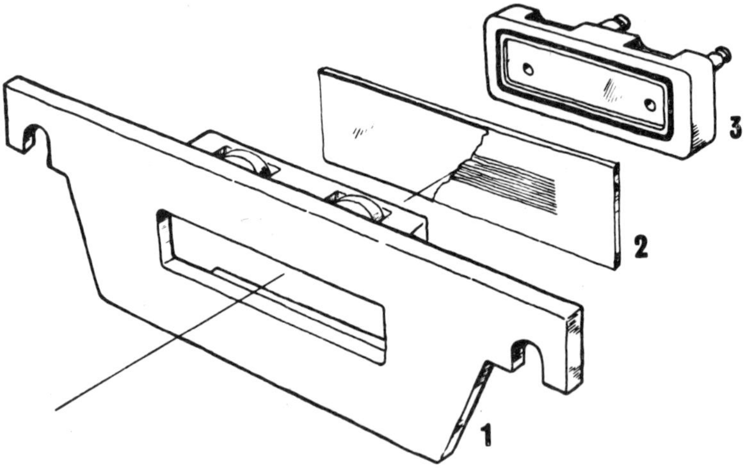

Fig. 5. External Target for Solid Materials:
1. Target mounting plate
2. Target plate carrying target material with target foil shown cut away
3. Target cooling block with 'O' ring

available; for instance, a thin metal foil (such as indium for ^{117}Sb production) may simply be clamped to the target plate, or a salt (such as NaBr for ^{81}Rb production) may be pressed firmly or melted into grooves cut in the plate. A typical target of this type is shown in Fig. 5, which also shows the method of applying water cooling to the back of the target plate.

Variations on the standard pattern of solid external targets are used in preparing radioactive gases incorporating ^{11}C or ^{13}N. Carbon-11 is produced by the reactions $^{10}B(d, n)^{11}C$ and $^{11}B(d, 2n)^{11}C$; the target material is B_2O_3, which is bombarded with deuterons of ~14 MeV in the target system shown in Fig. 6. A layer of B_2O_3 1–2 mm thick, is melted on to the wedge. The deuteron beam passes through the 0·050 mm aluminium window and strikes the surface of the B_2O_3 at grazing incidence. With a defocused beam of ~40 μA the power density is 10–20 watts cm^{-2} on the surface of the wedge, which is deliberately in poor thermal contact with the surrounding box.

This power density is sufficient to melt the B_2O_3, releasing ^{11}C-labelled gases

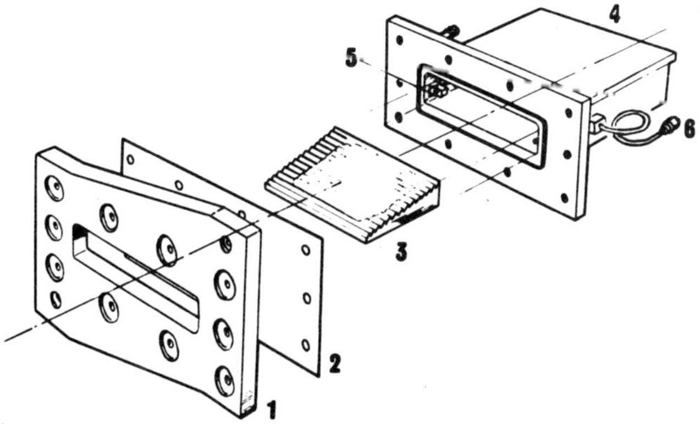

Fig. 6. External Target for ^{11}C *Production:*
1. Target mounting plate (water cooling not shown)
2. Foil window
3. Brass wedge supporting B_2O_3
4. Gas-tight target box
5. Wedge locating pins
6. Gas connections

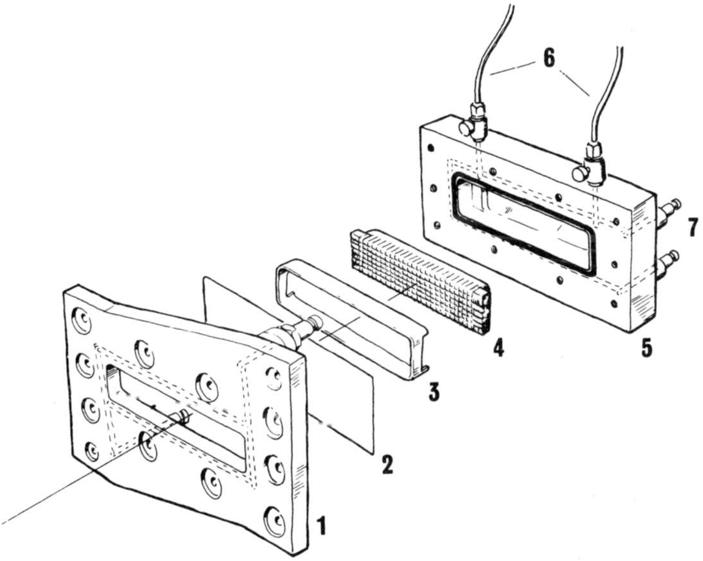

Fig. 7. External Target for ^{13}N *Production:*
1. Target mounting plate
2. Foil window
3. Spacer
4. Graphite block
5. Gas-tight target box
6. Gas connections
7. Cooling water connections

into a stream of carrier gas passing through the box at ~50 ml min^{-1}. When pure hydrogen is used as carrier gas ^{11}C appears as ^{11}CO and $^{11}CH_4$; when helium containing a little CO is used ^{11}CO, $^{11}CO_2$ and $^{13}N_2$ are produced.[6] The volume of the target box is kept small (~200 cm^3) to maximise the specific activity of the effluent gas.

The normal working pressure is about 0·1 kg cm^{-2} above atmospheric pressure. The ^{11}C target is used on an open-circuit system, a typical yield being 0·1 mCi ml^{-1} measured in a 100 ml bulb in an ionisation chamber approximately 100 m from the target. After about 12 h bombardment the B_2O_3 migrates from the area of the beam strike causing a loss of yield, necessitating re-coating of the wedge.

Nitrogen-13 is produced by the $^{12}C(d, n)^{13}N$ reaction in the target system shown in Fig. 7. Graphite is used as the target material when very high specific activity $^{13}N_2$ is required for making solutions of $^{13}N_2$ for physiological research.[7] Activated charcoal, the target material previously reported,[8] has proved to be unsuitable for this purpose owing to difficulties encountered in removing the

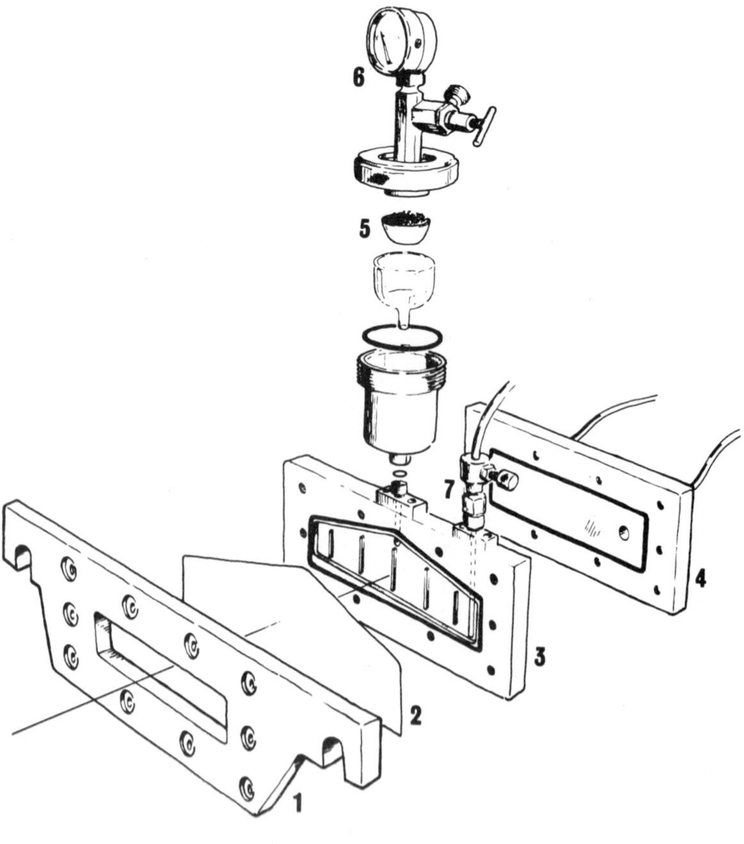

Fig. 8. External Target for ^{18}F Production
1. Target mounting plate
2. Foil window
3. Target plate with target water cavity
4. Target cooling block with 'O' ring
5. Catalyst in container
6. Pressure gauge
7. Tube and valve for filling/emptying cavity with target water

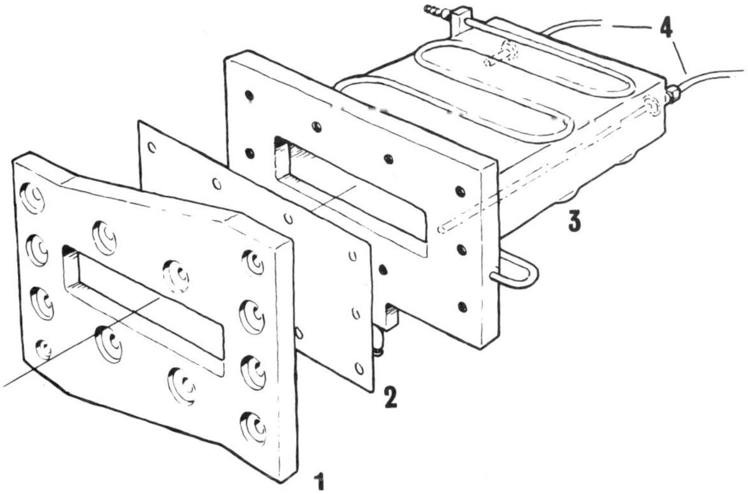

Fig. 9. External Target for ^{15}O Production
1. Target mounting plate
2. Foil window
3. Water-cooled gas-tight target box
4. Gas connections

adsorbed gases before irradiation. In an atmosphere of He, Ar or H_2 only small amounts of $^{13}N_2$ are released from the irradiated graphite. However, in the presence of carbon dioxide, the graphite (which is raised to a temperature above 1450°C by the beam power, typically about 100 W cm^{-2} at 50 μA) is eroded continuously as the following reaction occurs: $C + CO_2 \longrightarrow 2CO$. The purpose of cutting the graphite target material into a matrix, as depicted in Fig. 7, is to minimise the heat loss from the area struck by the beam. In this way the release of nitrogen-13 from the target has been greatly increased. The useful life of such a target is about 8 h.

This continuous erosion releases $^{13}N_2$ previously trapped in the graphite lattice. The yield from this target system is quite sensitive to power density; at high power densities erosion is more rapid. The target effluent, a mixture of CO_2 and CO of variable composition, is normally passed over heated copper oxide and the resulting CO_2 removed batch-wise. This leaves $^{13}N_2$ in the accumulated residual permanent gases found in the system. Typical specific activities achieved are ~7mCi ml^{-1}, which on mixing with physiological saline yields solutions of $^{13}N_2$ containing up to 150 μCi ml^{-1}. The major problem in using this type of target system has been to find a beam entry window which would retain its strength at elevated temperatures. Stainless steel foil (EN58B) 0·025 mm thick has proved to be very reliable at beam currents up to 70 μA.

(c) Liquids. In practice, the only liquid target material at Hammersmith is water, which is bombarded by the α-particle beam to produce ^{18}F through the reaction ^{16}O (α, pn). The target system is shown in Fig. 8. Its construction is similar to that for external solid targets in that a layer of water, thick enough to stop 30 MeV α-particles, is sandwiched between a thin air-cooled titanium foil (through which the beam enters) and a specially shaped water-cooled titanium target plate.[9]

The target is sealed, and during bombardment the pressure builds up to about 1·1 kg cm^{-2} above atmospheric due to the formation of radiolytic H_2 and O_2. This increases the boiling point of the target water, and thereby helps to retain it in the beam so that good yields of ^{18}F are achieved (normally in excess of 1 mCi μA^{-1} h^{-1} for a 1 h bombardment). A catalyst in the chamber above the level of the water recombines the radiolytic gases which would otherwise build up sufficient pressure to burst the target foil. This target will withstand beam currents up to 80–100 μA without risk, provided the beam is spread over an area not less than about 5 cm^2.

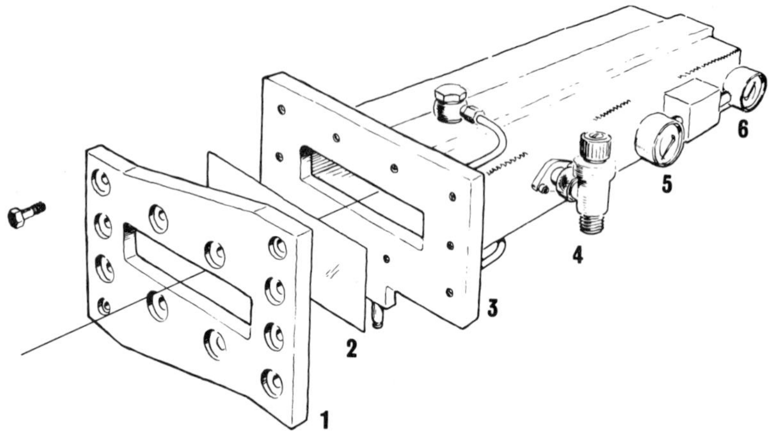

Fig. 10. External Target for ^{85m}Kr Production
1. Target mounting plate
2. Foil window
3. Water-cooled gas-tight target box
4. Gas connection
5. Pressure gauge
6. Vacuum gauge

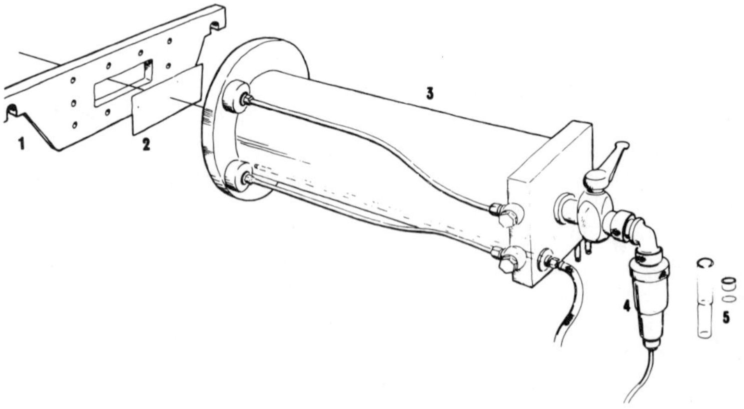

Fig. 11. External Target for ^{43}K Production
1. Target mounting plate
2. Foil window
3. Water-cooled gas-tight target box
4. Filter assembly
5. Sintered glass filter stick, glass-fibre filter paper and glass retaining ring

(d) Gases. Essentially, such a target is simply a water-cooled metal box filled with gas, and having a thin foil window for beam entry. The thickness of this window, its material, the target box dimensions and its operating pressure are all interdependent as discussed below.

Three such targets are in regular use. The first is shown in Fig. 9 and is used to prepare molecular oxygen labelled with ^{15}O. This is produced by the reaction ^{14}N (d, n) ^{15}O, using nitrogen as the target gas, to which 4% O_2 is added. The oxygen acts as a carrier for the ^{15}O; if it is absent the yield of ^{15}O as O_2 is very small. The target beam entry window is 1 mm magnesium, which reduces the beam energy from 15 to 5 MeV. This has the effect of reducing the contaminants otherwise produced by the reactions ^{14}N (d, nα) ^{11}C and ^{16}O (d, nα) ^{13}N.[8]

The length of the box is sufficient to stop the 5 MeV deuteron beam with the target gas at about 0·5 kg cm^{-2} above atmospheric pressure. Using 35 μA of 5 MeV deuterons and a target gas flow rate of 0·7 1 min^{-1}, the yield of $^{15}O_2$ is typically 120 mCi 1^{-1} measured in a 500 ml bulb in an ionisation chamber 90 m from the target.

A target similar in design (Fig. 10) is used to produce ^{85m}Kr by the ^{84}Kr(d, p) reaction. The 45 cm long target box is filled with krypton to a pressure of 0·7 kg cm^{-2} above atmospheric, ideally sufficient to stop the deuteron beam. This pressure rises to about 1·75 kg cm^{-2} during bombardment at 30 μA. The target box is therefore rigidly constructed, and is fitted with a 0·025 mm titanium foil window which readily withstands the pressure and temperature differential. After bombardemnt, krypton is recovered from the target by sorption on molecular sieve in a trap cooled by liquid nitrogen.

Finally, the target used for the production of ^{43}K is shown in Fig. 11. Argon is the target gas, and when this is bombarded in the α-particle beam ^{43}K results from the ^{40}Ar(α, p) reaction. The target box is filled to a pressure of 0·7 kg cm^{-2} above atmospheric, and is fitted with a foil window which reduces the incident beam energy to 15 MeV, which is approximately the threshold for the ^{40}Ar(α, pn) ^{42}K reaction. This foil window consists of 0·025 mm titanium which, as in the krypton target, has the required mechanical strength, backed by a 0·2 mm aluminium foil, to degrade the beam energy.[10]

During bombardment, the argon is recirculated through the target box at about 100 1 min^{-1} by oil-free diaphragm pumps. On leaving the target box the gas passes through a glass-fibre filter paper supported on a coarse sintered glass disc, which traps more than 70% of the ^{43}K produced. At the end of bombardment the activity is readily recovered by washing the filter with a few ml of dilute hydrochloric acid.

REFERENCES

1. Gallop, J. W., Vonberg, D. D., Post, R. J., Powell, W. B., Sharp, J., and Waterton, P. J., *Proc. I.E.E.* **104 B,** 452, (1957).
2. Vonberg, D. D. and Fowler, J. F., *Nature* **198,** 827, (1963).
3. Clark, J. C., Matthews, C. M. E., Silvester, D. J., and Vonberg, D. D., *Nucleonics* **25,** 54, (1967).
4. Glass, H. I. and Silvester, D. J., submitted to *Brit. J. Radiol.* (1969).
5. Silvester, D. J., Sugden, J., and Watson, I. A., *Radiochemical and Radioanalytical Letters,* in press (1969).
6. Clark, J. C. and Buckingham, P. D., submitted to *Radiochemical and Radioanalytical Letters* (1969).
7. Clark, J. C. and Buckingham, P. D., to be published.
8. Buckingham, P. D. and Forse, D. J., *Int. J. Appl. Radn. and Isotopes* **14,** 439, (1963).
9. Clark, J. C..and Silvester, D. J., *Int. J. Appl. Radn. and Isotopes* **17,** 151, (1966).
10. Clark, J. C., submitted to *Radiochemical and Radioanalytical Letters* (1969).

SESSION 6

Radiation Damage

Chairman: Dr. C. B. Amphlett

Cyclotron bombardment to simulate fast neutron damage in reactor materials

R. S. Nelson

Solid State Division, A.E.R.E., Harwell, Berks., England

ABSTRACT

The basic physical considerations underlying the production of radiation damage by both energetic charged particles (>1 MeV) and fast neutrons are discussed. The relevant formulae are derived in order to make direct comparisons between the primary recoil spectra and the total damage produced by such irradiations. As an example the damage produced in an iron sample is calculated in detail. Finally, the result of a typical simulation experiment is given to illustrate the theoretical conclusions.

INTRODUCTION

Particle accelerators, and in particular cyclotrons, have been used to study radiation damage in solids for many years. The early experiments of Eggleston,[1] and of Cooper *et al*,[2] paved the way for more sophisticated experiments, e.g. those of Barnes and Mazey.[3] These experiments were concerned with the basic nature of radiation damage and defects in metals; however, cyclotrons are now being used to simulate fast neutron damage in reactor materials to solve technologically-important problems (Mazey *et al*[4]).

In this paper we will outline the fundamental considerations underlying the production of radiation damage during high-energy charged-particle bombardment of solids, and then make some comparisons with a typical fast neutron irradiation. Finally, the results of a specific simulation experiment will be used to illustrate the usefulness of such an approach.

DAMAGE PRODUCTION WITH CHARGED PARTICLES

Energetic charged particles (i.e. >1 MeV) interact with the atoms of a solid in a rather complicated way. It has been the usual practice to separate collisions into two groups, viz. elastic collisions, where the total kinetic energy is conserved, and inelastic collisions which result in ionisation and electron excitation.

The energy loss per unit length, called the 'stopping power', can then be calculated for each process and finally combined and integrated to give the total range of the particles in the solid. In general, for energies above 1 MeV, and especially for light particles such as protons and carbon ions, inelastic losses dominate and almost exclusively determine the particle range. However, in metals atomic displacement can only result from elastic collisions, and it is these which will concern us in the present context.

We must first discuss the detailed nature of the interaction between the moving particle and the atoms of the solid. At the high relative velocities under consideration we can assume that the interaction results from the Coulomb repulsion between essentially unscreened nuclei. Under these conditions analytical treatment is relatively simple and leads to the well-known Rutherford scattering laws. The differential cross-section for transferring energy E_t to $E_t + \mathrm{d}E_t$ in such a Rutherford collision is

$$\mathrm{d}\sigma_p(E_t) = 4\pi a_B^2 Z_1^2 Z_2^2 \frac{M_1}{M_2} \frac{E_R^2}{E} \frac{\mathrm{d}E_t}{E_t^2} \tag{1}$$

where a_B is the Bohr radius for hydrogen, Z_1, Z_2, and M_1, M_2 refer to the atomic numbers and masses of the incident and struck particles respectively, E_R is the Rydberg energy, and E is the energy of the moving particle. The average number of recoils at energy E_t in the interval $\mathrm{d}E_t$ per particle of energy E is then

$$n_p(E_t)\mathrm{d}E_t = \mathrm{d}\sigma_p(E_t)S_p(E) \tag{2}$$

where $S_p(E)$ is the number of particles with energy E per unit area per particle, which in the case of a monoenergetic parallel beam of charged particles is numerically equal to unity. The total number of displacements per atom produced by N_p particles of energy E is then simply given by

$$N_d(E) = N_p \int_{E_d}^{E_t(\mathrm{max})} \mathrm{d}\sigma_p(E_t)S_p(E)n_d(E_t) \tag{3}$$

where E_d is the displacement energy, $E_t(\mathrm{max}) = 4M_1M_2E/(M_1 + M_2)^2$, and $n_d(E_t)$ is the number of displacements per primary recoil of energy E_t. In general, there is no simple expression for $n_d(E_t)$ and one has to make use of curves calculated for a specific case; for example, $n_d(E_t)$ has been plotted as a function of E_t for iron in Fig. 1 (Nelson[5]). However, in rough calculations it may suffice to use the old Kinchin-Pease[6] result $n_d(E_t) = E_t/2E_d$.

The depth distribution of radiation damage produced by a beam of charged particles can be determined by converting energy to depth from a knowledge of the stopping power $(\mathrm{d}E/\mathrm{d}x)_E$. For instance, the depth below the surface, x, corresponding to an energy E_x is given by

$$x(E_x) = \int_{E_x}^{E_i} \left[1/(\mathrm{d}E/\mathrm{d}x)_E \right] \mathrm{d}E \tag{4}$$

where E_i is the initial beam energy at the target surface. The number of displacements per atom at this depth is simply found by inserting E_x for E in Eqn (3)

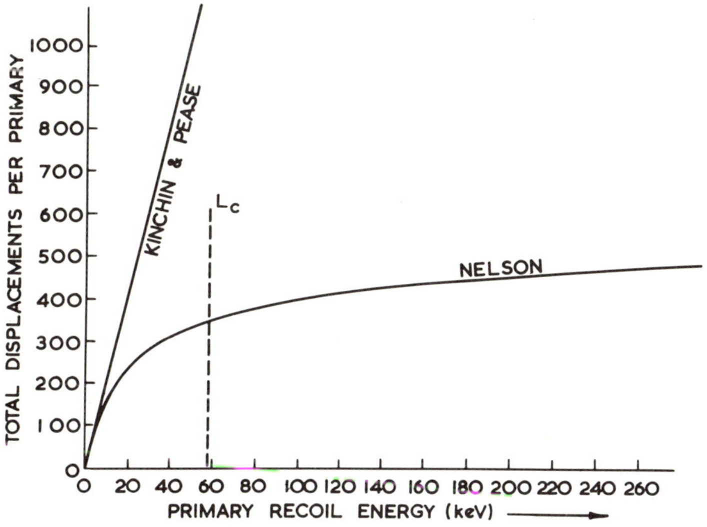

Fig. 1. The average total number of displacements per primary recoil in iron plotted as a function of primary recoil energy, where L_c *is the Kinchin and Pease limiting energy for ionisation*[6] *(from Nelson*[5]*)*

Further, the average total number produced per atom by N_p particles in slowing down from E_i to E_f is

$$N_d(E_i, E_f) = N_p \int_{E_f}^{E_i} \int_{E_d}^{E_t(\max)} \mathrm{d}\sigma_p(E_t) S_p(E) n_d(E_t) \mathrm{d}E \tag{5}$$

DAMAGE PRODUCTION WITH FAST NEUTRONS

Fast neutrons have no charge and collisions with lattice atoms are generally perfectly elastic. The scattering is assumed to be isotropic inasmuch as every possible energy transfer from zero to the maximum is equally probable, in other words the differential cross-section is simply

$$\mathrm{d}\sigma_n(E_t) = \frac{\sigma_n \mathrm{d}E_t}{E_t(\max)} \tag{6}$$

where σ_n is the total cross-section. The main difficulty in calculating the number of displacements is that the neutron energy spectrum $S_n(E_n)$ must be known. Assuming this, the number of recoils at energy E_t in the interval $\mathrm{d}E_t$ per atom per neutron is then

$$n_n(E_t)\mathrm{d}E_t = \int_{\frac{M_2 E_t}{4}}^{E_n(\max)} \mathrm{d}\sigma_n(E_t) S_n(E_n) \mathrm{d}E_n \Bigg/ \int_0^{E_n(\max)} S_n(E_n)\mathrm{d}E_n \tag{7}$$

and as $\quad E_t(\max) = 4E_n/M_2$

$$n_n(E_t)\mathrm{d}E_t = \frac{\sigma_n M_2}{4}\,\mathrm{d}E_t \int_{\frac{M_2 E_t}{4}}^{E_n(\max)} \frac{S_n(E_n)\mathrm{d}E_n}{E_n} \Bigg/ \int_0^{E_n(\max)} S_n(E_n)\mathrm{d}E_n \qquad (8)$$

The total number of displacements per atom produced by N_n neutrons is then

$$N_d = N_n \int_{E_d}^{E_t(\max)} n_n(E_t) n_p(E_t)\mathrm{d}E_t \qquad (9)$$

COMPARISON BETWEEN CHARGED PARTICLE AND NEUTRON IRRADIATION

For the purposes of comparison we must discuss a specific case which in this instance we take to be iron. The first comparison is illustrated in Fig. 2, which shows the recoil spectra per atom per MeV per particle for both cases,

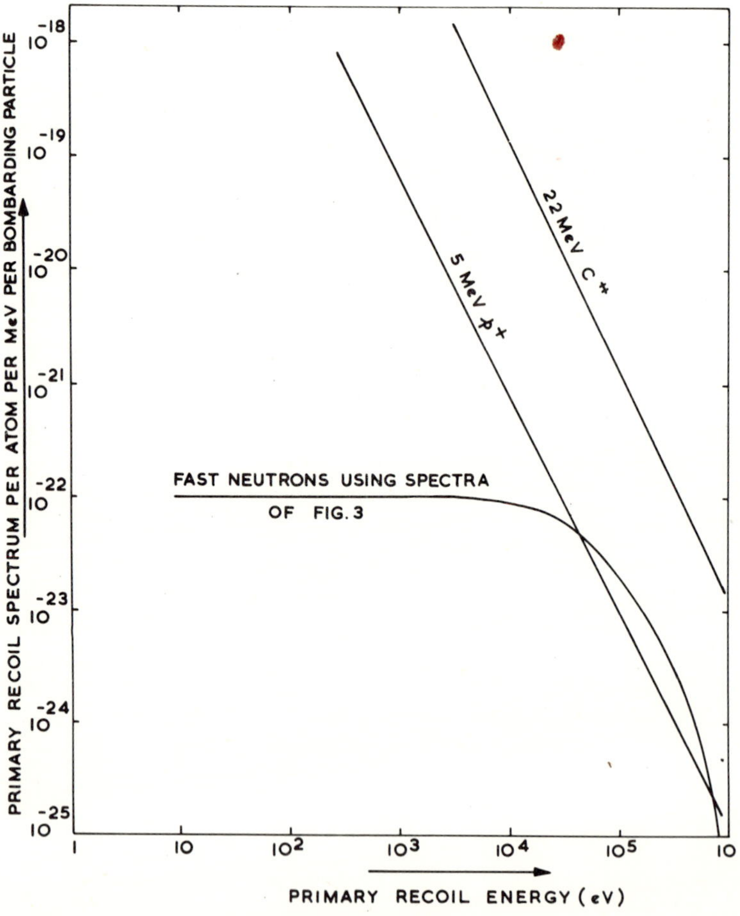

Fig. 2. Primary recoil spectrum for 22 MeV C^{++} *bombardment, 5 MeV proton bombardment and fast neutron bombardment*

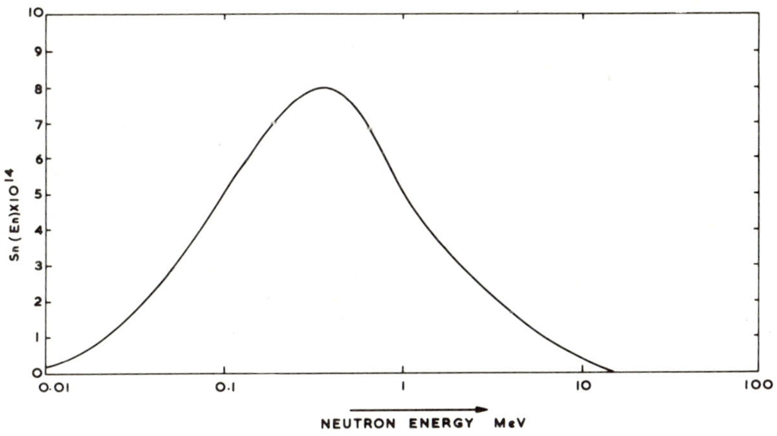

Fig. 3. A typical fast neutron spectrum

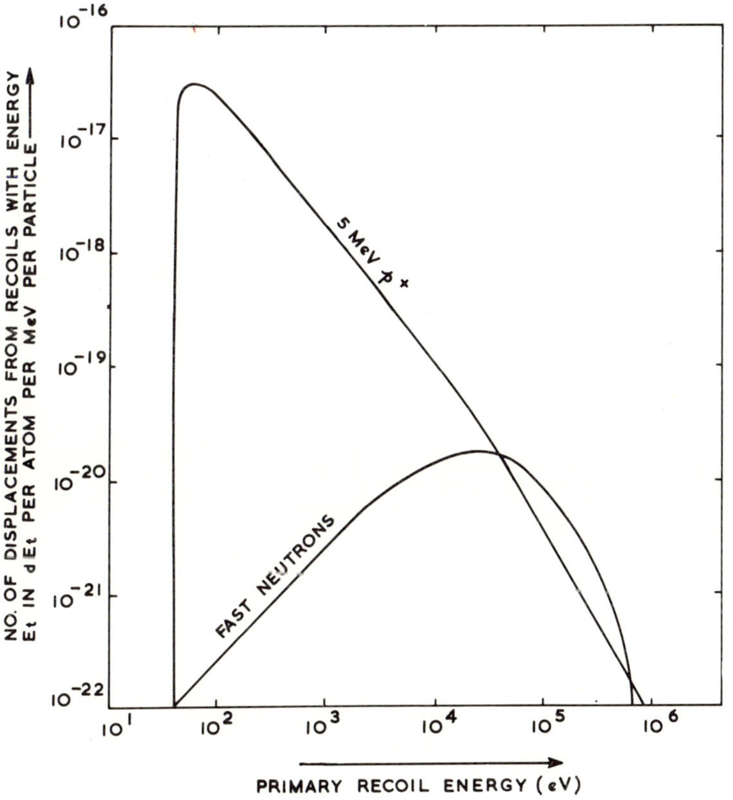

Fig. 4. Number of displacements produced by primary recoils with energy E_t in the interval dE_t per atom per MeV per particle plotted as a function of primary recoil energy for both 5 MeV protons and fast neutrons

as calculated from Eqns (2) and (8) respectively. A neutron spectrum typical of a fast reactor environment, as shown in Fig. 3, was used. From these curves it is quite clear that for recoil energies above ~10 keV the recoil spectra are qualitatively not too different; however, below 10 keV the charged-particle bombardment produces relatively many orders of magnitude more recoils at low energies. This is only to be expected, due to the preference for low-energy recoils resulting from Rutherford collisions.

Next, the number of displacements resulting from recoils at energy E_t in the interval dE_t per atom per MeV per incident particle is compared in Fig. 4. By summation under the curves in Fig. 4, the total number of displacements

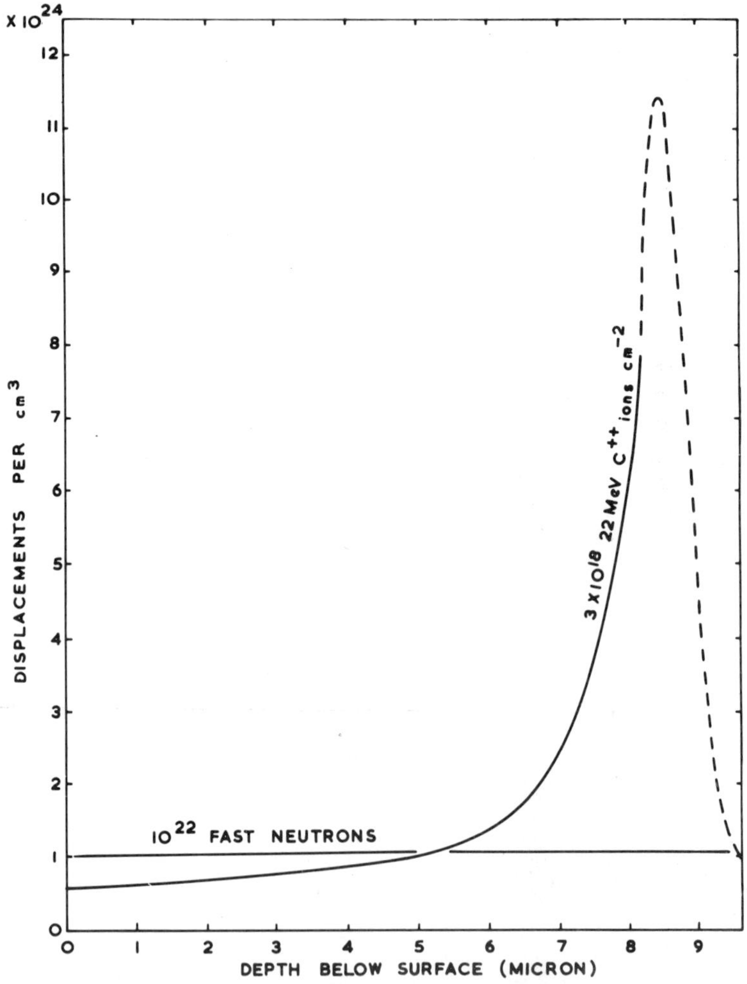

Fig. 5. The total number of displacements produced by 3×10^{18} 22 MeV C^{++} ions cm^{-2} as a function of depth below the surface. For comparison the horizontal line indicates the number of displacements produced by 10^{22} fast neutrons

per atom per ion or per neutron can be estimated for each case. Finally, the total number of displacements as a function of depth is shown in Fig. 5 for a 22 MeV C^{++} ion bombardment of iron to a dose which corresponds to a typical neutron irradiation of 10^{22}n cm^{-2}. The deepest part of the curve–shown dotted

–corresponds to the energy region where the Rutherford collision approximation is no longer valid and electron screening reduces the interaction. The curve shown is considered a reasonable approximation, and includes a correction for straggling. From such a plot we can see that during the first half of the particle's range the damage increases by only a factor of ~1·2, due to the slow variation in stopping power at high energies. It is in this first half of the specimen therefore, where the damage is approximately constant, that charged-particle damage can most usefully simulate that produced by fast neutrons. The average number of displacements produced in such a region can be calculated by using Eqn (5) and this can be compared with the number given by Eqn (9), in order to relate the charged particle dose (N_p) to the total neutron dose (N_n).

TYPICAL EXPERIMENT

It is well established that voids are formed in stainless steel after prolonged irradiation with fast neutrons at certain elevated temperatures.[7] However, the precise mechanism by which such voids are formed has been open to speculation, particularly from the viewpoint of the importance of the helium which is inevitably produced as a consequence of transmutation. In such reactor irradiations it is virtually impossible to isolate the individual effects of radiation damage and of helium. Furthermore, it is difficult and time-consuming to perform irradiations to high neutron doses at different dose-rates, and at accurately-controlled temperatures. However, in principle these limitations can be overcome by simulating the effects using charged particle accelerators.

The first requirement is to produce a uniform known concentration of helium within the specimen,[8] next, the sample must be irradiated with charged particles to produce a density of radiation damage equivalent to say 10^{22} neutrons cm^{-2} For this specific experiment 22 MeV C^{++} ions were chosen as a compromise between a reasonable penetration (~8 μm) and a sufficiently high damage cross-section. After irradiation, the samples were selectively electropolished and examined by transmission electron-microscopy. Fig. 6 is a photograph of a typical section, and clearly shows a high density of voids comparable to that observed after neutron irradiation.

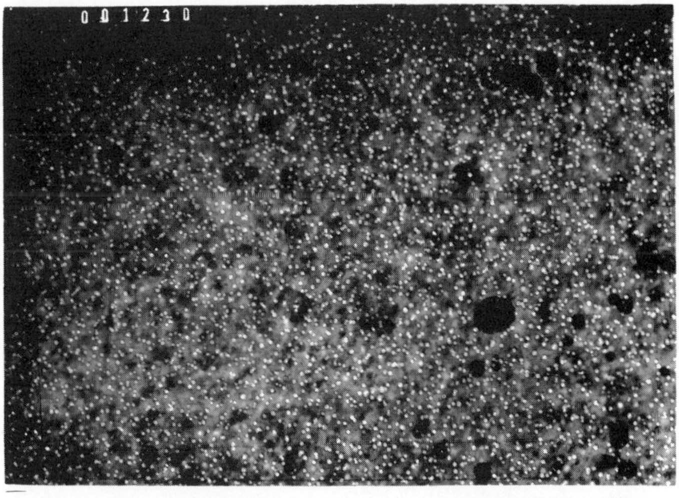

Fig. 6. A typical electron micrograph of a steel specimen irradiated with 10^{18} 22 MeV C^{++} ions at 550°C showing a high density of small voids

REFERENCES

1. Eggleston, R. R., *Acta Met.* 1, 679, (1963).
2. Cooper, H. G., Koehler, J. S., and Marx, J. W., *Phys. Rev.* 97, 599, (1965).
3. Barnes, R. S. and Mazey, D. J., *Phil. Mag.* 5, 1247, (1960).
4. Mazey, D. J., Nelson, R. S., Worth, J. H., and Clark, P. A., to be published (1969).
5. Nelson, R. S., *A.E.R.E.* R.6092 (1969).
6. Kinchin, G. H. and Pease, R. S., *Rep. Prog. Phys.* 18, 1, (1955).
7. Cawthorne, C. and Fulton, E. J., *Nature* 216, 575, (1967).
8. Worth, J. H., Proc. Int. Conf. on the Use of Cyclotrons in Chemistry, Metallurgy, and Biology, (1969).

DISCUSSION

R. T. King (Oak Ridge National Laboratory): Do you feel that dose-rate effects are important in this type of experiment?

R. S. Nelson: This depends on the particular case. For example, in processes involving heterogeneous nucleation of defects it is most likely that the total number of displacements is the important factor. In homogeneous nucleation, however, dose rate plays an important role.

J. W. Blue (N.A.S.A.): The discrepancy between neutron-produced displacement numbers and the calculations for 5 MeV protons in the low-energy region, i.e. low-energy displacements, is not as serious as is shown. This is because any reactor situation also includes γ-rays, and the Compton electrons from this source produce many low-energy displacements.

R. S. Nelson: The γ-ray component does in fact introduce a modification. However, as may be seen from Fig. 2, the number of recoils of 100 eV per particle produced by 5 MeV proton irradiation is nearly five orders of magnitude greater than for neutrons. I therefore think that to a first approximation the γ-ray component can be neglected.

R. L. Nelson (A.E.R.E.): I wish to comment on the diagram showing the considerable discrepancy at low energies between the primary recoil energy spectra from fast neutron and charged particle bombardment. In certain studies this discrepancy may not be as serious as might first appear. Although calculation of the total number of displacements requires the use of a displacement energy, E_d, of ~50 eV, the *effective* displacement energy may be much higher; this in turn puts less importance on the low-energy end of the recoil spectrum. We find in our work that *permanent* point defects are formed in collisions which involve energy transfer which is an order of magnitude greater than the normally-accepted value of E_d, although in other types of damage (e.g. disordering of alloys) the lower figure is relevant.

R. S. Nelson: Fig. 2 shows that even if E_d were 500 eV the number of recoils per 5 MeV proton per MeV is still 10^3 times that for neutrons. It is well established that in metals the displacement energy, defined as the energy required for permanent displacement stable against spontaneous recombination, is between 20 and 50 eV. It is only in some ionic materials that the displacement energy is greater, due to long-range Coulomb interaction between charged interstitials and vacancies.

K. Ehrlich (Karlsruhe): How did you modify the formula of Kinchin and Pease to calculate the number of displacements as a function of primary recoil energy? Did you substitute a kind of ionisation energy E_i for the transition energy E_t, and did you work with a fixed E_i (e.g. 50 keV for iron) or a variable E_i? In this case the displacement cross-section σ_d would be a function of E_i.

R. S. Nelson: Only that fraction of the recoil energy which is dissipated in elastic collisions is used. The component which is lost in inelastic processes can easily be determined from the work of Lindhard. In this approach elastic losses are treated rigorously and one does not simply assume a threshold for ionisation; for instance, it is well known that inelastic processes occur down to < 1 keV.

Techniques and apparatus for irradiating metallurgical specimens on the Harwell variable energy cyclotron

J. H. Worth

Metallurgy Division, A.E.R.E., Harwell, England

ABSTRACT

The Harwell Variable Energy Cyclotron has been used in a variety of metallurgical experiments. The build-up of nuclear reaction products that occurs in reactor structural and fuel materials has been simulated by implanting inert gas ions in metals and ceramics. Medium energy ions have been used to create radiation damage in many different materials to simulate fast neutron damage. Carbon ions have been implanted in molybdenum to study the effects of controlled amounts of interstitial impurities. Beams from the cyclotron will also be used to study the mechanism of ion channelling through single crystals.

The apparatus and techniques used for these experiments are discussed and methods of obtaining uniform concentrations of implanted ions throughout a volume are described. The factors governing the choice of beams for radiation damage experiments are also discussed.

INTRODUCTION

The Harwell Variable Energy Cyclotron (V.E.C.) was built to meet the needs of radiation chemists and reactor metallurgists,[1] who required an intense source of ions ranging in mass from hydrogen to xenon and in energy up to $\simeq$50 MeV/amu. This paper is concerned with the use of such beams for materials research and with the equipment that has been built by the author and his colleagues and used on the V.E.C.

Types of experiments. Experiments can be broadly grouped into three classes:

1. Ion implantation, where atoms are introduced into the atomic lattice of the target material. These ions may be inert gas atoms, so that gas bubble formation, behaviour and effects on mechanical properties can be studied. They can also be ions of a solid element, e.g. carbon may be implanted in pure metals to study the effects of interstitial impurities; donor or acceptor

atoms can be implanted in semi-conductors to make semi-conducting devices.

2. Radiation damage, where high-energy charged particles are used as projectiles to displace atoms from their sites in the crystal lattice of the material being irradiated. The damage rates obtainable from the use of charged particles are several orders of magnitude greater than can be obtained in nuclear reactors using epithermal or fast neutron fluxes. An inherent problem in the use of charged particles for creating damage is the non-uniformity of the damage along the track of the particle, the most intense damage occurring at the end of the track. This problem is discussed more fully in a later section of this paper.
3. Ion channelling, where a fine, well-collimated beam of ions is used to study the way ions are transmitted through single crystals and how the transmission is affected by different orientations of the crystal relative to the beam. Analyses of the distribution of the particles in the transmitted beam can give information on the inter-atomic forces and the location of impurity atoms in the crystal lattice. Ion channelling is dealt with more fully by Nelson and Thompson.[2]

EXPERIMENT DESIGN CONSIDERATIONS

The major problem facing the designer of target equipment for use on an accelerator such as the V.E.C. is dissipating the power of the beam. The V.E.C. has a design capability of accelerating 100 μA beams of 50 MeV protons, equivalent to 5 kW. When it is realised that this power can be focused on to an area of 1 cm^2 the magnitude of the design problem becomes apparent. However, the more usual power level when using beams of He, C and Ne ions and protons for metallurgical experiments is ~50 W with power intensities of 50 to 80 W cm^{-2}.

Cooling and temperature control. The most reliable and effective way of keeping the irradiation temperature of materials below 100°C is by cooling the back of the specimen with a jet of water. Heat-transfer rates as high as 1 kW cm^{-2} for a 79°C temperature rise have been achieved by direct cooling. The limitations of this technique are set by the need of the material being irradiated to contain the coolant from the accelerator vacuum. This precludes the use of jet cooling on thin or brittle materials. Foils and ceramic materials therefore have to be clamped to a heat sink, and this poses the problem of conducting heat across an interface that is in a vacuum. It has been found that reproducible results can be obtained only if a bonding medium is introduced into the interface. Two suitable bonding materials have been found after many trials: for low temperatures, i.e. 100°C, heat-resisting aluminium paint has been found very satisfactory; above 100°C a silver cement consisting of colloidal silver in an organic solvent (methyl butyl ketone) has been used. This high-conductivity paint is manufactured by Acheson Colloids under the trade name 'Dag dispersion 915'.

Heat-transfer tests have shown that a clamped interface in vacuum has a heat-transfer coefficient of 0·6 W cm^{-2} $°C^{-1}$. Aluminium paint improves this figure by a factor of 5 and the silver cement by a factor of 2. Aluminium paint gives a more reproducible and reliable bond than the silver Dag, probably because it never fully dries out, thus maintaining a flexible bond that is able to accommodate differential thermal expansion.

A problem allied to heat transfer is temperature measurement and control. Temperature measurement is made difficult by the small dimensions of typical targets and the steep temperature gradients arising from the high power intensities. Experience has shown that a 1 mm diam., mineral-insulated, isolated junction, metal-sheathed thermocouple is the most reliable method of temperature measurement. The siting of the junction relative to the irradiated area must be done carefully if serious errors are to be avoided.

Temperature control can be achieved by electrical resistance heating. The technique is to arrange the rate of heat transfer from the specimen being irradiated to be greater than the beam power at the desired working temperature, i.e. to over-cool the target. Electrical power is then added to bring the specimen up to temperature. For good control the electrical power needs to be 5 to 10 times greater than the beam power.

The heat from targets must eventually end up in a heat sink. It aids the design of experiments if this heat sink is operated at a low temperature. The heat sink used for the targets described in this paper is a 40-gallon tank of high-purity water chilled to 5°C by a 1 h.p. refrigerator.

Cryogenic targets. Some types of irradiation damage experiments must be carried out at temperatures approaching absolute zero, and these are usually performed by immersing the target material in liquid helium or liquid nitrogen. The liquid gas is contained in an insulated container (cryostat) and the particle beam enters the container through a window to reach the immersed target material. The author has built and used a liquid nitrogen cryostat for irradiating 2·3 mm diam. discs of single-crystal molybdenum with 20 MeV protons.

There are two main problems in designing such targets. Firstly, there is the difficulty of determining the maximum permissible power loading of the specimen if the desired irradiation temperature is not to be exceeded. Secondly, if liquid oxygen is allowed to accumulate in the cryostat there will be a build-up of ozone with the consequent hazard of explosions. The latter problem was overcome by periodically dumping part of the liquid nitrogen charge to limit the mean dose received by the liquid to $1{\cdot}48 \times 10^8$ rad l^{-1}. Published data were used to determine the limiting rate of heat transfer: Flynn *et al*[3] reported that the maximum allowable heat transfer from a horizontal tube was $5{\cdot}0$ W cm^{-2}, which gave a surface temperature 10°K higher than the bulk liquid. However, since the physical arrangement of the target differed from that of Flynn, it was decided to limit the target rating to $2{\cdot}0$ W cm^{-2}. Good reproducibility of results (Maher *et al*[4]) has justified this approach.

Dosimetry. Closely associated with the design problems of cooling, temperature measurement and control is the requirement for accurate dosimetry. Beam currents are generally in the range 1×10^{-9} to 1×10^{-5} A, and are measured by instruments with input impedances of the order of 1 MΩ. Target potentials as high as 10 V with respect to earth can therefore be encountered and accurate current measurements will be possible only if the insulation resistance is >100 MΩ. Insulation of this order is difficult to achieve, especially when water is used as a coolant. It has been found necessary to use demineralised water for cooling purposes, with the water circulating in a closed loop containing an ion-exchange column, which maintains the conductivity of the water at $\leqslant 2\ \mu\Omega^{-1}$.

Galvanic corrosion currents generated by dissimilar metals in impure water frequently pose a problem with water-cooled target systems. Our targets are of aluminium, stainless steel, copper, and bronze. With low-conductivity water these metals appear to be compatible, although nevertheless it is advisable to electrically separate copper and aluminium.

When a target incorporates a heater and thermocouple, the requirements for good dosimetry make it essential that the heater core is well insulated from its sheath and that the thermocouple junction is isolated from its surroundings.

A major source of error arises because the ions striking the target excite electrons near the surface of the material. The energy of these electrons can be as high as several hundred eV and they leave the bombarded surface. If they can be collected and returned to the target material there will be no error in measuring the ion beam current. However, if they escape from the target the measured beam current will be anomalously high.

Some of these electrons can be collected, and hence the accuracy of beam current measurement improved, if the target is incorporated in a Faraday cup. However, to be effective the cup has to be very deep because the emitted electrons have a cosine distribution about the beam axis.[5] Because the ion beam is very much 'stiffer' than the electron beam the latter can be bent preferentially by a weak magnetic field (50–100 gauss). Use of such a field around a target ensures that the electrons are collected by the Faraday cup. This technique cannot be used if ferromagnetic material is present in the target system. In this case it is necessary to use a negatively-biased electrode in front of the bombarded surface. This method is the one preferred by the author as it is simple, it keeps the target voltage low with respect to earth and it screens the target from stray electrons. Experience with 22 MeV carbon ions on stainless steel has shown that 100 V bias suppresses ~90% of the electron emission; 200 V brings the indicated target current down to a plateau where there is very little change until the suppressor potential exceeds 1000 V.

DESCRIPTION OF EQUIPMENT CURRENTLY IN USE

A typical arrangement has a beam-measuring box in the beam line immediately in front of the target chamber, which is used to determine the position and quality of the beam. The beam measuring box contains two pairs of slits, a Faraday cup, and a quartz scintillator. All these items can be moved into and out of the beam by remote control from the control room of the cyclotron. The two slits are mutually perpendicular and their widths are adjustable. With both slits in the beam, a small window is formed which can be moved over the cross-section of the beam. The beam transmitted through the window is collected by the Faraday cup and is plotted against slit position on an X-Y recorder. The scintillator is viewed by a television camera and mirror.

Universal target chamber. Mounted immediately behind the measuring box is a chamber with collimators at either end. To the rear of the chamber is bolted the holder containing the material to be irradiated. In the sides of the chamber are two 2 in diam. holes. Connected to one of these holes is a by-pass vacuum line; the other is used to introduce an energy-absorbing wedge or the electron suppression electrode.

Foil and pellet target holder. This general purpose, water-cooled holder (Fig. 1) was designed to be changed quickly in case of high activities and to reduce the time needed to change specimens when performing multiple irradiations. The two variants of the design permit strong metal foils to be cooled directly by a water jet or fragile cylindrical pellets to be clamped against a water-cooled face. A feature of this holder is the ability to locate a window in front of the specimen, in which case the latter has its own atmosphere, which may be an inert gas or vacuum.

Tensile blanks holder. This target was designed to hold plates 1½ in × 2½ in from which were stamped tensile test specimens. The material was 0·020 in

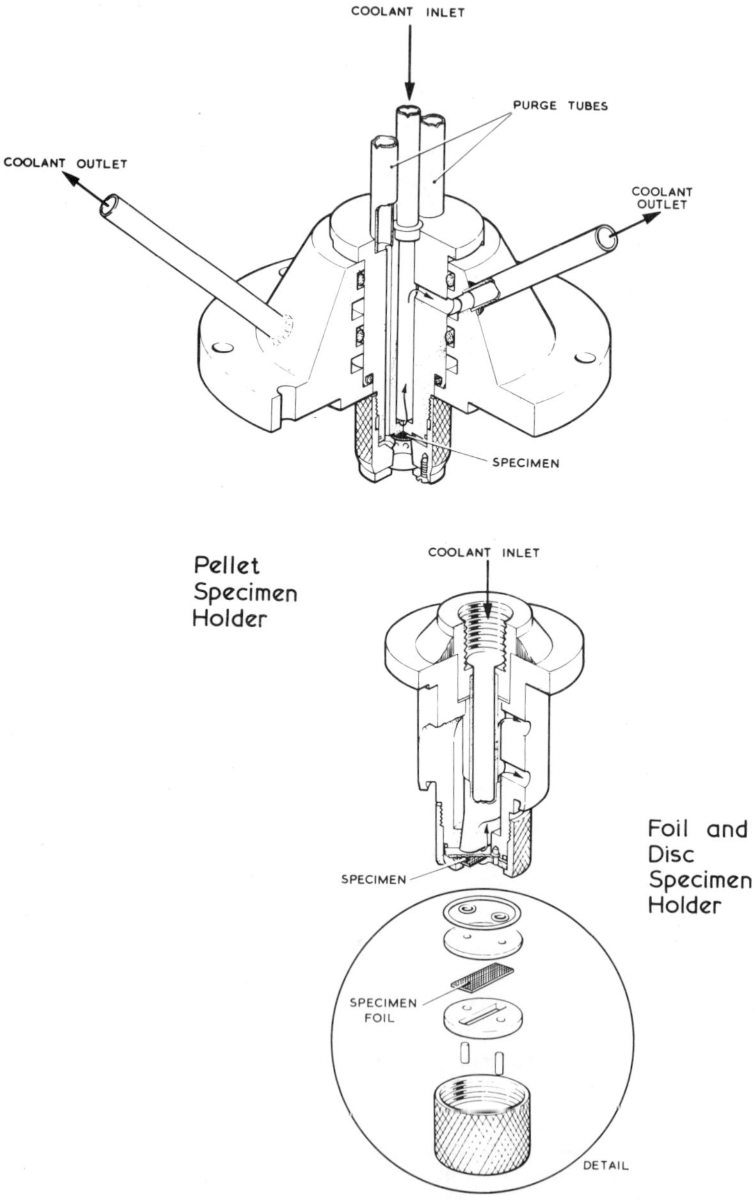

Fig. 1. General-purpose foil and pellet target holder

thick and could therefore be directly water-cooled. A feature of this target holder is the accurate registration of the plate relative to the final collimator (Fig. 2).

Liquid nitrogen cryostat. A drawing of this target is shown in Fig. 3. It was designed to irradiate molybdenum single crystals 1·5 in long, 0·125 in wide, and 0·001 in to 0·005 in thick. Alternatively it can irradiate 8 electron microscope

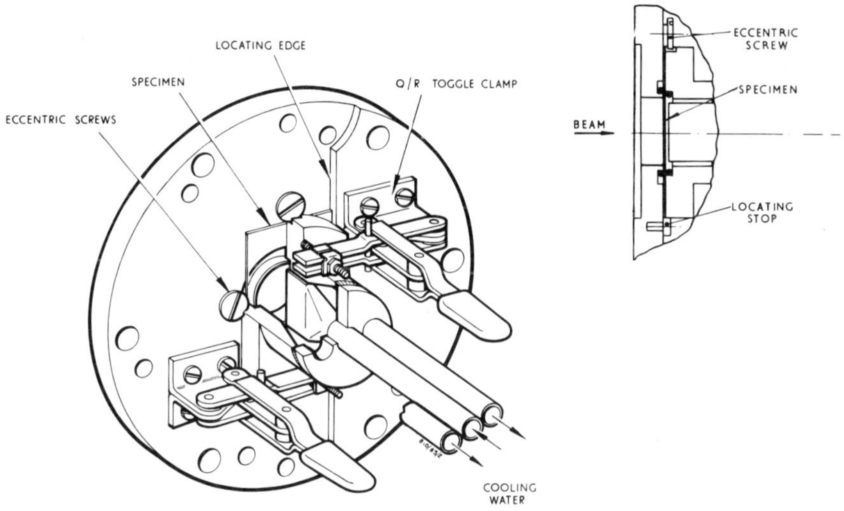

Fig. 2. Blank tensile specimen holder

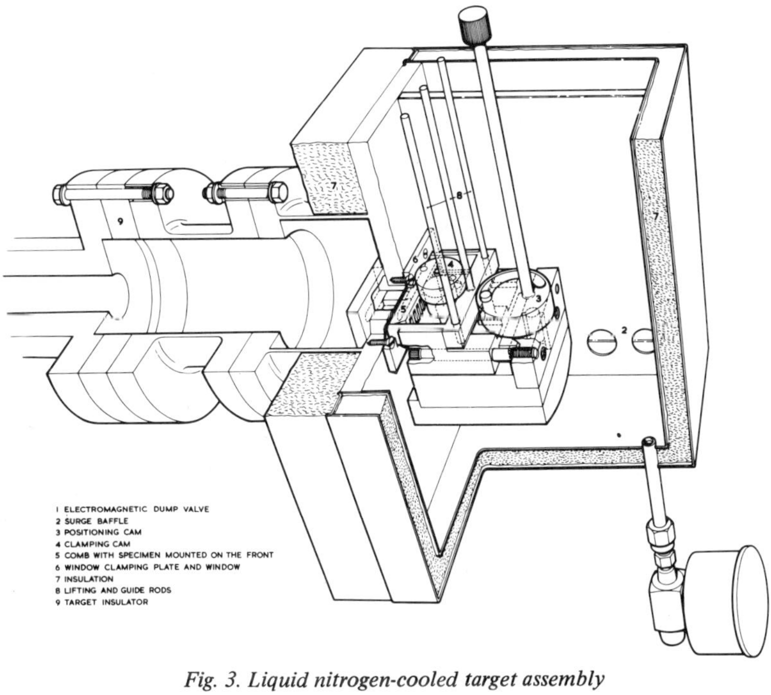

Fig. 3. Liquid nitrogen-cooled target assembly

size discs of 2·3 mm diam. It was designed for use with proton beams and electrons but could be used with beams of He^+ ions.

The level of liquid nitrogen is sensed by two gas bulbs operating pressure switches which control an electromagnetic valve in a transfer line connected to a pressurised Dewar. An electromagnetic dump valve in the bottom of the tank is used to periodically drain off a proportion of the liquid to prevent a dangerous build-up of ozone.

Controlled high-temperature target. This target is used to irradiate thin foils at temperatures up to 600°C. The heater is made from swaged coaxial, MgO-insulated cable, stainless-steel sheathed, that has been coiled and cast into a short copper cylinder. Heat is conducted away from the heater by a copper stem that is water cooled on its end. The foil to be irradiated is bonded to the end face of the heater by silver cement and clamped by a plate which has holes in it to define the irradiated areas. This plate carries a thermocouple in a groove, the thermocouple being held against the foil.

The signal from the thermocouple is taken to a recorder-controller which provides a three-term proportional control of the temperature. Heater power is controlled by a saturable reactor.

Ion channelling apparatus. This is in three parts:

(a) A collimator box containing a coarse and a fine collimator. This box is in the beam line close to where the beam enters the target room.

(b) A second pair of collimators and a chamber containing a 3-axis goniometer. This goniometer holds a target crystal in the path of the beam leaving the collimators. It is operated remotely by means of stepping motors and it is accurate to 0·01 degrees of arc.

(c) A camera chamber, designed to hold and expose X-ray film to the scattered beam leaving the crystal.

The size and separation of the collimators will give a beam with a divergence of less than 0·01°.

Energy absorbing wedge. To obtain uniform volume concentration of implanted ions it is necessary to modulate in some way the energy of the beam. A convenient way of achieving this modulation in beams of hydrogen and helium ions is to introduce a movable wedge into the beam close to the material to be bombarded. In general, because of the relationship between range and energy the wedge will be required to move non-linearly with respect to time. We have made several aluminium wedges, a typical arrangement being shown in Fig. 4. The wedges are moved by lead-screws driven by stepping motors, the drive pulses for which are obtained from a unit which reads timing instructions from punched paper tape. The latter is prepared on a computer by a programme which incorporates all the details of the experiment.[6] It can be seen from Fig. 4 that the wedge is double and that it has beam stopping blocks on either end, which enable the beam to be set up initially and also finally terminate the run.

Rotating target holder. Aluminium wedges can be used only for ions with atomic numbers up to 3. Above $Z = 3$ the high rate of energy loss calls for impossibly-thin wedges. However if the target material is rotated about an axis perpendicular to the axis of the beam the target material effectively becomes its own energy absorber. The angular movement required is not linear with time but the technique and apparatus used to move the energy-absorbing wedge can be

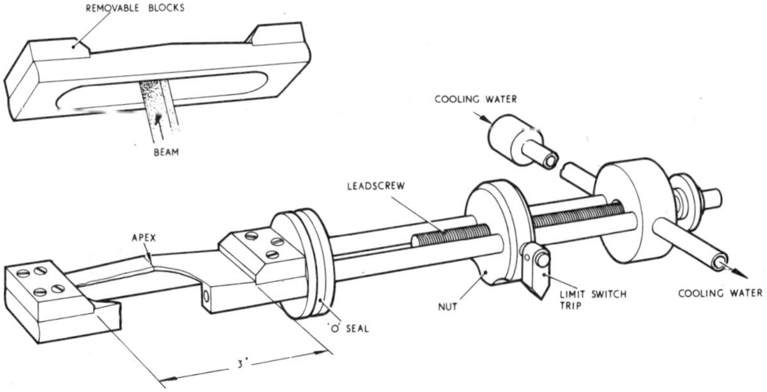

Fig. 4. Energy-absorbing wedge

adapted to rotate a target. The small angle necessary to implant ions near the surface means that a large fraction of the beam misses the target. This problem is discussed in more detail by Worth.[6]

Ancillary equipment. In addition to the equipment so far described there are the following pieces of apparatus.

Chilled water unit. This is a 40-gallon tank cooled by a 1 h.p. refrigerator. On the front of the tank is a circulating pump and a distribution manifold. Water from this unit is used to cool the target material and its electrical conductivity is kept low by an ion exchange column.

Beam scanning coils. Two sets of coils are fitted around the beam line before it enters the target room. The coil that deflects the beam vertically is placed around the beam line before the final pair of quadrupole focusing magnets and is excited by a 400 Hz sinusoidal current. The horizontally-deflecting coil is placed after the quadrupoles and is driven at 50 Hz. The 8:1 frequency ratio makes the beam describe a Lissajou figure over the irradiated area. Thus a uniform beam current distribution can be obtained over areas of 1–2 cm^2 without sacrificing as much of the beam current as would be necessary if the beam were merely defocused. The design of the coils and their positioning has been described by Barnfield.[7]

OPERATIONAL EXPERIENCE

The equipment described has been developed over a period of about four years. Experience has shown that the basic designs are correct, and high reliability has been achieved. Dosimetry has at times proved difficult and can be markedly influenced by surface contaminants. The use of an electron suppression electrode biased at –300 V has proved to be very effective.

Scanning the beam over the irradiated area by the use of coils is an efficient way to achieve uniform intensity; however, sinusoidal scanning coil currents are not ideal, since the dwell period at the peaks of the sine wave leads to current peaking at the ends of the sweep. A function generator and current amplifier is being built to overcome this difficulty.

The liquid nitrogen cryostat has been operated successfully for many runs and no trouble from ozone accumulation has been experienced.

Many metal foils have been irradiated at controlled temperatures to simulate fast neutron damage in reactor materials. With the techniques and materials currently in use temperatures are limited to ~650°C. Work has started to raise this limit to 800°C and eventually to 1200°C.

A large number of metal specimens has been implanted with hydrogen and helium, using the aluminium wedge to obtain a uniform concentration of ions within the material. Concentrations in the range 10^{-8} to 10^{-3} gas atoms/metal atom have been achieved.

RADIATION DAMAGE EXPERIMENTS

The variety of particles available from the V.E.C. makes it a very attractive tool for studying radiation damage effects. However, the choice of 'projectile' for any particular experiment is not always an easy one to make, as the following considerations will show. Firstly, an ion must be chosen such that if it comes to rest within the material its presence will not interfere with the phenomenon being investigated. For instance, it is pointless measuring the change in electrical resistance in semi-conductors as obtain a measure of damage if the semi-conductor is bombarded by ions that can serve as donors or acceptors. Therefore the best choice of ion is an ion of an element that is already present in the material, e.g. a carbon ion to bombard stainless steel. Secondly, an ion should be chosen that will cause the maximum number of displacements per unit track length. This requirement ensures that the maximum amount of displacement damage is created for the minimum amount of beam power and hence heating of the specimen.

Following Kinchin and Pease,[10] an energetic ion loses energy along the first part of its track in a solid mainly by electronic excitation and an occasional Rutherford collision with the nuclei of the target. Enough energy is transferred to the initially stationary atom for it to cause further collisions leading to displacement of the lattice atoms. The cross-section for these primary collisions is:

$$\sigma_d = \left(\frac{M_2}{M_1}\right) \frac{Z_1^2 Z_2^2 e^4}{E E_d}$$

where M_2 = mass of the ion
M_1 = mass of the material
Z_2 = atomic number of the ion
Z_1 = atomic number of the material
E = energy of the ion
E_d = energy required to create a vacancy-interstitial pair
e = electronic charge

It can be seen from this formula that the cross-section varies inversely as the energy. However, as the energy of the ion decreases it begins to pick up electrons and thus tends to become neutral. The effects of the electrons screening the nucleus are then felt, and the collisions are no longer Rutherford in character but tend towards the hard-sphere type. Kinchin and Pease considered

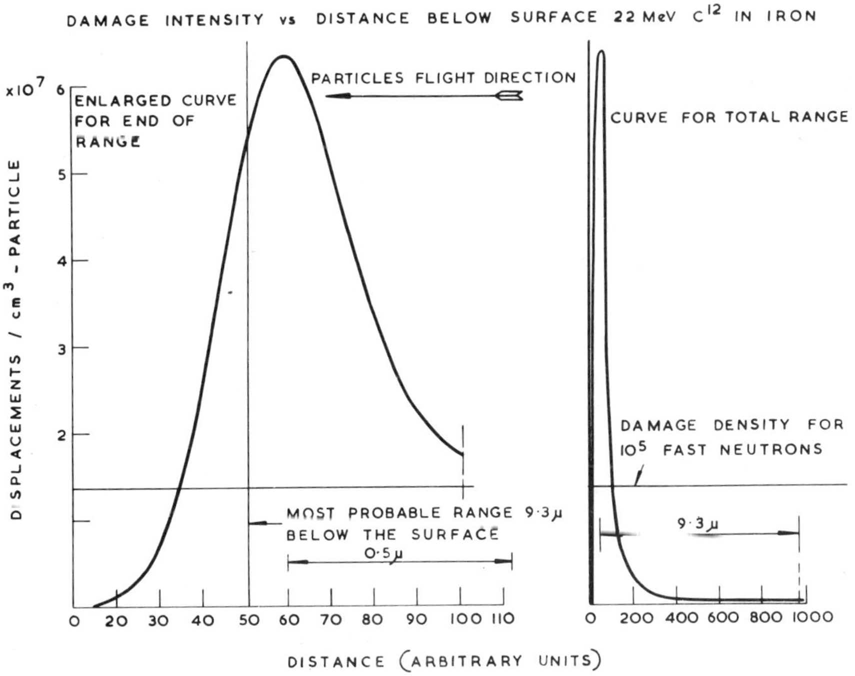

Fig. 5. Damage intensity vs distance below surface for 22 MeV ^{12}C *ions in iron*

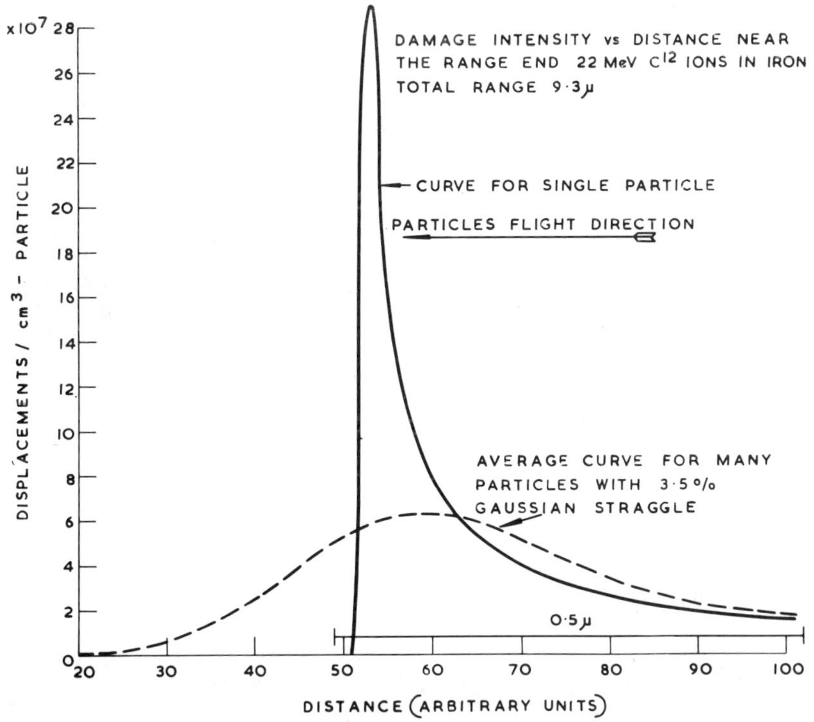

Fig. 6. Damage intensity vs distance near the range end, showing the effect of range straggle on the distribution of damage. 22 MeV ^{12}C *ions in iron*

that if the ion energy was greater than a certain value L_c energy was lost by electronic excitation, but that for energies less than L_c energy loss was by hard-sphere collisions; $L_c \cong M_1$ keV.

The author has combined the formulae of Kinchin and Pease with the range-energy formulae of Lindhard *et al*[8] to give a measure of the damage intensity along the track of the particle. Fig. 5 shows how the damage intensity varies along the track of a carbon ion in iron. The basic curve of damage intensity vs distance is drastically altered in shape when straggling effects are taken into consideration (Fig. 6). The modified curve is for straggling appropriate to a 22 MeV carbon ion and the straggling is assumed to be Gaussian in character; the percentage straggling was evaluated from the data of Caldwell.[9]

Consideration of the intensity of damage along a track will confirm whether the choice of particle from the criterion of chemical compatibility is suitable for a particular experiment. Although the cross-section for displacements is much lower for low atomic weight ions, this is sometimes more than compensated for by the higher particle fluxes available. This is the third factor governing the choice of particle for damage experiments.

In general the over-riding consideration is the chemical or physical effects of the ion when it comes to rest within the target lattice. Having thus chosen an element for the projectile, the next thing to consider is the rate of damage along the particle track to assess whether or not a sufficient volume of material will be sufficiently damaged for the particular experiment. The energy of the ion must then be chosen and a compromise made between beam current available for the various energies, the minimum acceptable penetration range and the ability of the target to dissipate the power of the beam.

CONCLUSION

A cyclotron such as the Harwell Variable Energy Cyclotron can be used in a variety of experiments associated with reactor metallurgy and solid state physics. The wide range of ions and energies available from the V.E.C. makes it extremely versatile, although the particle used for a particular experiment must be carefully chosen if the desired result is to be achieved. Target equipment is generally simple, but considerable experimentation is advisable to determine the true operating conditions; this is especially true for controlled-temperature apparatus.

ACKNOWLEDGEMENTS

The equipment described in this paper results from the efforts of many people. In particular the author wishes to thank Mr. P. A. E. Clark for his help in developing much of the apparatus. Most of the design work was done by Mr. J. White and Mr. A. R. Ball, whose help is gratefully acknowledged. Thanks are also due to Dr. R. S. Nelson and Mr. D. J. Mazey for many helpful discussions.

REFERENCES

1. Lawson, J. D., Report NIRL/R/85 (1965).
2. Nelson, R. S. and Thompson, M. W., *Phil. Mag.* **8**, 1977, (1965).
3. Flynn, T. M., Draper, J. W., and Roos, J. J., *Advances in Cryogenic Engineering* **7**, paper L-6, 539, (1961).
4. Clark, P. A., Campbell, D., Maher, D., and Worth, J. H. AERE, Report to be published.
5. Abbott, R. C. and Berry, H. W., *J. Appl. Phys.* **30**, 871, (1959).
6. Worth, J. H., Report A.E.R.E.-R5704 (1968).
7. Barnfield, R. W., Report A.E.R.E.-R5846.
8. Lindhard, J., Scharff, M., and Schiott, H., *Kgl. Danske Vid. Selsk, Mat-Fys. Medd.* **33**, 14 (1963).
9. Caldwell, D. A., *Phys. Rev.* **88**, 131, (1952).
10. Kinchin, G. G. and Pease, R. S., *Reports on Progress in Physics* **18**, 1, (1955).

DISCUSSION

W. L. Rautenbach (N.P.R.L., Pretoria): What are the advantages of a V.E.C. for metallurgical research compared to a tandem generator?
J. H. Worth: High-energy particles are required for deep penetration, so that damage zones are removed from possible surface effects. In general tandem generators do not produce sufficient currents of high-energy ions and, more important, they are limited in the variety of ions which they can accelerate.

Cyclotron simulation of neutron-transmutation produced gases in reactor cladding and structural materials

R. T. King

Metals and Ceramics Division, Oak Ridge National Laboratory, Oak Ridge, Tennessee

ABSTRACT

Cyclotrons have been used to simulate the nearly uniform concentrations of helium produced during neutron irradiation of metals by (n, α) reactions. Gas was injected into flat sheet specimens by degrading monoenergetic ion beams with either a stationary or a rotating energy moderator of continually-varying thickness. Target specimens were scanned in one and two dimensions, respectively, through these variable-energy beams, producing nearly uniform concentrations of gas through the specimens. Thermal monitoring and beam current detection systems were employed throughout the bombardments.

The effects of cyclotron-injected helium on the mechanical properties of Type 304 stainless steel, a potential cladding material for Liquid Metal Fast Breeder Reactors, have been investigated. Helium concentrations ranging from 0·083 to 20·0 atomic ppm severely reduced the ductility of Type 304 stainless steel in both creep and tensile tests over the ranges of strain rate and temperature for which the material fails intergranularly. It has been experimentally demonstrated, in agreement with theoretical calculations, that simply passing similar currents of α-particles completely through stainless steel does not produce sufficient displacement damage to cause significant changes in ductility. The resistance of a 0·2% Ti-Type 304L stainless steel to helium embrittlement has been demonstrated by cyclotron bombardment techniques.

The role of helium in low-ductility failures of aluminium reactor components at ~150°C after fast neutron irradiation has also been investigated. Tensile data for aluminium alloys containing 5·0 atomic ppm He have shown that helium generated by ^{27}Al(n, α) reactions reduces the elevated-temperature ductility of the 1100 and X8001 aluminium alloys, while the properties of the 6061 and 895 SAP alloys are not affected.

* Research sponsored by the U.S. Atomic Energy Commission under contract with the Union Carbide Corporation.

The gas-injection technique has thus been demonstrated to be a useful, relatively inexpensive tool in isolating particular neutron irradiation damage mechanisms.

INTRODUCTION

Recognition of the importance of neutron transmutation reactions has given rise to a new class of materials problems in reactor fuel cladding and structural elements. Although heavier elements are also produced, the hydrogen and helium created by (n, p) and (n, α) reactions are of great concern, since the irradiation-produced concentrations of these transmutation-generated gases can far exceed their solubility limits in metals, as well as the concentrations which can be introduced by normal metallurgical techniques. The distribution of gases produced in this manner depends upon the distribution of the parent atom species. Gas atoms produced by minor impurities which are strongly segregated at crystal-line imperfections or in second-phase particles are segregated at loci one recoil distance from the parent atoms; this is frequently the case for the reaction products of ^{10}B(n, α) and ^{6}Li(n, α) reactions, which have high thermal neutron cross-sections.[1] However, the cross-sections for fast neutron (n, p) and (n, α) reactions with most metals are relatively high.[2] Gas atoms produced by these matrix atom reactions are initially uniformly distributed throughout the irradiated metal, in the absence of segregation effects. In order to simulate such uniform gas distributions, and to study their effects on the properties of materials, we have developed cyclotron bombardment techniques for injecting nearly uniform concentrations of gas into metal specimens.

This work deals with an important problem in two materials in current reactor technology, the irradiation-induced loss of ductility (embrittlement) of stainless steel and aluminium alloys at high temperatures.

Stainless steel is the prime candidate fuel cladding material for the Liquid Metal Fast Breeder Reactor (LMFBR). It will be subjected to fast neutron fluences ($E > 1$ MeV) in excess of 1×10^{23} neutrons/cm^2 at service temperatures up to ~650°C while under stress. Current estimates[3] of the (n, α) and (n, p) cross-sections indicate that about 2×10^{-3} atom fraction (2000 atomic ppm) of hydrogen and 1×10^{-4} atom fraction (100 atomic ppm) of helium will be produced during irradiation. Because of its high diffusivity, much of the hydrogen may diffuse from the cladding in service; however, most of the helium is expected to remain in the cladding. Previous thermal neutron irradiations of stainless steels have demonstrated a loss of post-irradiation ductility in creep and tensile tests above ~500°C.[4,5] This embrittlement has been ascribed to helium generated from the ^{10}B(n, α) thermal neutron reaction, since boron is a normal impurity in stainless steel. Higgins and Roberts[6] have shown that the bombardment of flat stainless steel specimens with monoenergetic α-particles embrittled stainless steel, whereas a similar distribution of helium did not. Their technique produced a non-uniform concentration of helium at the centre of the specimens. The work described below is directed at demonstrating the effects of nearly uniform concentrations of helium on standard stainless steels, and provides evidence that the addition of 0·2% Ti improved the resistance of stainless steels to helium embrittlement. The latter class of alloys was developed at the Oak Ridge National Laboratory, and has been demonstrated to be superior to standard alloys under thermal neutron irradiation conditions.[7]

Neutron irradiation fluences between 10^{21} and 10^{22} neutrons/cm^2

($E > 1$ MeV) have recently been found to cause severe embrittlement of aluminium alloys in the range 100°–300°C.[8] This problem is significant because of the current extensive use of aluminium in test and isotope production reactors. Several possible causes have been suggested, including the hydrogen and helium from $^{27}Al(n, p)$ and $^{27}Al(n, \alpha)$ fast neutron reactions. About 5 ppm He and 35 ppm H result from a 1×10^{22} neutrons/cm^2 fluence. In the case of aluminium alloys, however, it is possible that neither gas species diffuses out at reactor operating temperatures, since the aluminium oxide normally present as a surface film may be an effective barrier to hydrogen effusion.[9] Experiments have therefore been directed at ascertaining the effects of these gases on mechanical properties and determining which alloys are most suitable for in-reactor use from this standpoint.

EXPERIMENTAL TECHNIQUES

Specimen preparation. Two lots of stainless steel were investigated, one being a commercial Type 304 stainless steel and the other a special vacuum-melted heat of 0·2% Ti-Type 304L stainless steel. The compositions of these materials are listed in Table 1.

Table 1. COMPOSITION OF STAINLESS STEELS BOMBARDED BY α-PARTICLES

	Type 304 SS	*0·2% Ti-Type 304L SS*
C	0·046	0·019
Mn	1·77	1·05
P	0·023	0·006
S	0·010	0·017
Si	0·62	0·046
Cr	18·15	19·23
Ni	9·60	11·22
Co	0·25	–
Ti	–	0·22
Mo	0·33	–
Co	0·03	–

The aluminium alloys include 1100, X8001, 6061, and 895 Sintered Aluminium Product (SAP). The compositions of the commercial products are given in Table 2.

In all the experiments to be described, a standardised materials handling technique was employed. Regardless of the starting stock size, the materials were cold-rolled to flat sheets and given a series of finish cold-rolling passes to a 50% final reduction of area. The finish sheet thicknesses were either 0·009–0·010 in or 0·019–0·020 in. Tensile specimens having 0·75 in gauge lengths and 0·25 in gauge widths were then machined from the sheets. The specimens were given a final annealing treatment before ion bombardment or testing: stainless steel specimens were annealed in 1 atm of high-purity argon for 1 h at 925°C or 1038°C, while the aluminium specimens were annealed in a 50% KNO_3–$NaNO_3$ salt bath at times and temperatures suggested by the Alcoa Aluminium Handbook.[10]

Table. 2. COMPOSITION OF ALUMINIUM ALLOYS

Alloy	*Composition* (*wt %*)								
	Ti	Ni	Zn	Si	Fe	Mn	Mg	Cu	Cr	Al_2O_3
1100	0·025	<0·02	0·1	0·11	0·55	0·023	0·005	0·1	<0·03	–
6061	0·031	<0·02	0·1	0·69	0·28	0·024	0·90	0·2	0·07	–
X8001	<0·01	0·95	≤0·1	0·12	0·51	0·017	0·003	0·02	0·05	–
895 SAP	–	–	–	–	–	–	–	–	–	8·95

Ion injection techniques. The experimental techniques for injecting nearly uniform concentrations of gas into flat metal specimens evolved in two stages. The common basis of both techniques is the well-known slowing of ions (α-particles and protons) by electronic excitation and ionisation of the medium through which they are passing.[11] Ignoring any scattering effects, the statistical nature of the energy transfer causes a spread in ion energy on passing through the medium. Thus the ion deposition frequency profile exhibits some standard deviation (σ) about the mean penetration distance, or range (R), while the mean particle pathlength (P) is greater than the range. The standard deviation, pathlength, and range increase with increasing particle energy. The ranges of protons and α-particles in stainless steel, graphite, and aluminium used for calculations in these experiments were taken from the calculated values of Janni[12] (protons) and the range scaling formulae of Barkas and Berger[13] and of Northcliffe[14] (α-particles).

The first bombardment technique depends upon producing a uniformly intense, stable current of collimated monoenergetic ions which are degraded in energy by a stationary graphite wedge (Fig. 1). A linearly-tapered wedge of maximum thickness greater than the range of the ion (after passing through the

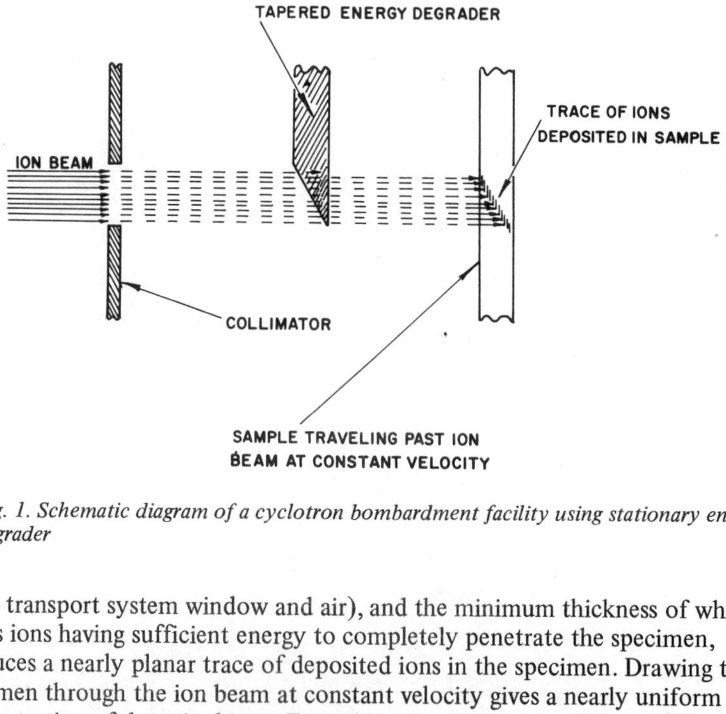

Fig. 1. Schematic diagram of a cyclotron bombardment facility using stationary energy degrader

beam transport system window and air), and the minimum thickness of which yields ions having sufficient energy to completely penetrate the specimen, produces a nearly planar trace of deposited ions in the specimen. Drawing the specimen through the ion beam at constant velocity gives a nearly uniform concentration of deposited ions. Forced-air circulation is used to cool both sides of the specimens and the temperature of a dummy specimen is monitored by thin chromel-alumel thermocouples spot-welded to the front and back of the specimen. This technique is no longer favoured for several reasons: (1) in our experience, over 90% of the beam is lost in collimation to obtain a nearly uniformly intense beam over the 0·25 in square target normally used in our experiments; (2) even under these conditions, the ratio of maximum to minimum beam current over the target can be as high as two or three; (3) too much operating time is required to adjust the beam focusing to obtain even these

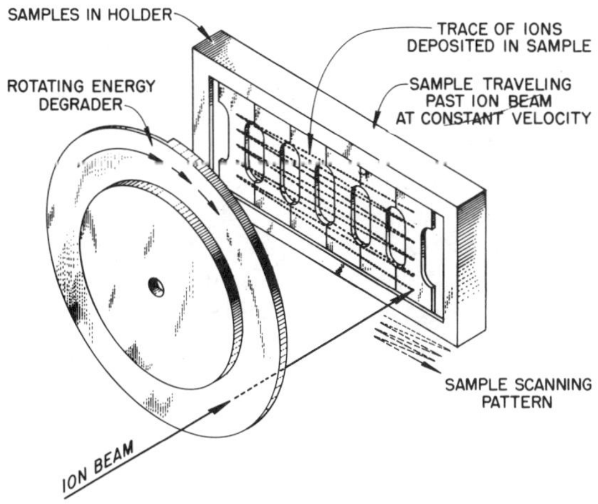

Fig. 2. Schematic diagram of a cyclotron bombardment facility using rotating energy degrader and sample scanning device

non-ideal conditions; (4) cyclotron beams are unstable in time, so that the concentration of ions in the deposited trace varies. Multiple scattering effects discussed below produce non-uniformities in the deposited beam from the front to the back of the specimens.

The second bombardment technique (Fig. 2) utilises a rapidly-spinning energy degrader wheel (usually aluminium), whose edge thickness is proportional to distance around the circumference. The time-variation produced in energy sweeps the trace of deposited ions from the front to the back of the specimens. A rack of specimens is scanned slowly in a fine zig-zag pattern so that all specimens are exposed to all areas of the collimated beam in small spatial increments. Repeated scanning has the effect of averaging ion current fluctuations in time and producing very homogeneous concentrations of deposited ions in the specimens. Forced-air circulation is used to cool the front side of the specimens, while the backs of the specimens rest against a water-cooled aluminium block. A dummy specimen with attached thermocouples is employed to monitor temperature on successive scanning passes. An independent Faraday cup mounted on a remotely-operated dolly behind the specimen assembly can be moved into the plane of the specimens periodically to monitor the total ion current. The net current on an insulated beam transport system end window is calibrated against the Faraday cup to provide a continuous beam current monitor while bombardments are in progress; this measurement is recorded continuously and integrated with time to determine the concentration of ions deposited in the specimens. In these experiments the specimens are staggered in three planes normal to the beam, so that a continuous series of specimen gauge lengths is presented to the beam.

A major problem of using energy-degrading techniques is that multiple scattering of ions by Coulombic interactions causes trajectory deflections of ions from the beam and consequent missing of the target specimens. Ball[15] has calculated the RMS multiple-scattering angles for various particles and particle energies as a function of scatterer thickness for various materials. Because of the high initial particle energies, small thicknesses and scattering distances involved, the beam transport system end window and approximately 0·75 in window-to-degrader air path losses are negligible. However, scattering in the degrader is significant, as is the scattering of low-energy particles in the ~1 in degrader to target air path. Using Ball's results, we have estimated that in the stationary energy degrader system ~15–65% of a 50 MeV α-particle beam is lost due to degrader scattering for the most and least energetic ions emerging from the degrader, respectively. This causes a corresponding inhomogeniety in the ions deposited from the back to the front of the specimens. However, the losses due to scattering by the rotating energy degrader are not greater than 2% for a 35-specimen target due to the large target-area to beam-area ratio employed, which is well within the uncertainty in the determination of the total ion current. The scattering losses of low-energy ions in the specimen-to-air path are difficult to estimate because the slowing of low-energy ions is not as well-characterised as that of high-energy ions. However, multiple scattering must reduce the ion concentration implanted in the front of the specimens to values below the average concentration.

EXPERIMENTAL RESULTS

A series of ion injection experiments has been performed at the Oak Ridge Isochronous Cyclotron and the Argonne National Laboratory Building 211 Cyclotron. For the sake of brevity, the important experimental parameters of target material, target history, injection technique, beam energy, beam current, collimated beam size, maximum specimen temperature, calculated injected gas concentration, and measured gas concentration are presented in Table 3. The aims and principal results of each experiment are put into perspective below in the short description of each experiment and its results, given in chronological order.

First experiment. The earliest experiment was performed on Type 304 stainless steel in a standard well-annealed condition in order to assess the effects of small concentrations of helium on the tensile properties of the material. In addition to injecting mean helium concentrations of 0·14 and 0·083 atomic ppm into specimens bombarded with the graphite degrader in place, one set of specimens was bombarded for the time required to inject 0·14 ppm with the degrader removed from the system. In the latter case, the ions entered and left the specimens at mean energies of 77·2 and 45·2 MeV respectively.

Each set of specimens, together with a set of unbombarded control specimens, was tensile tested at temperatures from 500° to 900°C. At least three control specimens were tested at each temperature to determine the variations which might be expected in the test results. The yield strength, ultimate tensile strength, and uniform elongation data showed that bombardment had no appreciable effects on any of these properties. One possibly significant exception is the apparent reduction of uniform strain at 900°C of the helium-bearing specimens, which is apparently due to the total elongation of the bombarded specimens being less than the total elongation of the control specimens. However, the total elongation to failure of specimens containing cyclotron-injected helium was significantly reduced between 700°C and 900°C. The total elongation results are plotted in Fig. 3, while the other properties are given in Table 4.

On the other hand, the specimens bombarded without the degrader in place exhibited ductilities similar to the unbombarded control specimens except at the 900°C test temperature. This reduced ductility at 900°C is probably explained by the occurrence of a few high-angle multiple-scattering events which

Table 3. EXPERIMENTAL ION BOMBARDMENT AND SPECIMEN CONDITIONS FOR VARIOUS INJECTION EXPERIMENTS

Exp. No.	*Material*	*Annealing temp. (°C)*	*Experiment degrader type*	*Beam energy* (MeV)	*Collimated beam size* (in)	*Maximum specimen temp.* (°C)	*Calculated mean helium content (atomic ppm)*	*Measured helium content (atomic ppm)*
1.[3]	304 S.S.[8]	1038	Stationary C	80·0 ± 0·5	0·27 X 0·27	210	0·14[5]	—
	304 S.S.	1038	Stationary C	80·0 ± 0·5	square	210	0·083[5]	—
	304 S.S.	1038	None	80·0 ± 0·5		210	0	—
2.[3]	304 S.S.[9]	1038	Rotating Al	80·0 ± 0·5	0·5 diam.	465	0·83[6]	0·53[1]
		925	Rotating Al	80·0 ± 0·5	circle	465	0·83[6]	
	0·2% Ti-[9]	1038	Rotating Al	80·0 ± 0·5		465	0·83[6]	0·90[1]
	304L S.S.	925	Rotating Al	80·0 ± 0·5		465	0·83[6]	
3.[4]	304 S.S.[9]	1038	Rotating Al	46·0 ± 0·5	1·0 diam.	325	20·0[6]	18·0, 11·0, 17·0[2]
		925	Rotating Al	46·0 ± 0·5	circle	325	20·0[6]	22·6[1]
	0·2% Ti-	1038	Rotating Al	46·0 ± 0·5		325	20·0[6]	
	304L S.S.[9]	925	Rotating Al	46·0 ± 0·5		325	20·0[6]	22·0[1]
4.[4]	1100-0 Al[8]	325	Rotating Al	46·0 ± 0·5	1·0 diam.	120	5·0[6]	—
	X8001-0 Al[8]	325	Rotating Al	46·0 ± 0·5	circle	120	5·0[6]	—
	6061-0 Al[8]	413[7]	Rotating Al	46·0 ± 0·5		120	5·0[6]	—
	895 SAP[8]		Rotating Al	46·0 ± 0·5		120	5·0[6]	—

[1] Analysis by Atomics International.
[2] Analysis by Oak Ridge National Laboratory.
[3] Performed at Oak Ridge Isochronous Cyclotron.
[4] Performed at Argonne National Laboratory Building 211 Cyclotron.
[5] Known uncertainties 50% of values quoted.
[6] Known uncertainties 15% of values quoted.
[7] Cooled at 28°C per hour to 260°C, annealing time 2 h.
[8] Specimen thickness 0·19–0·20 in.
[9] Specimen thickness 0·09–0·095 in.

Table 4. EFFECT OF α-PARTICLE BOMBARDMENT ON THE TENSILE PROPERTIES OF TYPE 304 STAINLESS STEEL

Sample	*Test Temp.* (°C)	*Yield Strength* (1000 psi)	*Ultimate Tensile Strength* (1000 psi)	*Uniform Elongation* (%)
Control	500	17·0–18·0	68·8–71·6	44·1–45·6
	600	12·4–14·4	48·0–52·4	35·5–39·5
	700	12·8–13·6	29·6–33·2	21·9–25·2
	800	11·0–14·4	18·3–23·2	9·36–17·1
	900	8·6– 8·8	10·1–10·8	11·0–16·3
Bombarded with Degrader in place	500	15·0	59·0	43·2
	600	15·6	52·0	33·6
	700	15·0	31·6	20·6
	800	11·2	17·0	11·6
	900	8·4	9·4	6·5
Bombarded without Degrader	500	14·4	60·8	44·1
	600	12·8	51·8	40·8
	700	14·2	43·0	30·3
	800	12·4	21·4	17·9
	900	8·6	11·0	12·2

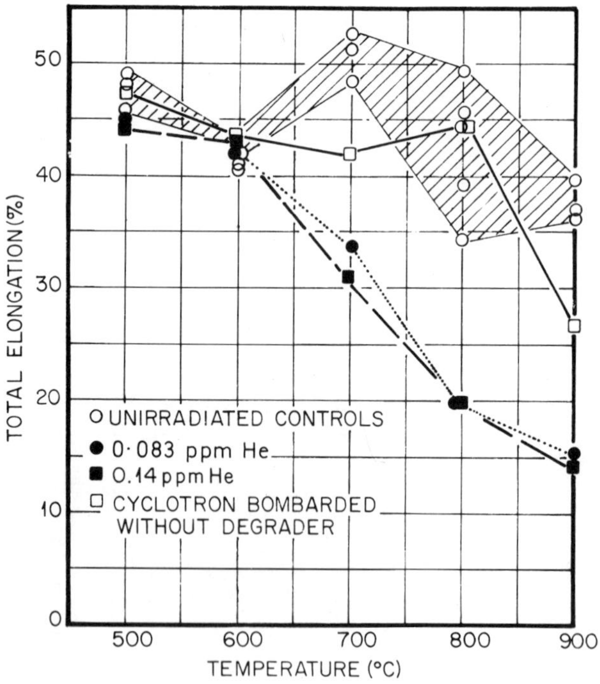

Fig. 3. Effect of cyclotron-injected helium and bombardment on the tensile ductility of Type 304 stainless steel annealed for 1 h at 1038°C

deposit a few atoms of helium in the specimen, and the effect is only noticed at 900°C where the effect of helium on ductility is greatest.

Second experiment. After ascertaining that cyclotron-injected helium did reduce the elevated-temperature ductility of Type 304 stainless steel, we compared the tensile properties of standard Type 304 stainless steel with a 0·2% Ti-Type 304L stainless steel, which is resistant to thermal neutron irradiation damage. The calculated injected helium concentration of 0·83 ppm agrees well with one analytical gas determination, although the second analytical result is unexplainably low (see Table 3).

As in the preceding experiment, sets of bombarded specimens were tensile tested from 500° to 900° C together with control specimens, at a strain rate of 0·026 min^{-1}. No significant changes were found in the yield strength, ultimate tensile strength, or uniform elongations of the specimens as a result of bombardment. The effects of injected helium on the total strain of the Type 304 stainless steel are shown in Figs 4 and 5, while the results for the 0·2% Ti-Type 304L stainless steel are presented in Figs 6 and 7. Within the limits of experimental error, the results demonstrate that, although 0·83 atomic ppm of helium has no effect on the tensile ductility of Type 304 stainless steel at 500° and 600°C, severe reductions in ductility occur over the 700°–900°C test temperature range. The absolute loss of ductility is greater for the specimens annealed for 1 h at 1038°C than for those annealed for 1 h at 925°C. Although

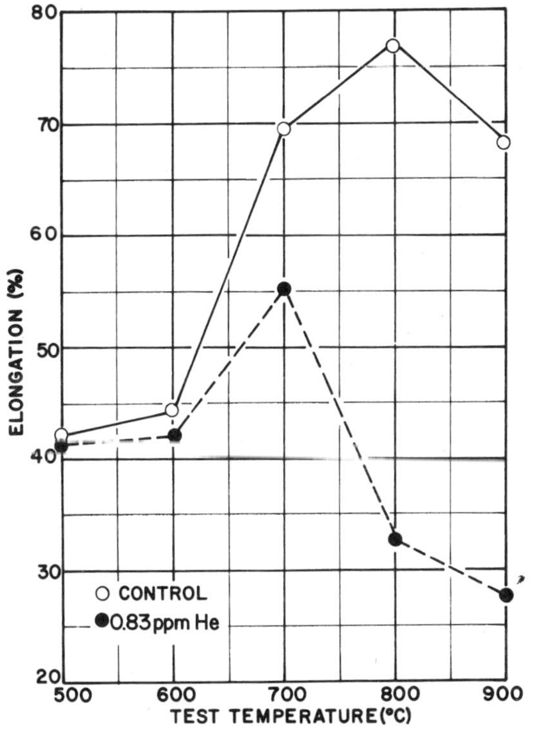

Fig. 4. Ductility of Type 304 stainless steel annealed 1 h at 925°C and tensile tested at 0·026 min^{-1} strain rate

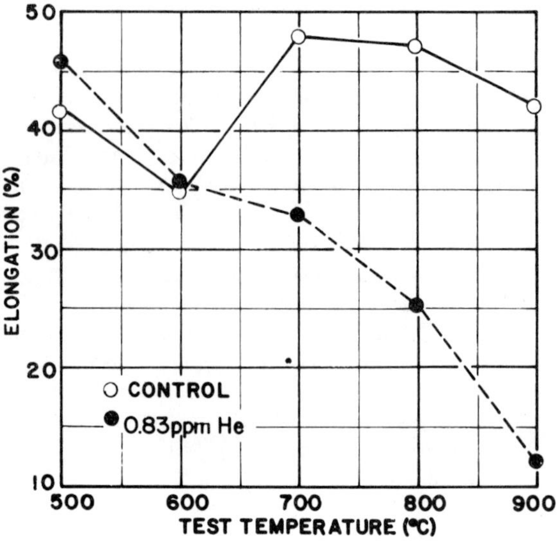

Fig. 5. Ductility of Type 304 stainless steel annealed 1 h at 1038°C and tensile tested at 0·026 min^{-1} strain rate

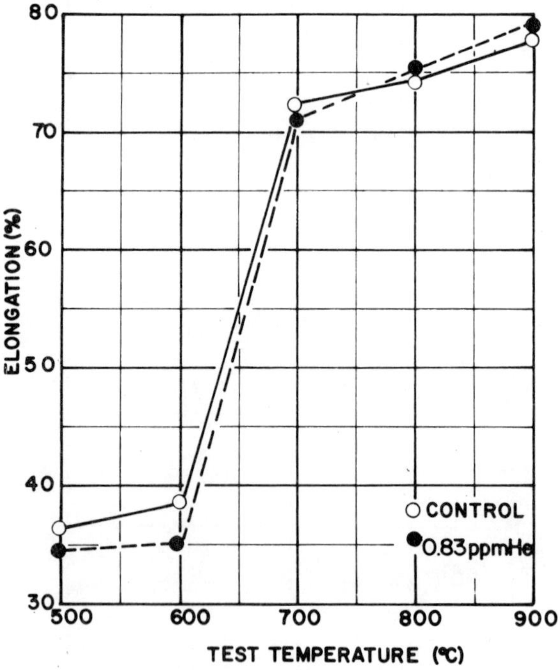

Fig. 6. Ductility of 0·2% Ti-Type 304L stainless steel annealed for 1 h at 925°C and tensile tested at 0·026 min^{-1} strain rate

the ductility of the 0·2% Ti-Type 304L stainless steel annealed for 1 h at 1038°C is qualitatively similar to that of the Type 304 stainless steel, the titanium-modified alloy annealed for 1 h at 925°C showed no loss of ductility over the entire test temperature range.

As-bombarded specimens were subsequently thinned and examined by transmission electron microscopy in both diffraction and absorption contrast to determine whether any observable displacement damage effects or helium bubbles might have formed during the bombardment. No evidence for either was found.

Tensile-tested specimens were examined metallographically to determine the probable mode of fracture. A strong tendency was found for Type 304 stainless steel to fracture in an intergranular manner at the higher test temperatures. The 0·2% Ti-Type 304L stainless steel, on the other hand, had far fewer intergranular cracks formed at any given test condition. Helium-bearing specimens of both materials generally tended to form more intergranular cracks than comparable control specimens. A comparison between the fracture appearance of the Type 304 and 0·2% Ti-Type 304L stainless steels is shown in Fig. 8.

Third experiment. Since the potential applications of stainless steel in reactors may involve stresses and temperatures which will cause creep deformation, a bombardment experiment was performed to provide specimens for creep-rupture testing. The materials and annealing conditions were selected to be the same as those used in the previous experiment for purposes of comparison, although the higher helium content of 20 atomic ppm was selected as being more representative of the helium contents expected in stainless steel after several months' exposure in a fast reactor having a neutron flux of 1×10^{15} neutrons cm^{-2} s^{-1}.

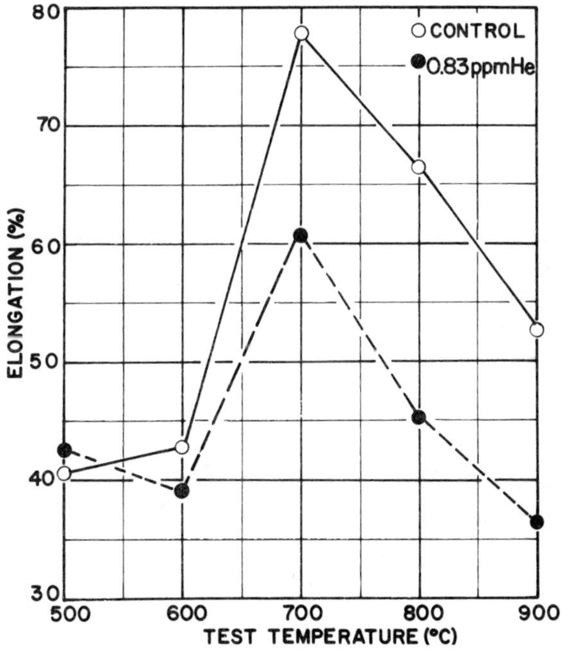

Fig. 7. Ductility of 0·2% Ti-Type 304L stainless steel annealed 1 h at 925°C and tensile tested at 0·026 min^{-1} strain rate

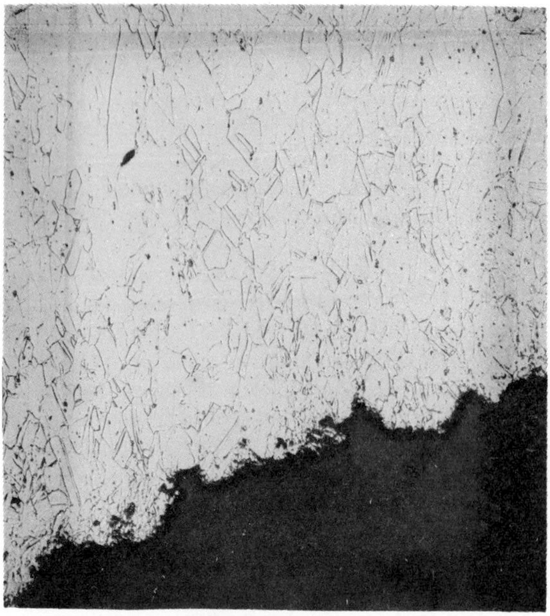

(a)

(b)

Fig. 8. Microstructures of (a) 0·2% Ti-Type 304L and (b) Type 304 stainless steels tensile tested at 700°C. Both steels were annealed at 1038°C for 1 h and contain 0·83 ppm He. Note the extensive intergranular cracking in the Type 304 stainless steel which is not observed in the titanium-modified alloy (× 100)

The creep-rupture tests were performed at 600°C and 700°C, temperatures near the peak operating temperatures expected in the Liquid Metal Fast Breeder Reactor. The results of these tests are plotted in Figs 9–12 as graphs of total rupture strain versus time to rupture. The applied stress for each datum is given on the graphs, so that rupture life effects may be compared directly at a given stress level.

At 600°C the Type 304 stainless steel control specimens annealed at 925°C exhibit consistently higher ductility than those annealed at 1038°C, although the rupture lives are nearly alike (Fig. 9). At 700°C the Type 304 stainless steel control specimens annealed at 925°C have consistently lower rupture life but higher ductility than those annealed at 1038°C (Fig. 10). The cyclotron-injected

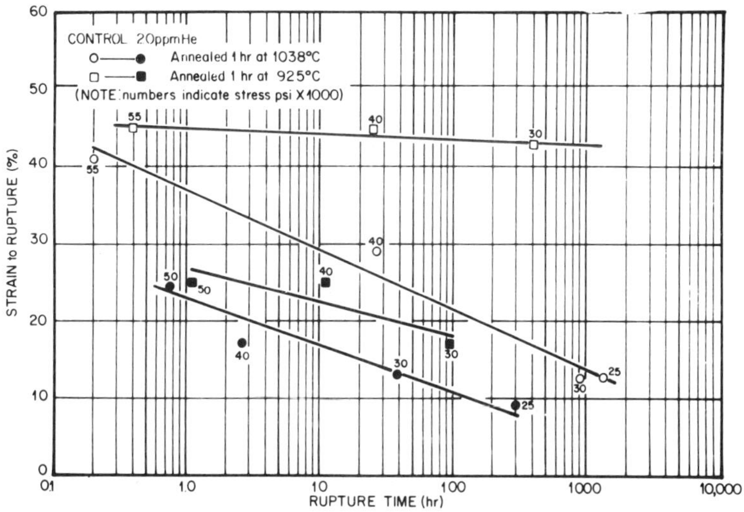

Fig. 9. Strain versus rupture time of Type 304 stainless steel at 600°C

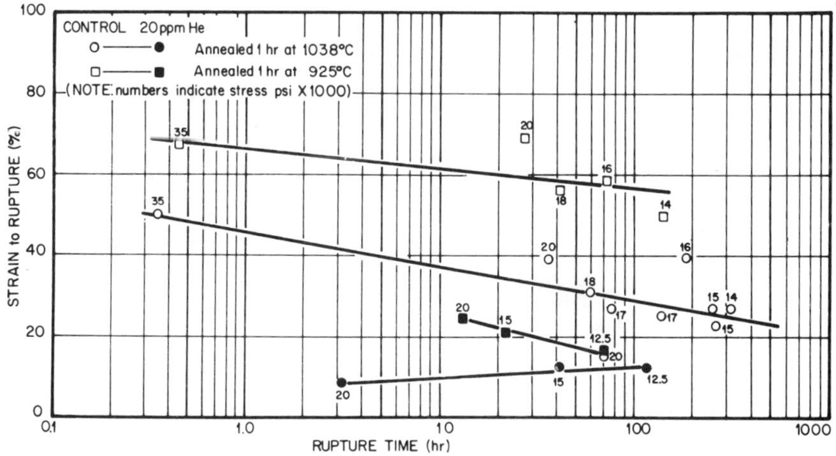

Fig. 10. Strain versus rupture time of Type 304 stainless steel at 700°C

20 ppm of helium reduces the ductility of the alloy as well as its rupture life, particularly at low applied stress levels.

The behaviour of the control specimens of the 0·2% Ti-Type 304L alloy is qualitatively similar to that of Type 304 stainless steel, except that at high stresses there is little effect of annealing temperature on rupture life at 700°C. The rupture life of the titanium-modified alloy is only slightly reduced by 20 ppm of helium at either test temperature. Furthermore, the ductility of this alloy tested at high strain rates at 600°C is not altered by 20 ppm of helium. At lower stresses and strain rates however, the ductility is reduced, though not as significantly as that of Type 304 stainless steel. The ductility loss in the 0·2% Ti-Type 304L alloy due to helium is greater at 700°C than at 600°C, but the

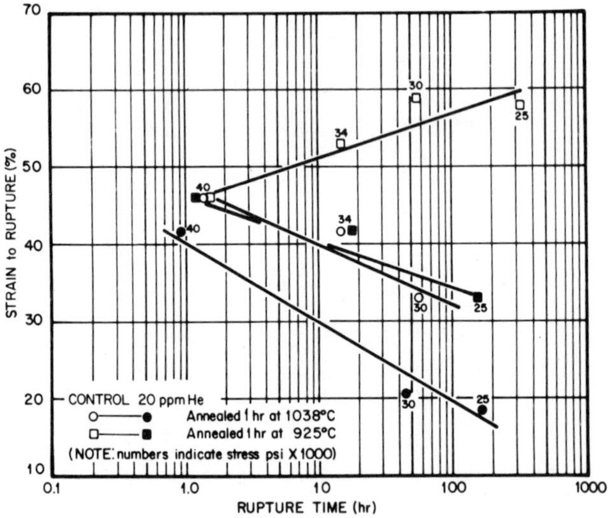

Fig. 11. Strain versus rupture time of 0·2% Ti-304L stainless steel at 600°C

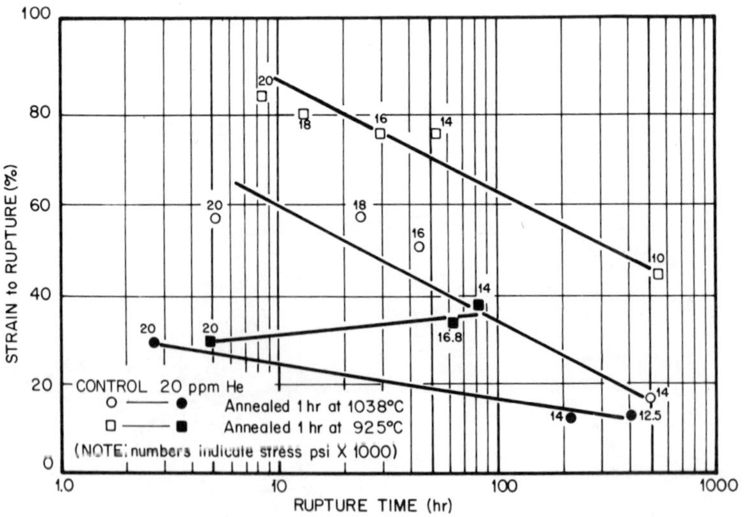

Fig. 12. Strain versus rupture time of 0·2% Ti-304L stainless steel at 700°C

Table 5. EFFECT OF 5 ATOMIC PPM CYCLOTRON-INJECTED HELIUM ON THE TOTAL ENGINEERING ELONGATION OF SELECTED ALUMINIUM ALLOYS

Test Temperature (°C)	*Total Engineering Elongation, %*							
	1100-0 *Aluminium*		X8001-0 *Aluminium*		6061-0 *Aluminium*		*SAP* 895	
	Control	5 ppm He	*Control*	5 ppm He	*Control*	5 ppm He	*Control*	5 ppm He
21	46·9, 35·3		53·2		34·8		15·8	
38							13·4	
93	55·7		52·1	55·5	29·0	28·6	8·1	9·7
149	56·9	52·9	51·2	45·5	31·0		7·2	8·1
204	53·6	51·2	48·5	36·1	31·9	28·5	9·1	> 5·4
260	54·8	32·7	48·4	30·5	41·3	43·4	9·5	7·5
316	43·4, 66·3, 59·6	27·5	54·3	28·4	40·7	40·6	6·3	5·8
371	53·9	25·9	52·9		33·5	39·9	4·5	4·1
482	66·7	31·5	65·3	55·2	40·2	51·7	4·8	4·6

ductility of specimens annealed for 1 h at 925°C remains significantly higher than that of Type 304 stainless steel, and does not vary greatly with stress in tests lasting up to 100 h.

Transmission electron microscopy on all specimens tested at 600°C and 30 000 psi stress failed to disclose any evidence of gas bubble formation during testing, either on grain boundaries or structure imperfections, or in the matrix. The structures observed could be related to the total strain: the structures of specimens exhibiting lower rupture strains were dominated by dislocation tangles, whereas well-developed subcells of 0·5 to 2·5 μm were common in specimens which deformed over 30%. No other major differences in deformation mode were observed.

Fourth experiment. In order to determine the effects of helium on the tensile properties of aluminium alloys, sets of 1100-0, X8001-0, 6061-0, and 895 SAP specimens were bombarded to produce 5 atomic ppm of helium. The selection of materials was based on current reactor usage except in the case of 895 SAP alloy, which was selected to assess whether a fine, stable second phase dispersion produces resistance to the anticipated effects of helium.

These specimens, together with suitable control specimens, were tensile tested at temperatures from 38°C (100°F) to 427°C (800°F). As in the bombarded stainless steels, no change was observed in any tensile properties except the total fracture strain as a result of bombardment. The fracture strain data for 0·0026 min^{-1} strain rate testing of the four materials are presented in Table 5. Above the 149°–204°C test temperatures the cyclotron-injected helium reduced the tensile ductility of the 1100-0 and X8001-0 alloys, whereas the 6061-0 and 895 SAP alloy ductilities were not affected by helium over the range of temperatures investigated.

In order to determine whether the fracture mode was affected by the presence of helium, specimens of X8001 and 1100 aluminium tested at 316°C and 371°C were examined by optical metallography. It was difficult to ascertain whether the fracture mode was transgranular or intergranular, but no evidence of intergranular cracking near the failure was observed. Since helium bubbles should form more readily during testing at higher temperatures if they form at all, the bombarded specimens of 6061-0 and 1100-0 aluminium tested at 316°C and 371°C respectively were examined by transmission electron microscopy. No gas bubbles were found.

DISCUSSION

One of the major difficulties in interpreting the results of neutron irradiation experiments is that several types of damage usually occur simultaneously, and the experimenter is often at a loss to decide unequivocally which irradiation-caused event is responsible for observed property changes in materials. Other methods exist for introducing helium into metals, but they too involve complicating uncertainties. Diffusion techniques do not permit exceeding the solid solubility of the gas at given partial pressure and temperature, and probably produce inhomogeneous gas distributions. Transmutation-produced gases obtained by doping with an element of high thermal neutron cross-section may be segregated, and can only be produced by changing the composition of the alloy in question. Apart from inelastic collision events, ion injection produces only one important complicating factor, viz. the displacement damage produced by the ion through Coulombic nuclear interactions. In order to calculate the magnitude of this effect, we have used the equations of Kinchin and Pease[16] for the number of displaced atoms produced by a moving ion as a function of transferred energy, assuming a sharp ionisation threshold; in this manner we have estimated $\sigma_D(E)$, the total displacement cross-section for stainless steel and aluminium as a function of ion energy E. By using the tabulated values of Janni[12] of dE/dx for protons, and making suitable conversions for α-particles, we have calculated the number of displaced atoms produced by an ion slowing from energy E_1 to energy E_2, $N(E_1, E_2)$[17] to be

$$N(E_1, E_2) = \int_{E_2}^{E_1} \frac{\sigma_D\,(E)}{dE/dx}\, dE \qquad (1)$$

The approximate results of these calculations are that α-particles of ~50 MeV energy produce ~5 10^{-7} displacements atom^{-1} μA^{-1} cm^2, a value which increases with decreasing ion energy to values of ~3 10^{-5} displacements atom^{-1} μA^{-1} cm^2 for 800 keV ions. Damage production rates are only about half as great in aluminium. These results lead to the following conclusions: (1) in a degraded beam experiment, displacement damage is produced non-uniformly from the front to the back of the specimen; (2) significantly fewer displacements per atom are produced in attaining any given helium level during ion bombardment than during fast neutron irradiation. The latter point may be demonstrated by estimating the ratio, m, of displaced atoms to helium generation during fast neutron irradiation as

$$m = \frac{\sigma_D \theta}{\sigma(n, \alpha)} \qquad (2)$$

If we assume[16] $\sigma_D \sim 1$ barn, and that θ, the number of secondary displaced atoms produced per primary knock-on event, is ~500[16], while the (n, α) cross-section for iron, $\sigma(n, \alpha)$ is ~0·003 barns,[3] then $m \cong 1{\cdot}7 \times 10^5$ displaced atoms/helium atom. On the other hand, the maximum number of displacements produced by slowing a 50 MeV α-particle to rest in stainless steel is ~150, the upper limit on m for a cyclotron experiment.

The first cyclotron experiment in which Type 304 stainless steel was bombarded to mean helium contents of 0·14 and 0·083 ppm demonstrates that even these very low concentrations of helium have a substantial effect on the elevated temperature ductility of stainless steel. The results of Martin and Weir[7] for Type 304 stainless steel irradiated to 2×10^{20} neutrons/cm^2 (thermal) show a qualitatively similar decrease in ductility with increasing test temperature over the 500°-900°C range; the results cannot be compared quantitatively because of the differences in material compositions, fabrication history and helium distribution. The amounts of displacement damage produced by bombarding specimens with a degraded and undegraded beam differ significantly, since the rate of displacement damage production in the absence of the degrader was only ~5 10^{-7} displacements atom^{-1} μA^{-1} s^{-1} cm^2, while with the degrader in place the less energetic ions produced more displaced atoms per unit time. However, the lack of change in strength properties at the test temperatures clearly indicates that the displacement damage had been annealed out during heating to the test temperature, so that the loss of ductility in the helium-bearing specimens may clearly be ascribed to the injected gas. The failure of electron microscopy to find evidence of displacement damage in later experiments further substantiates the negligible role of this form of damage in our ion bombardments.

The mechanism by which helium causes embrittlement has been discussed by several authors.[2,7,18] Although there is some disagreement on the details of the mechanism, it is generally agreed that helium is very insoluble in metals, and may either be transported to grain boundaries by diffusion or swept there by dislocations. Once at the boundaries, it forms small gas bubbles which either act as crack nuclei that hasten intergranular failure, or grow under an applied stress,[18] link up, and cause failure. On the basis of this hypothesis Martin and Weir[7] have proposed several methods of alleviating the problem, based on either reducing the amount of helium present at grain boundaries, reducing the stress concentrations which build up at grain boundaries by shortening the effective lengths of sliding boundaries, altering the surface tension which plays a role in bubble growth, or changing the concentration of tramp impurities present at grain boundaries. Their methods have been successful under thermal neutron irradiation conditions.

The second experiment was intended to demonstrate whether the proposed solutions to irradiation damage proposed by Martin and Weir were useful, given an initial uniform distribution of helium throughout stainless steel. Type 304 stainless steel annealed for 1 h at 925°C has an ASTM 6 grain size, whereas annealing for 1 h at 1038°C produces the coarser ASTM 5 grain size, with very few precipitate particles in either structure. The first striking aspect of this experiment, viz. the observation that the finer-grained material is the less severely embrittled, supports either of two hypotheses: (1) that reducing the concentration of helium at grain boundaries improves ductility, since if all helium were swept to grain boundaries the finer-grained material would have the lower grain boundary helium concentration, and (2) that reducing the effective sliding length of grain boundary decreases the stresses which operate in growing helium bubbles. Although the grain size of the titanium-modified alloy tended to be one ASTM size finer in either condition than the Type 304 stainless steel, the same effect of grain size was observed in it. The second important result of this experiment is that the tensile ductility of the titanium-modified stainless steel, annealed for 1 h at 925°C to produce a fine grain size, is not affected by as much as 0·83 ppm of helium. This result supports any of the proposed solutions to the helium-induced embrittlement problem, and does not allow us to distinguish which might be important. The marked difference in behaviour between the two alloys cannot be ascribed to any difference in precipitate structure.

It is noteworthy, however, that the effect of helium in all of the tensile tests described above was greatest at higher temperatures, where the intergranular mode of fracture becomes dominant. Thus, all these results are in keeping with the proposals that the role of helium is one of accelerating intergranular fracture.

The apparent difference in the degree to which Type 304 and 0·2% Ti-Type 304L stainless steels tend to fracture intergranularly in the absence of helium seems to be closely related to their relative abilities to resist high-temperature embrittlement.

The results of constant-load creep experiments are not strictly comparable with constant strain-rate tensile tests, since the times and stresses involved are dissimilar. However, we may make certain correlations by noting that under the slower strain rate conditions of long-time creep tests, the intergranular fracture mode becomes more important at lower temperatures, so that the microstructure of commercial Type 304 stainless steel creep-tested at 600°C contains a large number of intergranular cracks. Hence, it is not surprising that Type 304 stainless steel, which was not affected by helium in tensile tests in 600°C, exhibits a sharp reduction in both rupture life (at lower stresses) and ductility in creep tests at 600°C after injecting 20 atomic ppm of helium (third experiment, Fig. 9). At 700°C, the creep ductility and rupture life are greatly reduced by that amount of helium at all stresses investigated(Fig. 10). The strength of the alloy, as indicated by the minimum creep rate, was unaffected by the cyclotron-injected helium. As before, the finer-grained specimens consistently exhibit higher ductility.

The behaviour of 0·2% Ti-Type 304L stainless steel differs from that of Type 304 material only in the magnitude of the effects observed (Figs 11 and 12). Reductions of rupture life were observed, but were generally less severe than in the commercial alloy. The rupture life of titanium-modified Type 304L stainless steel was consistently lower than that of Type 304 stainless steel, due to higher creep rates caused by reduced solid-solution strengthening and precipitate formation during test, as a result of the lower carbon content. The significant point from the standpoint of eventual use, however, is the consistently higher ductility of the titanium-modified alloy, which is desirable for reactor applications.

The fracture behaviour of aluminium alloys has been investigated by Mullendore and Grant,[19] who found that at intermediate temperatures around 260°-371°C pure Mg-Al alloys containing up to 5·1% Mg tend to fracture intergranularly in creep tests lasting from one to several hundred hours. Grain boundary sliding and intergranular fracture are more likely to occur with increasing Mg content, indicating that the magnitude of grain boundary effects is sensitive to composition. Grant[20] has also found similar behaviour in other aluminium alloys. The results are also sensitive to strain rate, since at fast strain rates grain boundary sliding does not occur readily. At sufficiently high temperatures grain boundaries are highly mobile and stress concentrations do not build up along boundaries, whereas at very low temperatures sliding does not occur. Neither of these conditions, therefore, leads to intergranular cracking.

From the standpoint of Grant's work, therefore, we might suspect that cyclotron-injected helium hastens intergranular fracture in the 1100 and X8001 aluminium alloys by the mechanisms cited above for stainless steel. Below 149°C for 1100 aluminium, and below 204°C for X8001 aluminium, grain boundary sliding probably does not play an important role in tensile deformation, and helium is unimportant in the transgranular fractures which occur. At the 427°C test temperature, where the reduction of ductility for both alloys is less severe than in the range 260°-371°C, one might presume that grain boundary mobility increases sufficiently to reduce the tendency for intergranular cracking. However, this picture is not certain, due to the apparent lack of grain boundary cracks in the microstructure of tested specimens. It is not inconceivable that the first grain boundary crack to form intensifies the stress locally at its tip to a level sufficient to promote transgranular fracture across the remainder of the specimen. However, this event would be in agreement both with the observations and the proposed explanation. Further evidence from tests performed at varying strain rates may help to clarify the situation.

If the proposed explanation for the helium-induced embrittlement of aluminium is correct, then we may speculate that the tests to date have involved strain rates and temperatures at which 6061-T6 and 895 SAP alloys do not fracture intergranularly, so that cyclotron-injected helium can play no important role in their fracture. The reasons for this behaviour are not clearly understood, although the same arguments used to explain the beneficial effects of titanium on stainless steel may be equally well applied to the aluminium alloys which are not embrittled by helium. From the standpoint of developing improved materials having ductility after irradiation suitable for reactor service, however, these tests have provided a starting point which may permit more rapid alloy development.

It should be noted that in all experiments of this kind the object is not to supplant the testing of neutron-irradiated materials, but rather to aid in understanding the results of neutron irradiation experiments. Because of the vast differences in irradiation times, temperatures, and events occurring in neutron irradiation experiments and ion bombardment experiments, ion bombardments can never really duplicate the effects of neutron irradiation. However, in a few cases like those presented above, the effects of certain types of irradiation damage can be isolated and characterised.

SUMMARY

(1) By using suitable ion energies, degrader thicknesses, and geometrical configurations, uniform concentrations of gases may be injected into flat metal specimens to simulate fast neutron transmutation produced gas.

(2) The elevated temperature tensile and creep ductility of Type 304 stainless steel is severely reduced by cyclotron-injected helium in a manner analogous to the effects of thermal neutron produced helium. As little as 0·083 atomic ppm of helium has a pronounced effect.

(3) 0·2% Ti-Type 304L stainless steel is less susceptible to the loss of high-temperature ductility caused by cyclotron-injected helium.

(4) The role of helium in reducing the elevated-temperature ductility of stainless steels involves increasing the rate at which intergranular fractures start or propagate, under stress and temperature conditions for which these materials normally tend to fracture in an intergranular manner.

(5) Both X8001 and 1100 aluminium alloys in the fully-annealed condition are also susceptible to helium embrittlement at the 5 ppm He level in tensile tests over a limited temperature range. This range may correspond with test conditions under which grain boundary sliding and intergranular fracture tend to occur.

(6) 6061 aluminium alloy in the –0 temper condition and 895 SAP alloy containing 5 ppm He do not exhibit any loss of ductility over the range of temperatures for which the 1100 and X8001 aluminium alloys are embrittled.

REFERENCES

1. Woodford, D. A., Smith, J. D., and Moteff, J., *J. Nucl. Mater.* **24**, 118, (1967).
2. Weir, J. R., Jr., *Science* **156**, 1689, (1967).
3. Alter, H. and Weber, C. E., *J. Nucl. Mater.* **16**, 68, (1965).
4. Hinkle, N. E., *Am. Soc. Testing Mater. Soc. Tech. Publ.* **341**, 344, (1963).
5. Robertshaw, F. C., Moteff, J., Kingsbury, F. D., and Pugact, M. A., *Am. Soc. Testing Mater. Spec. Tech. Publ.* **341**, 372, (1963).
6. Higgins, P. R. B. and Roberts, A. C., *Nature* **206**, 1249, (1965).

7. Martin, W. R. and Weir, J. R., Jr., *Am. Soc. Testing Mater. Spec. Tech. Publ.* 426, 440, (1968).
8. Cunningham, J. E., ORNL-TM-2138 (1968).
9. Smithells, C. and Ransley, C., *Proc. Roy. Soc.* **152A,** 706, (1953).
10. Alcoa Aluminium Handbook, Aluminium Co. of America, Pittsburgh, Pa. (1967).
11. Livingston, M. and Bethe, H. A., *Rev. Mod. Phys.* **9,** 245, (1937).
12. Janni, J. F. Technical Report No. AFWL-TR-65-150 (1966).
13. Barkas, W. H. and Berger, M. J., National Academy of Sciences–National Research Council Publication 1133, 103, (1964).
14. Northcliffe, L. C., *ibid,* 173.
15. Ball, J. B., ORNL-3311 (1962).
16. Kinchin, G. S. and Pease, R. S., *Rept. on Prog. in Phys.* **18,** 1, (1955).
17. King, R. T., to be presented at AIME Fall Meeting, Philadelphia, Pa. (1969).
18. Barnes, R. S. and Nelson, R. S., AERE-R4952, Harwell (1965).
19. Mullendore, A. W. and Grant, N. J., *J. of Metals,* 973, (1954).
20. Grant, N. J., private communication.

DISCUSSION

K. Ehrlich (Karlsruhe): I would like to comment on the simulation of He embrittlement of cladding materials. We have used this method to study the influence of He on different vanadium-base alloys. Small tensile specimens were irradiated to concentrations of 10^{-6} He atoms/atom over the whole thickness of the samples. On testing at temperatures $\leqslant 750^{\circ}C$ we found no influence on the ductility in comparison with unirradiated specimens. Tensile tests at $850^{\circ}C$ and $950^{\circ}C$ showed a remarkable loss in ductility of the irradiated specimens. This means that at least some of the B.C.C. vanadium alloys show high-temperature He embrittlement. The advantage of this method is that the irradiation time to produce a sufficiently high He concentration is much less than for neutron irradiation. It is also possible to study the specific influence of He on the high-temperature behaviour of alloys, whereas in irradiation experiments at high fast neutron doses leading to the same He concentrations the ductility can be influenced by other effects such as the production of very stable dislocation networks.

ACKNOWLEDGEMENTS

We wish to thank E. Bolling, K. Boling, and H. Kline in preparing, bombarding and testing specimens. The operating staff of the Oak Ridge Isochronous Cyclotron, particularly A. Riikola, and the Argonne National Laboratory Building 211 Cyclotron, particularly M. Oselka, are thanked for their many suggestions and care in producing the ion beams. We acknowledge the assistance of W. Ghormley in the optical metallographic work, and J. O. Stiegler and C. K. H. Dubose for electron microscopy investigations. The many helpful contributors through discussions include W. R. Martin, J. R. Weir, E. E. Bloom, J. E. Cunningham, R. Jones, O. E. Oen, C. B. Fulmer, and E. Newman. We also wish to thank G. R. Love, E. E. Bloom, H. E. McCoy, and J. R. Weir for reviewing this manuscript.

SESSION 7

Activation Analysis

Chairman: Dr. T. B. Pierce

Influence of energy distribution in charged particle activation analysis

J. Giroux, J. P. Thomas, M. D. Tran, and J. Tousset
Institut de Physique Nucléaire, Université de Lyon,
43 Bd. du 11 Novembre 1918, 69–Villeurbanne (France)

ABSTRACT

This paper investigates the influence of energy distribution upon the activation curves of the reaction producing ^{18}F from oxygen with α-particles. Experiments were performed with mylar and mica in different beams: one at 35 MeV incident energy with good resolution and the other at 54 MeV with different widths, the origins and importance of which are discussed. Noticeable discrepancies between these activation curves have been established according to the different distribution widths, which involve variations between the values for the 'equivalent thickness' in the calibration. Moreover, the energy distribution exaggerates the difficulty in eliminating interferences caused by parasitic reactions with thresholds occurring in the matrix. All these terms can be evaluated if the excitation functions of the useful reactions are expressed in a simple analytical form; for an exponential or gaussian function, shape factors are calculated to take account of the slopes of excitation functions and of the widths of energy distributions.

INTRODUCTION

A charged particle beam always shows a certain energy distribution around probable physical values. The conditions of irradiation and operation as well as the nature of the accelerator itself may modify to a large extent the shape and width of this distribution. Among other problems arising in activation analysis we particularly studied[1,2] the straggling and energy degradation of charged particles. However, the nominal resolution of the incident beam is often most important.

These variations in the energy distribution frequently involve modifications to the shape of activation curves,[3,4,5] i.e. to the activities of the successive slices of a thick sample, which change the value of the 'equivalent thickness' used in Engelmann's method. The mean range of the particles in the matrix (Ricci's method) is also altered. The discrimination of activities induced by two competing reactions by means of energy thresholds becomes more difficult as this distribution increases in width.

Our purpose is to investigate these points, taking as our examples well-known reactions, e.g. ^{18}F formation from oxygen by irradiation with α-particles. Differences, sometimes important, between the various curves may be interprete and we give a few values of this effect when excitation functions can be expresse by simple analytical forms.

EXPERIMENTAL

The activation curves plotted in Fig. 1 were obtained by stacking mica foils in two irradiation positions of the Synchrocyclotron of the University of Lyon. The internal target, located on the maximum energy acceleration radius, presents

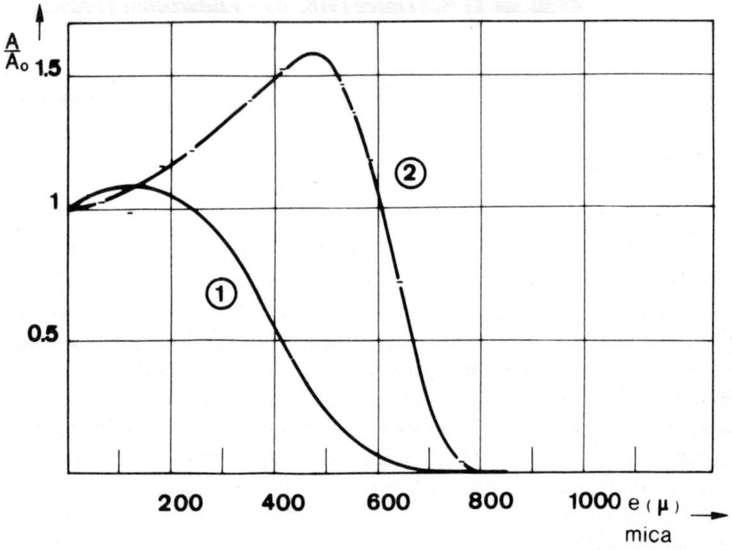

Fig. 1. $O \xrightarrow{\alpha} {}^{18}F$ *activation curves obtained by irradiating mica stacks in the internal (1) and semi-extracted beams*

a very broad energy distribution, and so the corresponding curve (1) is given here just to illustrate an extreme case. The second position makes use of a semi-extracted beam which still shows a broad distribution (cf. below) around 54 MeV.

Fig. 2 shows results for mylar irradiated in the second position. Within the bounds of range-energy relationships[2] and experimental error, the two curves corresponding to mylar and mica irradiated in the same beam, sensibly agree. Curve b of Fig. 2 illustrates the third irradiation position of the accelerator, which is associated with a pneumatic transfer system.

This external beam is characterised by a narrow energy distribution around a mean value which can vary along the dispersion of the whole extracted beam according to the adjustment of the deflecting magnet.

We also give activation curves for the reaction $^{19}F \xrightarrow{\alpha} {}^{18}F$ investigated with stacked teflon foils for the two last beams (Fig. 3).

A comparison of two oxygen activation curves performed with two accelerators delivering different incident energy beams (E_o = 54 and 35 MeV) is shown in Fig. 4.

Diaphragms of different diameters used for the seco[illegible] of the synchrocyclotron may play a prominent par[illegible]

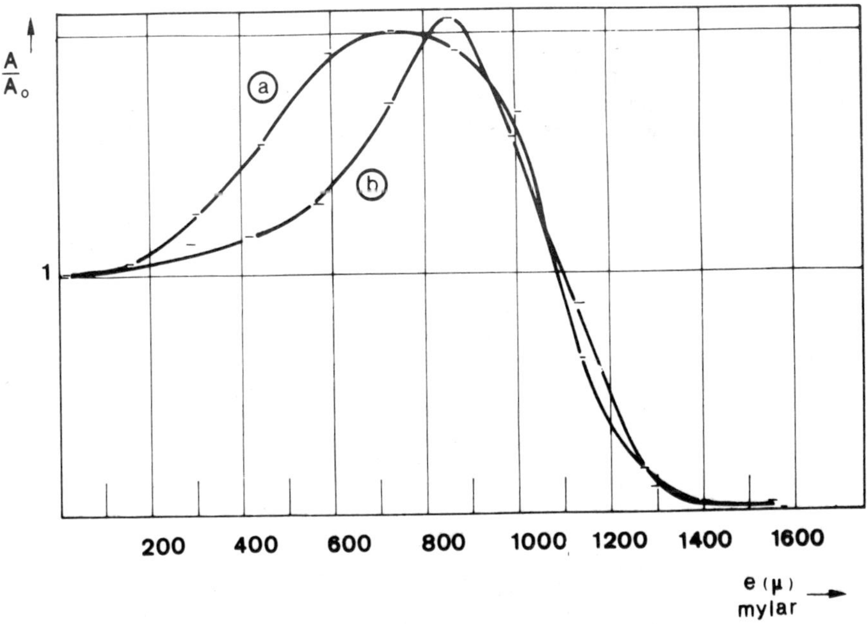

Fig. 2. $O \xrightarrow{\alpha} {}^{18}F$ *activation curves obtained by irradiating mylar stacks in the semi-extracted and extracted beams*

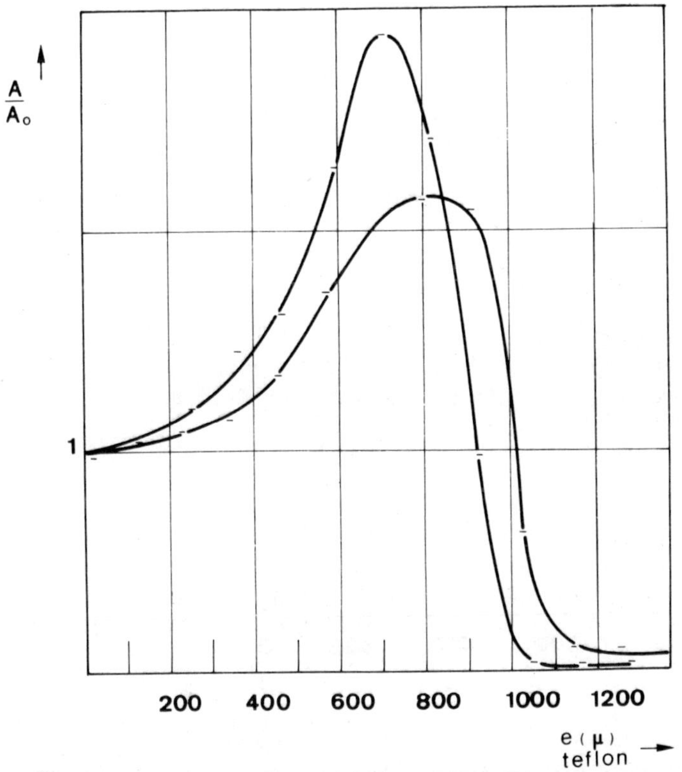

Fig. 3. ${}^{19}F \xrightarrow{\alpha} {}^{18}F$ *activation curves obtained by irradiating teflon stacks in the semi-extracted (a) and extracted (b) beams*

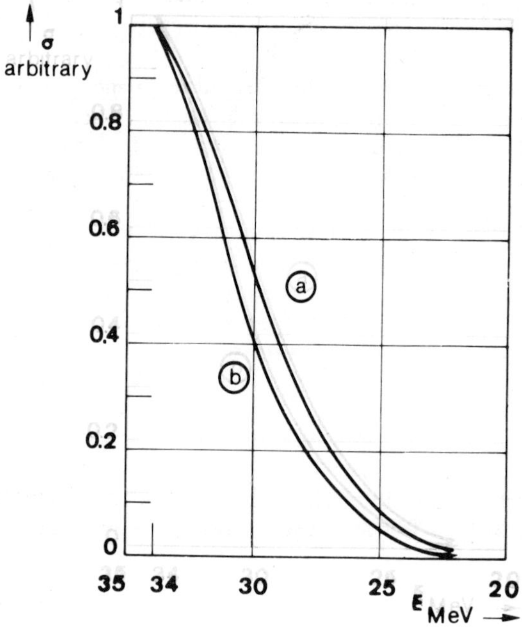

Fig. 4. $O \xrightarrow{\alpha} {}^{18}F$ *activation cross-sections below 35 MeV, with two energy distributions from two different accelerators. (a)* $E_0 = 54$ *MeV*
(b) $E_0 = 35$ *MeV normalised*

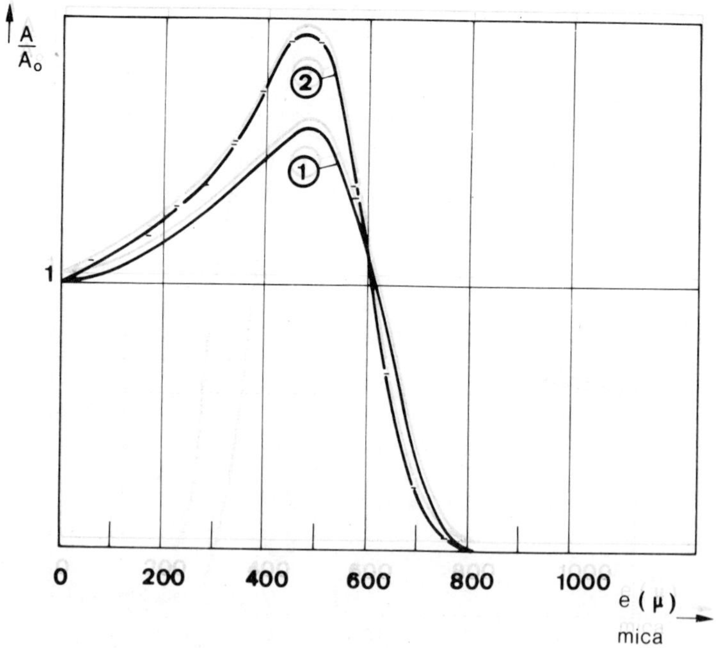

Fig. 5. Influence of diaphragm diameter: curve 1:16 mm, curve 2:6 mm ($O \xrightarrow{\alpha} {}^{18}F$ *reaction with semi-extracted beam)*

DISCUSSION

Importance of the energy distribution. The distribution in energy arises principally from two causes: a nominal dispersion due to the accelerator itself and to the irradiation position, and the dispersion caused by the degradation of the particle energy through absorbers and samples. From different studies of straggling it appears that the width of the distribution is proportional to the energy loss (except for a small energy loss), and the factor of proportionality does not in practice depend upon the nature of the absorber. The variances of the two distributions approximated to gaussian ones are added, but their relative importance is very variable. Straggling being an expected and known phenomenon, danger arises when one compares activation curves, mainly from the initial distribution of the particles. The requirement of high intensities often leads to irradiations with unanalysed beams and the use of large aperture diaphragms still further exaggerates the effect of dispersion.

For instance, if we consider the second irradiation position of the synchrocyclotron of the University of Lyon, the full width at half maximum of the distribution reaches 3 MeV for a mean value of the beam of 54 MeV. It is easily reduced, with an evident loss of intensity, by magnetic analysis and use of small diameter diaphragms, to <1 MeV at the third irradiation position. It follows that, on account of straggling, this last beam, slowed down to 30 MeV, presents an energy distribution with a standard deviation of ~1·8 MeV. It is clear that such a dispersion can affect an activation curve in parts where the excitation function has sharp slopes.

We have not yet demonstrated significantly the influence upon the energy distribution of the operating conditions of the accelerator (e.g. ion source position, acceleration voltage) but we are continuing this study.

Influence of these distributions. The effect of dispersion, which is obvious from the previous figures, may be estimated in advance by a quick computation

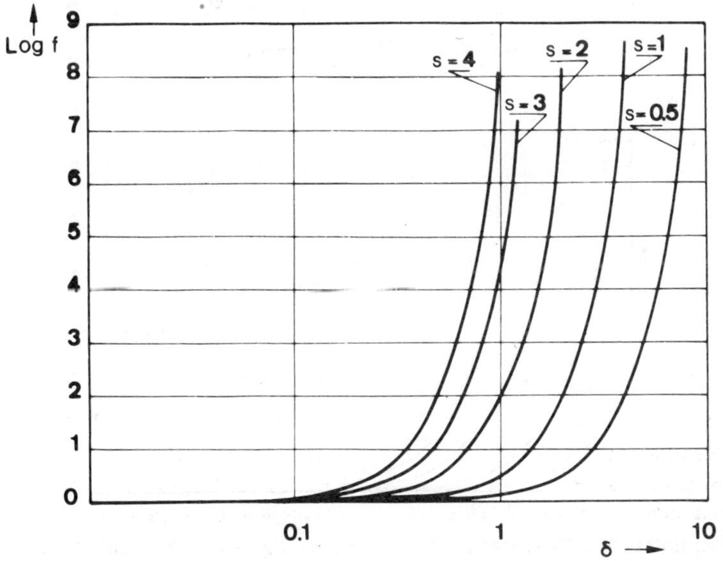

Fig. 6. Variations of the shape factor vs the coefficient δ for a few values of the standard deviation s

when the excitation function can be expressed in a simple analytical form. For an exponential excitation function, a common case which is well interpreted by the evaporation theory, a shape factor f, defined[6] as the ratio of the average activation cross-section measured in the target and the cross-section corresponding to the mean energy $\bar{E}$ of the particles striking the target, is given by the exponential expression:

$$e^{\frac{\delta^2 s^2}{2}}$$

δ expresses the exponential variation of the excitation function ($\sigma_E = \sigma_{\bar{E}}\, e^{\delta(E-\bar{E})}$) and s is the standard deviation of the distribution. Fig. 6 shows the variation of f with δ calculated for a few values of s.

In the particular case of the excitation function of the reaction $O \xrightarrow{\alpha} {}^{18}F$, the curve obtained below 36 MeV has a nearly gaussian shape, as shown by the representation in Fig. 7 of the curve obtained from a 35 MeV beam with good resolution. We then have a shape factor which may be expressed as follows:

$$f = \frac{d}{\sqrt{(s^2 + d^2)}} \exp\left[\frac{(E_m - \bar{E})^2 s^2}{2d^2(s^2 + d^2)}\right]$$

where E_m is the energy corresponding to the extrapolated maximum of the gaussian curve and d is the standard deviation.

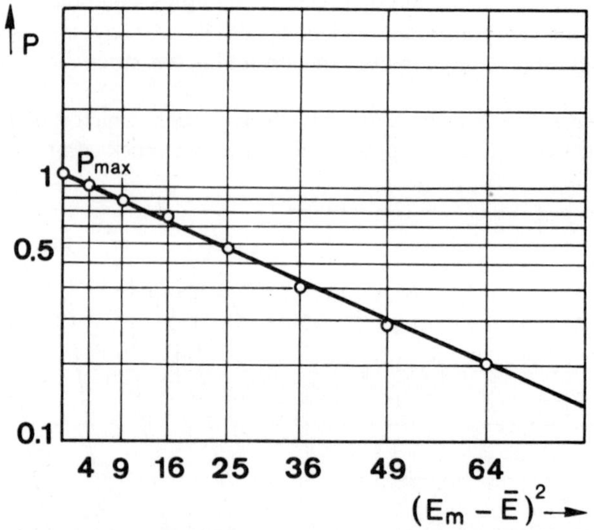

Fig. 7. Below 35 MeV the $O \xrightarrow{\alpha} {}^{18}F$ *excitation function is assimilated to gaussian*

Fig. 8 summarises variations in the shape factor f vs the difference $(E_m - \bar{E})$ for a few values of s, in the particular case where $d = 4{\cdot}25$, which corresponds to the example investigated. This shape factor applied to oxygen activation curves obtained from broad distribution beams (Fig. 4a or 9a) leads to a corrected curve (Fig. 9b) which agrees satisfactorily with the curve plotted with a well-resolved beam (Fig. 10). On the other hand, this ratio obviously

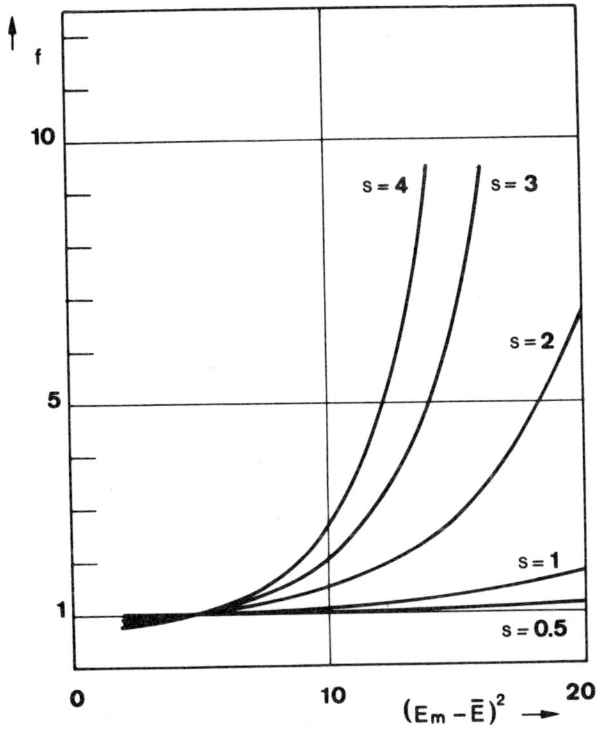

Fig. 8. Variations of the shape factor vs $(E_m - \bar{E})$ for a few values of the standard deviation s

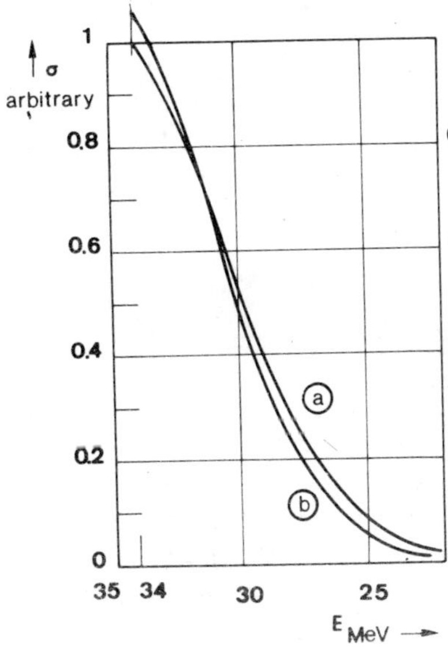

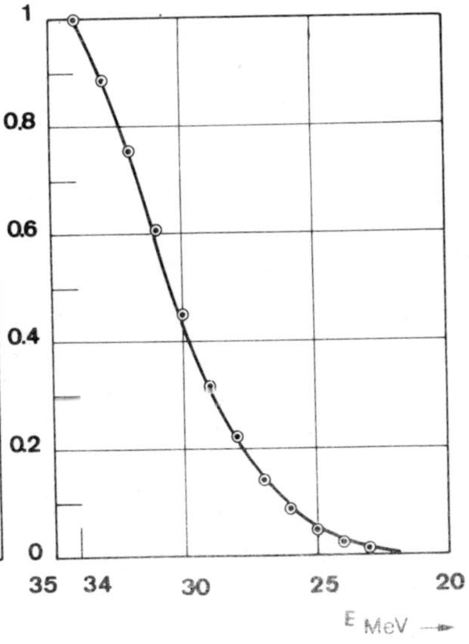

Fig. 9. Curve (b) was obtained by multiplying (a) by the shape factor calculated with s constant and equal to 2 MeV

Fig. 10. Full line: curve corresponding to a narrow distribution. Encircled points: plotting [illegible] from a curve corresponding to a broad distribution (s = 2 MeV) equals unity if the cross-sections either do not vary or vary in a linear manner, except in the neighbourhood of the threshold energy as shown below.

The values of the 'equivalent thickness' derived from the experimental activation curves thus show significant differences, the origin of which is clear (Table 1).

Table 1

E MeV	$e(\mu)$ $s_0 = 1{\cdot}3$: $E_0 = 54$ MeV	$e(\mu)$ $E_0 = 35$ MeV
34	110	93
32	65	51
30	32	23
27	9	5·5

The values of equivalent thickness in silicon for $O \xrightarrow{\alpha} {}^{18}F$ reactions from activation curves provided with two different beams do not agree.

Another, more important consequence lies in the well-marked shift towards low energies of the influence on a reaction with threshold energy. We give as an example the case of oxygen analysis in silicon by α-activation, for which a parasitic reaction[7] clearly appears above 40 MeV. With a broad distribution, the analysis is disturbed far below this value to ~36 MeV.

Fig. 11 shows the schematic behaviour of experimental activation cross-sections in the neighbourhood of the threshold when the excitation function has a constant

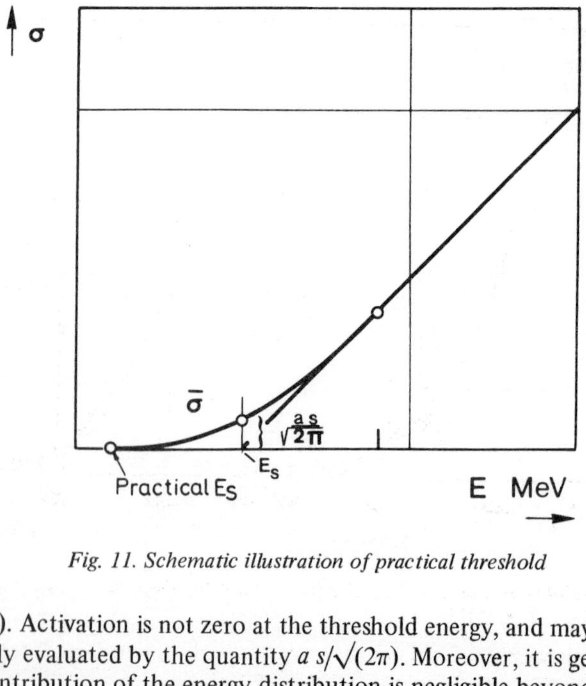

Fig. 11. Schematic illustration of practical threshold

gradient (a). Activation is not zero at the threshold energy, and may be theoretically evaluated by the quantity $a\,s/\sqrt{(2\pi)}$. Moreover, it is generally true that the contribution of the energy distribution is negligible beyond 2*s*.

CONCLUSION

The existence of an energy distribution of variable width is not a difficulty in activation analysis as long as one is certain of its reproducibility. However, frequent calibrations (mainly activation curves or direct measurements of the width) are required for any irradiation position of the accelerator. Slight corrections are necessary in the reciprocal transpositions of excitation functions and activation curves and a practical threshold is introduced with an energy which is lower than the calculated value for a parasitic endoergic reaction.

REFERENCES

1. Lacroix, M. J., Tran Minh Duc, and Tousset, J., Euratom Symposium 'Practical aspects of activation analysis with charged particules'–Liége (Belgium). EUR 3896, 351 (1968).
2. Tran Minh Duc and Tousset, J., Modern trends in activation analysis, Gaithersburg (U.S.A.), Vol. II, 754, (1968).
3. Engelmann, Ch., Rapport C.E.A. 2559 (1964).
4. Barrandon, J. N., Debrun, J. L. and Albert, Ph., Euratom Symposium 'Practical aspects of activation analysis with charged particules'–Liège (Belgium). EUR 3896, 277 (1968).
5. Rook, H. L., Schweikert, E. A. and Wainerdi, R. E., Modern trends in activation analysis, Gaithersburg (U.S.A.), Vol. II, 768 (1968).
6. Tran Minh Duc, Thesis, Université de Lyon (1968).
7. To be published.

DISCUSSION

H. D. Maccabee (M.R.C., Harwell): Have you tried comparing measured activation curves with theoretical activation curves? I think that you will find that the measured curves are broader than the predictions, if the straggling is accounted for by assuming a gaussian energy distribution with variance given by Bohr's theory. Our experiments have shown that the variance of straggling distributions in thick absorbers increases faster than Bohr's prediction.
M. D. Tran: We have already measured the energy-distribution broadening caused by straggling phenomena, and we have confirmed in previous work that the straggling width is much greater than that predicted by Bohr's assumptions. We have, of course, taken the experimental widths for correcting activation curves.

Contribution to the study of Be, B, C, N, O and F determination by p, ^{3}He and α activation

Ch. Engelmann

Département d'Etudes Métallurgiques, C. E. N. Saclay
Boite Postale n°2, 91–Gif sur Yvette, France

ABSTRACT

From activation curves obtained by irradiating thick targets containing Be, B, C, N, O and F with p, ^{3}He and α-particles, it is shown that these elements are measurable at concentrations of 10^{-10} to 10^{-9} g/g using 10 μA beams, assuming that there are no interferences from other impurities. B, N, O (by the $^{16}O(p,\alpha)^{13}N$ reaction, which proves to be much more sensitive than the reaction on ^{18}O) and F may be determined by activation with 15 MeV protons. Be, B, C and O may be determined by activation with 20 MeV ^{3}He particles, and Be, B, C, N, O and F by activation with 42 MeV α-particles.

The curves can also be used to plot graphs which give, as a function of the charged-particle energy, the magnitude of certain interferences attributable to the formation of the same active isotope from one or several other neighbouring elements. From these we deduce the relative selectivity of p, ^{3}He or α-particles for the determination of one or other of the above elements.

The possibilities are summarised in a table showing, for several energies of each of the three particles, the minimum concentrations which are measurable.

The criterion adopted in the calculation of these values is referred to a relatively high limiting activity, with the result that in practice it should be possible in certain cases to determine smaller concentrations than those indicated.

In addition, the activation curves can be used to predict the production yield of isotopes such as ^{11}C, ^{13}N, and ^{18}F obtained by irradiating thick targets of unspecified composition by weight.

INTRODUCTION

During the last few years charged particles have been used more and more to determine by activation analysis [illegible] light elements in extremely pure materials. We give here experimental results which contribute to knowledge of the practical possibilities of using p, ^3He and α-particles to determine low concentrations of Be, B, C, N, O and F. Unlike other authors in the field,[1,2] we wished to consider in principle the actual conditions prevailing when an analysis is made; this involves the determination of the specific activities induced in the elements to be analysed, contained in a matrix which absorbs the whole beam.

The results of this experimental study are presented as curves which give, as a function of the incident particle energy, the activities induced in the impurity investigated, per unit mass concentration, for a constant beam intensity. These curves can then be used to consider the possibilities of using the three particles in question for the determination of a given light element, taking into account the intrinsic sensitivities obtainable, but also allowing for competitive nuclear reactions on neighbouring elements, some of which were also studied in the course of our experiments.

From the same integrated activation curves it is also possible to calculate the yields of isotopes such as ^{11}C, ^{13}N, and ^{18}F which would be obtained by irradiating thick samples.

EXPERIMENTAL

The activation curves were all obtained by irradiating targets thicker than the range of the charged particles. The materials used were beryllium metal, H_3BO_3, graphite, AlN, SiO_2, and LiF.

In order to be able to standardise the results with respect to a reference activity, ^{13}N for protons, ^{18}F for ^3He and α-particles, all the experiments were carried out at fixed energy. The variation in energy was obtained using aluminium absorbers. The beam was monitored with thin mica discs (a few mg/cm^2), covered on both faces with 25 μ aluminium sheets to stop recoil nuclei, placed in front of the screen and target assembly and consequently always irradiated at the same energy. The respective reference activities chosen for the mica monitors are as follows:

10^7 dpm/mg of mica for ^{13}N at the end of a 1 min irradiation for protons

$$[E_p \simeq 15\,\text{MeV};\ ^{16}O(p, \alpha)^{13}N];$$

10^6 dpm/mg of mica for ^{18}F at the end of a 1 min irradiation for ^3He particles

$$[E_{^3He} \simeq 21\,\text{MeV};\ ^{16}O(^3He, p)^{18}F + {}^{16}O(^3He, n)^{18}Ne \xrightarrow{\beta+} {}^{18}F]$$

and α-particles

$$[E_\alpha \simeq 42\,\text{MeV};\ ^{16}O(\alpha, \text{p n and d})^{18}F + {}^{16}O(\alpha, 2n)^{18}Ne \xrightarrow{\beta+} {}^{18}F]$$

These reference values for the mica can be used to calculate the true beam intensities, using the cross-sections published in the literature, e.g. by taking $\sigma = 25$ mb for the $^{16}O(p, \alpha)^{13}N$ reaction at 15 MeV,[3] $\sigma = 50$ mb for the series of reactions

$$^{16}O(^3He, p)^{18}F \text{ and } {}^{16}O(^3He, n)^{18}Ne \xrightarrow{\beta+} {}^{18}F \text{ at } 21\,\text{MeV},^4$$

Table 1. LIST OF NUCLEAR REACTIONS FOR WHICH THE ACTIVATION CURVES HAVE BEEN DETERMINED AND OTHER PRINCIPAL REACTIONS POSSIBLE WITH THE ELEMENTS IRRADIATED UNDER THE EXPERIMENTAL CONDITIONS DESCRIBED

Element irradiated	*Nuclear reactions dealt with in our experiments*	*Threshold in* MeV	*Other reactions possible*	*Threshold in* MeV	*Other references*
Be	$^{9}Be(^{3}He, n)^{11}C$	Q>0			
	$^{9}Be(\alpha, 2n)^{11}C$	18·8			
B	$^{11}B(p, n)^{11}C$	3			
	$^{10}B(^{3}He, d)^{11}C$	Q>0	$^{11}B(^{3}He, n)^{13}N$	Q>0	Ricci et Hahn[2]
	$^{11}B(^{3}He, t)^{11}C$	2·5			
	$^{10}B(\alpha, n)^{13}N$	Q>0	$^{10}B(\alpha, t)^{11}C$	15·6	
	$^{11}B(\alpha, 2n)^{13}N$	14·2	$^{11}B(\alpha, tn)^{11}C$	34	
C	$^{12}C(p, \gamma)^{13}N$	Q>0			
	$^{13}C(p, n)^{13}N$	3·2			
	$^{12}C(^{3}He, \alpha)^{11}C$	Q>0	$^{12}C(^{3}He, d)^{13}N$	4·4	Ricci et Hahn[2]
	$^{12}C(\alpha, \alpha n)^{11}C$	25			
N	$^{14}N(p, \alpha)^{11}C$	3·1	$^{14}N(p, d)^{13}N$	8·9	
			$^{14}N(p, pn)^{13}N$	11·3	
	$^{14}N(^{3}He, \alpha)^{13}N$	Q>0			
	$^{14}N(\alpha, \alpha n)^{13}N$	13·6	$^{15}N(\alpha, n)^{18}F$	8·1	
O	$^{16}O(p, \alpha)^{13}N$	5·5			
	$^{18}O(p, n)^{18}F$	2·6			
	$^{16}O(^{3}He, p)^{18}F$	Q>0	$^{16}O(^{3}He, 2\alpha)^{11}C$	6·3	Debrun, Barrandon, and Albert[1]
	$^{16}O(^{3}He, n)^{18}Ne \xrightarrow{\beta+} {}^{18}F$	3·8			
	$^{16}O(\alpha, d)^{18}F$	20·4	$^{16}O(\alpha, 2\alpha n)^{11}C$	32·5	id.
	$^{16}O(\alpha, pn)^{18}F$	23·2			
	$^{16}O(\alpha, 2n)^{18}Ne \xrightarrow{\beta+} {}^{18}F$	29·7			
F	$^{19}F(p, d)^{18}F$	8·6			
	$^{19}F(p, pn)^{18}F$	11			
	$^{19}F(^{3}He, \alpha)^{18}F$	Q>0			
	$^{19}F(\alpha, \alpha n)^{18}F$	12·6			

and $\sigma = 220$ mb for the series of reactions

$$^{16}O(\alpha, p\ n\ \text{and}\ d)^{18}F + {}^{16}O(\alpha, 2n)^{18}Ne \xrightarrow{\beta+} {}^{18}F \text{ at } 42\,\text{MeV}$$

(in fact the value 220 mb is given for 40 MeV α's, but from the shape of the excitation function there would seem to be little difference at 42 MeV), we find the following intensities:

I_p = 0·85 μA
$I_{^3He}$ = 0·94 μA
I_α = 0·25 μA (assuming the above hypothesis to be valid).

RESULTS

The activation curves (Figs. 1–10) correspond to the nuclear reactions of Table 1. They show, as a function of the charged particle energy, the activities obtained in disintegrations per minute at the end of irradiation, where the light element concerned is present at a mass concentration of 10^{-6} in the target, irradiated for one minute in a beam of intensity corresponding to the values shown above.

The ranges of the three charged particles in the target materials are similar,[6] which means that the various curves can quite justifiably be compared with one another. Obviously, for a matrix of different stopping power the activities induced for a given composition by weight are proportional to the respective range. For this reason the results shown are extrapolated to all target compositions.

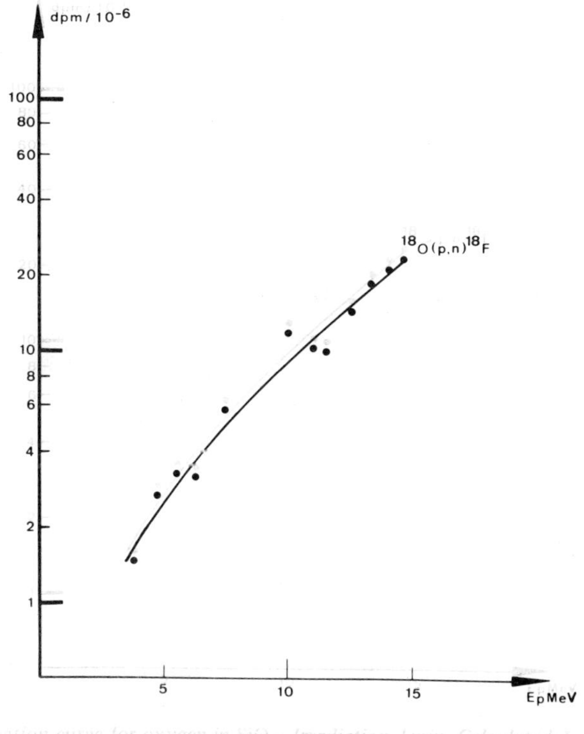

Fig. 1. Activation curve for oxygen in SiO_2*: Irradiation 1 min. Calculated* $I_p = 0{\cdot}85\ \mu A$

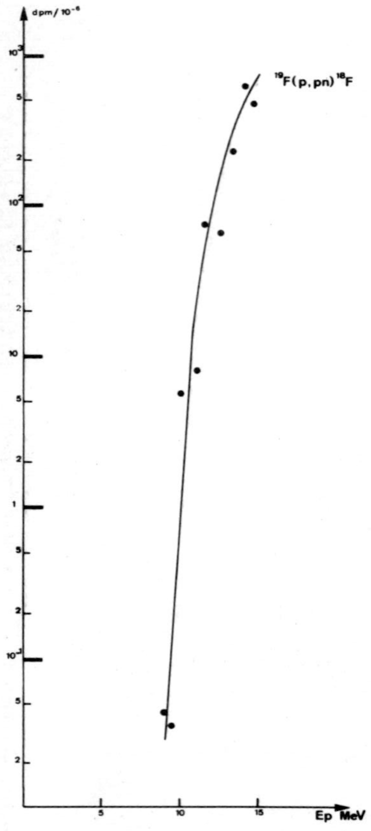

Fig. 2. Activation curve for fluorine in LiF: *Irradiation 1 min. Calculated* $I_p = 0{\cdot}85\ \mu A$

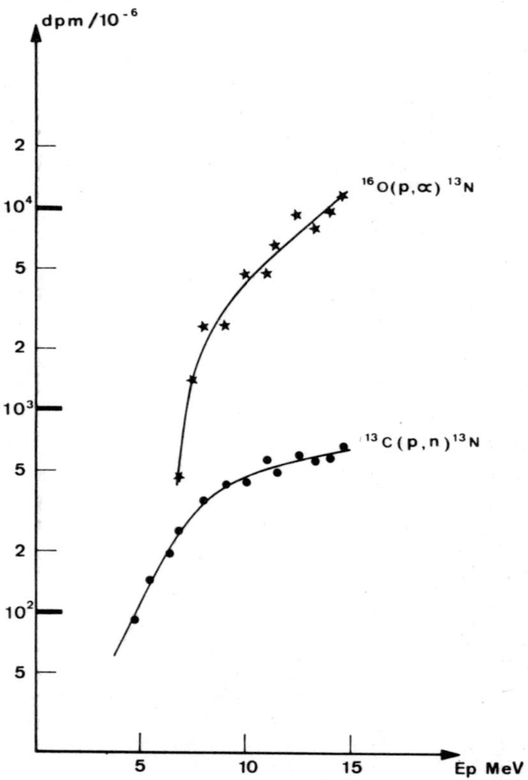

Fig. 3. Activation curves for carbon as graphite and oxygen in SiO_2: *Irradiation 1 min. Calculated* $I_p = 0{\cdot}85\ \mu A$

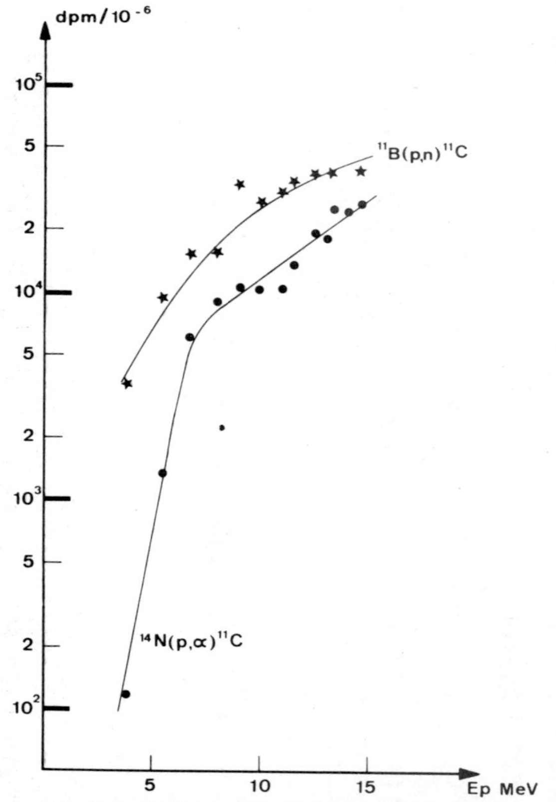

Fig. 4. Activation curves for boron in H_3BO_3 *and nitrogen in* AlN: *Irradiation 1 min. Calculated* $I_p = 0{\cdot}85\ \mu A$

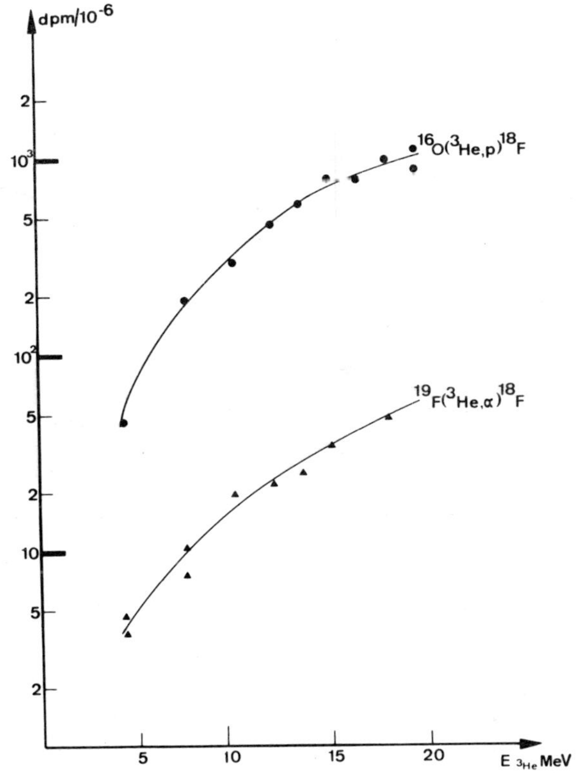

Fig. 5. Activation curves for oxygen in SiO_2 *and fluorine in* LiF: *Irradiation 1 min. Calculated* $I_{3He} = 0.94\ \mu A$

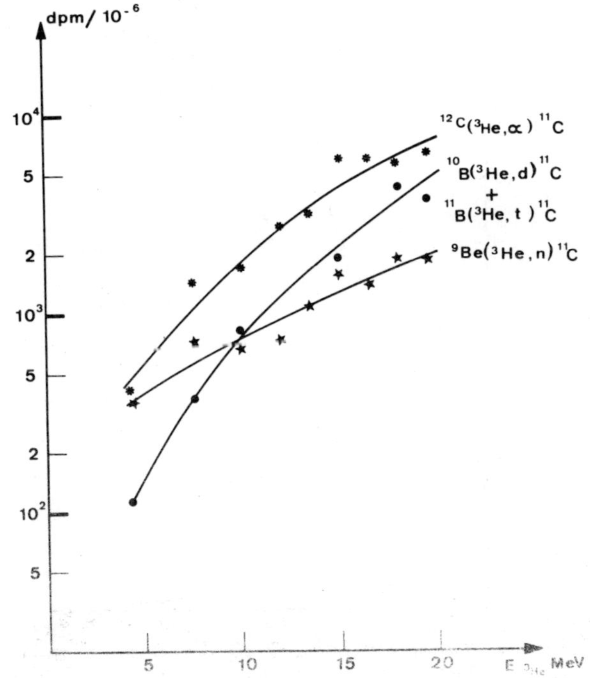

Fig. [illegible] Activation curves for [illegible]rbon as graphite, [illegible]

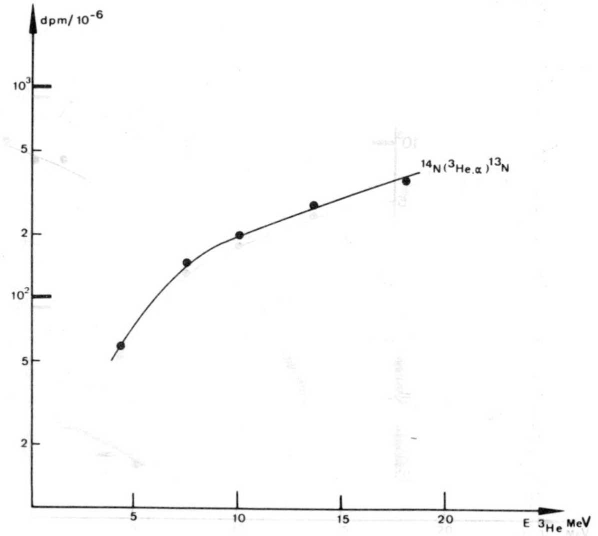

Fig. 7. Activation curve for nitrogen in AlN: *Irradiation 1 min. Calculated* $I_{3He} = 0{\cdot}94\ \mu A$

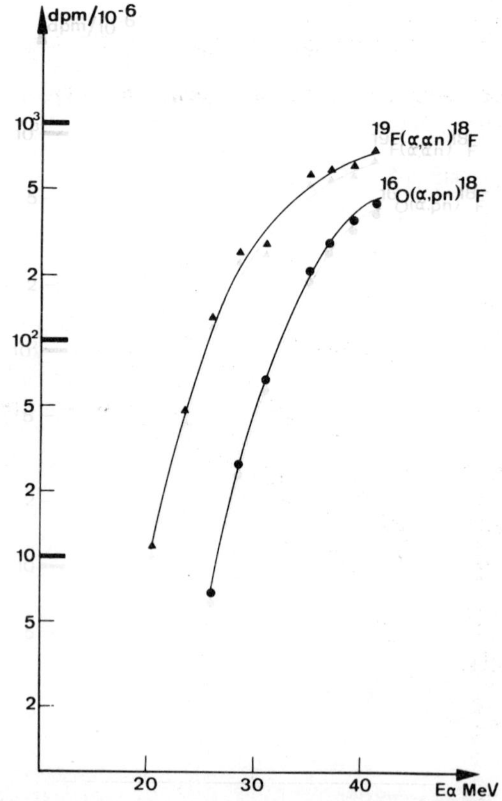

Fig. 8. Activation curves for oxygen in SiO_2 *and fluorine in* LiF: *Irradiation 1 min. Calculated* $I_\alpha = 0{\cdot}25\ \mu A$

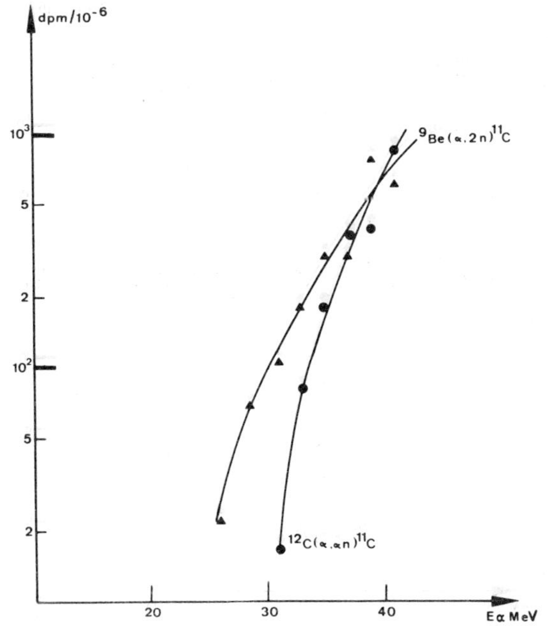

Fig. 9. Activation curves for carbon as graphite and beryllium as metal: Irradiation 1 min. Calculated I_{α} = *0·25 μA*

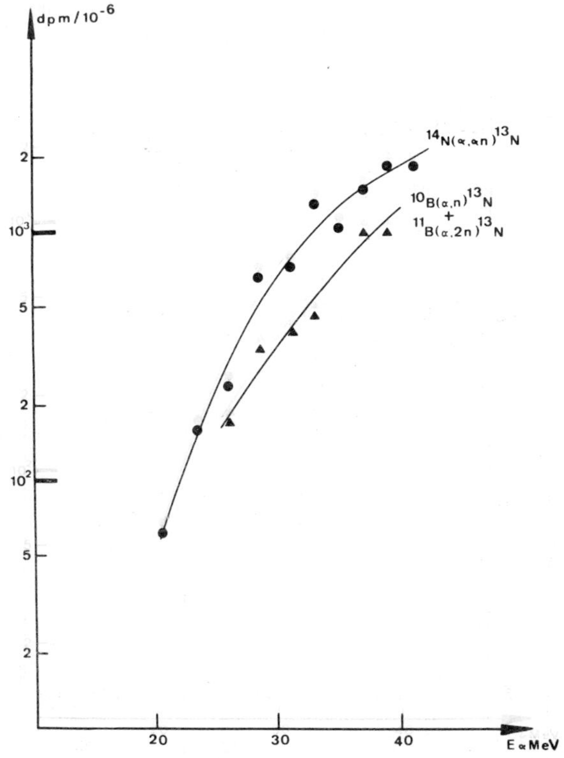

Fig. 10. Activation curves for nitrogen in AlN *and boron in* H_3BO_3*: Irradiation 1 min. Calculated* I_{α} = *0·25 μA*

Table 2. MINIMUM CONCENTRATIONS MEASURABLE IN 10^{-9} g/g IF THE CRITERION ADOPTED IS 1000 dpm AT THE END OF AN IRRADIATION LASTING ONE HALF-LIFE IN A 10 μA BEAM

	E, MeV	Be	B	C	N	O		F
			[^{11}C]	[^{13}N]	[^{11}C]	[^{13}N]	[^{18}F]	[^{18}F]
	5		0·92	110	9·2		430	
	7·5		0·40	40	0·83	8	216	
ρ	10		0·24	27	0·50	3	114	
	12·5		0·17	22	0·33	1·6	68	8·3
	15		0·13	18	0·22	1	43	1·6
		[^{11}C]	[^{11}C]	[^{11}C]	[^{13}N]		[^{18}F]	[^{18}F]
	5	16·5	44	12	175		17	240
	10	8·8	8·2	3·6	66		3·7	70
^{3}He	15	5	3	1·5	44		1·7	34
	20	3·3	1·3	0·85	29		1·1	20
		[^{11}C]	[^{13}N]	[^{11}C]	[^{13}N]		[^{18}F]	[^{18}F]
	26	70	20		12		45	2·9
	30	17·5	10		5·4		6·4	1·1
α	34	8	6	16	3·2		2·1	0·7
	38	3·9	3·5	4·4	2·2		1	0·5
	42	2		1·75	1·65		0·7	0·4

Production of ^{11}C, ^{13}N *and* ^{18}F *by irradiation of thick targets.* The possibilities of the three particles as a function of energy, within the range investigated, are calculated directly from the experimental curves.

Determination of Be, B, C, N, O *and* F *by* p, ^{3}He *and α-particle activation.* By defining the minimum quantity measurable as the mass concentration of the above elements giving an activity of 1000 dpm at the end of irradiation, we can draw up Table 2 which summarises all the possibilities if competitive nuclear reactions are neglected. The values given were calculated on the assumption that the sample to be analysed is irradiated in a 10 μA beam for a period equal to the half-life of the radioelement formed and measured for analysis purposes. As before, we also assume that the range of the charged particles in the sample analysed is similar to that in aluminium. When heavy materials are analysed the sensitivity of the measurement is improved in proportion to the range.

DISCUSSION

Other activation curves could be plotted. Some of the corresponding nuclear reactions are shown in Table 1, and quantitative data concerning them can be found in the publications already mentioned.[1,2] Those presented here relate to

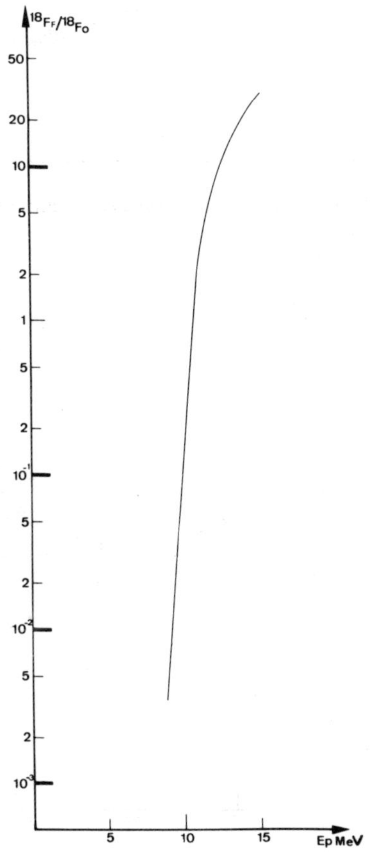

Fig. 11. Variation of the mutual fluorine-oxygen interference as a function of the proton energy

the main activities obtained and measured by irradiation of the targets used. Thus, not all the possible competitive nuclear reactions are taken into consideration here, and consequently the following discussion is limited to those dealt with in our own experiments.

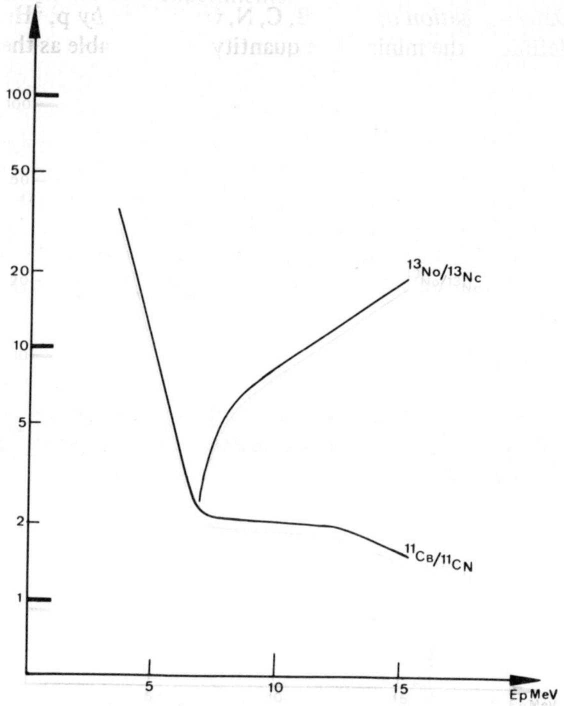

Fig. 12. Variations of the mutual boron-nitrogen and oxygen-carbon interferences as a function of the proton energy

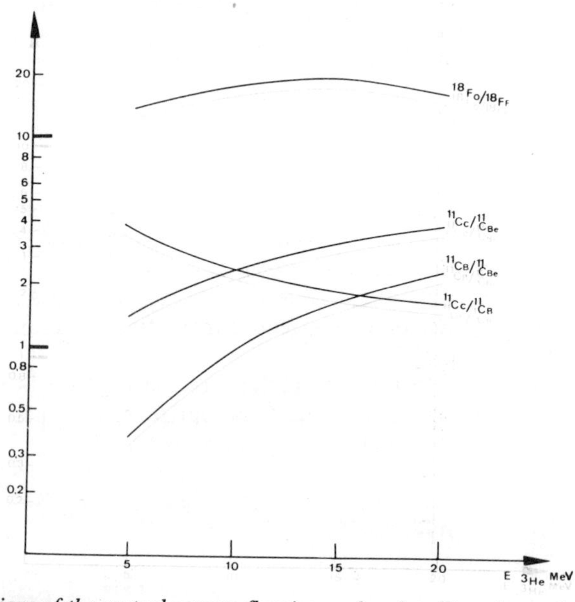

Fig. 13. Variations of the mutual oxygen-fluorine, carbon-beryllium, boron-beryllium, and carbon-boron interferences as a function of the ^{3}He *energy*

So far as sensitivity is concerned, the results in Table 2 require no further comment. From the viewpoint of the influence of competitive nuclear reactions, however, an analysis of the activation curves is necessary. For this purpose we have used them to obtain graphs showing, as a function of energy, how certain of the elements investigated can interfere in the determination of another. The quantities plotted as ordinates represent, for each energy, the ratios of the integrated activities of two elements assumed to be at the same concentration in the sample and giving the same active isotope (Figs 11–14). We should

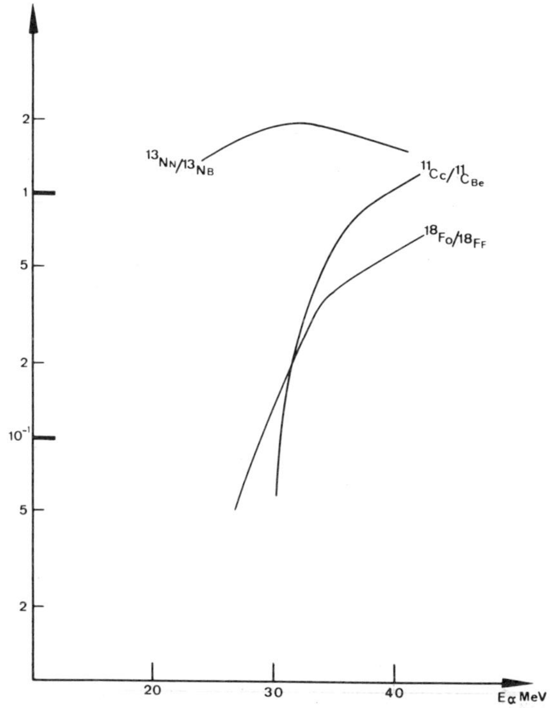

Fig. 14. Variations of the mutual oxygen-fluorine, carbon-beryllium, and nitrogen-boron interferences as a function of the α-particle energy

point out that the curves relating to C and O determination by ^{3}He and α-activation (Be and F interference) (cf. Figs. 13 and 14) give results similar to those of Debrun, Barrandon and Albert,[1] who used the stacked foil method.

The curves thus obtained need no explanation. They show clearly the importance of certain competitive nuclear reactions, and how in some cases it is possible to avoid them or at least reduce their effect by a suitable choice of particle or particle energy.

In any case the values in Table 2 and the curves in Figs. 11–14 provide a useful basis on which to choose the most appropriate charged particle and energy for the particular analytical problem involved.

ACKNOWLEDGEMENTS

We extend our particular thanks to Mlle. M. Loeuillet and M. R. Chazot who helped us in this work, and to the operators of the two cyclotrons at CEN, Saclay.

REFERENCES

1. Debrun, J. L., Barrandon, J. N., and Albert, Ph., *Bull. Soc. Chim. France n° 3* 1011, (1969).
2. Ricci, E. and Hahn, R. L., *Anal. Chem.* **40**, 54, (1968).
3. Hill, H. A., Haase, E. L., and Knudsen, D. B., *Phys Rev.* **123**, 1301, (1961).
4. Markowitz, S. S. and Mahony, J. D., *Anal. Chem.* **34**, 329, (1962).
5. Furukawa, M. and Tanaka, S., *J. Phys. Soc.* Japan **16**, 129, (1961).
6. Williamson, C. F., Boujot, J. P., and Picard, J., *Rapport CEA* R 3042 (1966).

Experimental data on the yields of ^{11}C, ^{13}N, ^{18}F isotopes used for the detection of carbon, nitrogen, oxygen and neighbouring light element impurities by means of activation analysis with different charged particles (p, d, ^3He, α)

N. N. Krasnov, P. P. Dmitriyev, S. P. Dmitriyeva, I. O. Konstantinov, and G. A. Molin

Institute of Physics and Power Engineering, Obninsk, U.S.S.R.

ABSTRACT

Experimental values are given of the yields of ^{11}C, ^{13}N, and ^{18}F isotopes produced by the irradiation of carbon, nitrogen, oxygen, and neighbouring light elements (Be, B, F, Na, Mg, and Al) by protons, deuterons, ^3He ions, and α-particles. Experimental data are presented in the form of curves giving the dependence of isotope yields on particle energy (E_p = 22 MeV, E_d = 22 MeV, E_{3He} = 30 MeV, E_α = 44 MeV). On the basis of these data various methods of detecting carbon, nitrogen, and oxygen impurities by bombardment with the above four types of particles are considered.

INTRODUCTION

One of the promising applications of charged particle activation analysis is the determination of carbon, nitrogen and oxygen impurities in different materials. It is convenient to determine these impurities by measuring the radioactivity of such isotopes as ^{11}C, ^{13}N, and ^{18}F.

To choose the best conditions for analysis, i.e. to choose the type of particles and particle energy, it is necessary to have experimental data on the isotopic yields or reaction cross-sections. We have shown[1] that in practice it is much more convenient to use values of isotopic yields rather than values of reaction cross-sections. In this case the radioactivity of a certain isotope *k*, produced from the impurity element *i* which is present in the material *m*, is determined by a simple expression:

$$^kN_{im} = {}^kB_i\eta_{im}\frac{P_m}{P_i}I\left(\frac{1-e^{-\lambda_k t}}{\lambda_k}\right) \tag{1}$$

where $^kN_{im}$ is the activity of isotope k in disintegrations per sec at the end of exposure; ^kB_i (dps/μAh) is the k isotopic yield for the thick target consisting completely of the i element; η_{im} is the relative weight content of the i element in the m material; P_m and P_i are coefficients proportional to the charged-particle range in the material m and the element i (values of these coefficients for all elements are given in ref. (1)). I is the beam current in μA, t is the exposure time in h and λ_k is the k isotope decay constant.

This expression is the basic formula for charged-particle activation analysis and leads[1] to simple expressions for the determination of impurities by the comparator method both in thick targets and in thin foils. Eqn (1) also gives equations for simultaneous detection of impurities of some elements, using the activity of one isotope produced from these elements. The use of P_m and P_i coefficients proportional to charged-particle range allows us to eliminate completely the necessity of calculating charged-particle ranges in the materials investigated, and simplifies the procedure for calculating the amount of impurities. Any substance with any known content of the element to be determined may be used as a standard. On the basis of Eqn (1) impurities can be determined by the absolute method if the values of the isotopic yields are known.

The purpose of the present study was to determine the yields of ^{11}C, ^{13}N and ^{18}F isotopes from carbon, nitrogen and oxygen by irradiation with protons, deuterons, ^3He and ^4He ions. Since the radioactive isotopes ^{11}C, ^{13}N and ^{18}F may also be formed from other light elements, the yields of these isotopes were measured by irradiating Be, B, F, Na, Mg, and Al with corresponding particles.

All the experimental data are given in the form of curves showing the dependence of isotopic yields on particle energy. These data allow us to choose the optimum conditions of irradiation (type and energy of particles) for each problem in order to evaluate easily the sensitivity of the analysis and to use the absolute method for determination of impurities.

The technique of measuring isotope yields. Samples were irradiated on the cyclotron at the Institute of Physics and Power Engineering,[2] which accelerates protons to ~22 MeV, deuterons to ~22 MeV, ^3He ions to ~30 MeV, and ^4He ions to ~44 MeV. The samples were exposed to the external cyclotron beam in a special target assembly which allowed three samples to be exposed simultaneously. The samples were cylindrical, of diam. 10 mm and thickness larger than the particle range. The dose was measured by means of copper monitoring foils ~20 mg/cm^2 thick which were mounted in front of the samples. The activity of these monitoring foils was compared with that of a standard foil irradiated by a fixed beam current. To prevent isotopic loss during irradiation the samples were placed in a well-cooled target-holder and were irradiated at a low current (0·1 μA). The exposure time was usually 10 min. The particle energy was varied by means of copper stopping foils. The dependence of average particle energy on the foil thickness was calculated according to references 3 and 4.

The activity of the samples was measured by a one-channel scintillation γ-spectrometer calibrated with ^{85}Sr standard sources. The dependence of activity on time (the decay curve) was measured for every sample. The activity of specific isotopes was determined by analysis of the decay curve and the values of the yields were calculated from this activity with the aid of Eqn (1).

The elements, radioactive isotopes produced from these elements and materials which were subjected to irradiation are given in Table 1. The values

Table 1

Elements	*Isotopes formed*				*Target*	$\eta_{im} \frac{P_m}{P_i}$
	22 MeV protons	22 MeV deuterons	30 MeV ^{3}He ions	44 MeV α-particles		
Be	–	–	^{11}C	^{11}C	Be	1
B	^{11}C	^{11}C	^{11}C, ^{13}N	^{11}C, ^{13}N	B	1
C	^{11}C, ^{13}N	^{11}C, ^{13}N	^{11}C, ^{13}N	^{11}C, ^{13}N	C	1
N	^{11}C, ^{13}N	^{11}C, ^{13}N	^{11}C, ^{13}N	^{11}C, ^{13}N, ^{18}F	AlN	0·38
O	^{13}N, ^{18}F	^{13}N, ^{18}F	^{11}C, ^{18}F	^{18}F	Al_2O_3	0·51
F	^{18}F	^{18}F	^{18}F	^{18}F	$(CF_2)_n$	0·74
Na	–	–	^{18}F	^{18}F	NaCl	0·42
Mg	–	–	^{18}F	–	Mg	1
Al	–	–	^{18}F	–	Al	1

of the yields, which are given by the curves below, are for thick targets consisting completely of the elements indicated.

The error in determining the isotopic yields is ±10%, and results from errors in measuring exposure quantity and absolute isotopic activity. The points shown on the curves are average values obtained from the treatment of the experimental data. The error limits (±10%) are deliberately not shown on the curves, because where there are two curves in one figure the error limits frequently overlap. The initial energy of the particles is shown within ±2%. When the particle energy is close to the reaction threshold, experimental values of the yields may be over-estimated because of a spread in particle range (straggling), which is a well-known defect of the stacked-foil technique. The origin of the isotopic yield curves was therefore determined by theoretical calculation of the reaction threshold.

YIELDS OF ^{11}C AND ^{13}N PRODUCED BY IRRADIATION OF CARBON

Experimental values of the yields of ^{11}C and ^{13}N produced by irradiation of carbon are given in Fig. 1. The yields are given in dps per microampere-hour (dps/μAh). There are no direct published data on these yields, but there are papers in which excitation functions of reactions were measured by bombarding carbon with charged particles, and by integrating an excitation function it is possible to derive values of isotopic yields. For comparison with our data we have integrated the following excitation functions taken from published data: $^{12}C(^{3}He, \alpha)^{11}C$,[5] $^{12}C(^{3}He, pn)^{13}N$ and $^{12}C(^{3}He, \alpha)^{11}C$,[6] $^{12}C(d, n)^{13}N$,[7] $^{12}C(d, t)^{11}C$ and $^{12}C(d, n)^{13}N$.[8] The yields calculated in this way differed from our data: in ref. (5) by + 5% (at 30 MeV), in (6) by + 10% (at 24 MeV), in (7) by + 20% (at 20 MeV), and in (8) by + 100% (at 22 MeV).

Consideration of the curves in Fig. 1 shows that in principle all four particles may be used to determine carbon. With low particle energies, which are desirable to minimise activation of the matrix, the greatest yields and consequently high sensitivity are achieved by irradiation with deuterons (^{13}N)

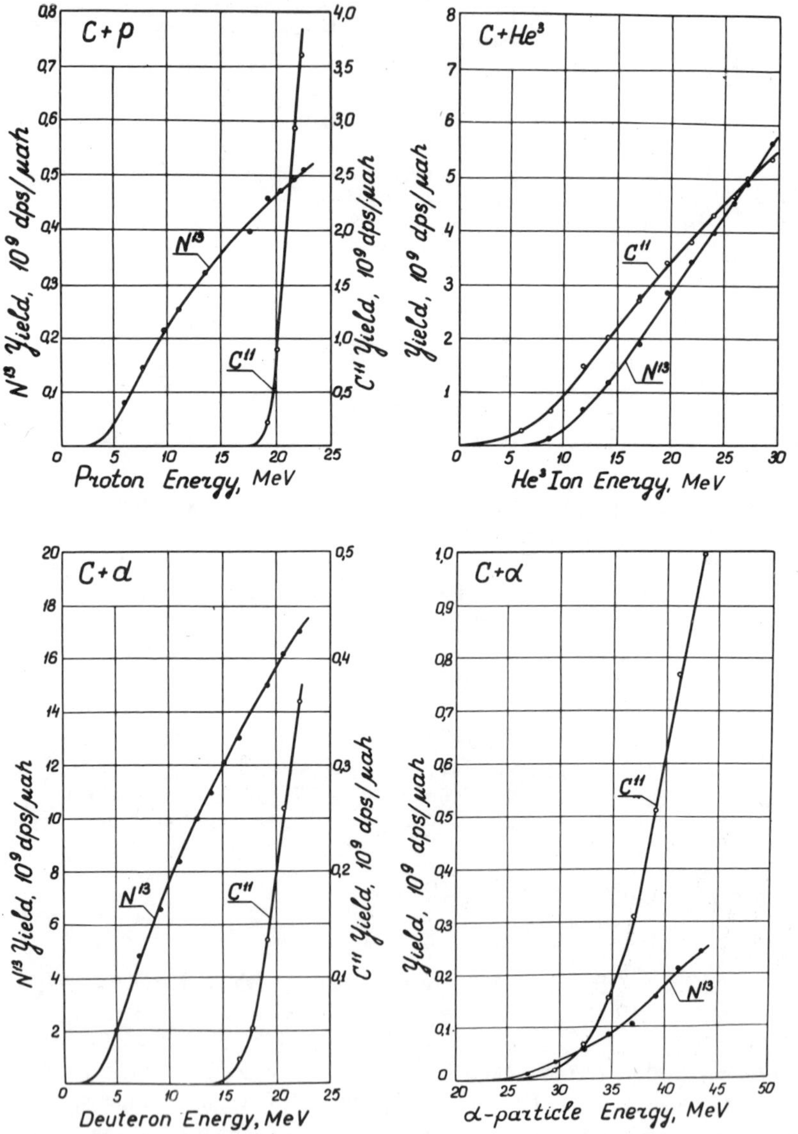

Fig. 1. Yields of ^{11}C *and* ^{13}N *formed by irradiation of carbon with protons, deuterons,* ^{3}He *and* ^{4}He *ions*

and ^{3}He ions (^{11}C). These two methods are therefore recommended for determination of carbon impurities.

YIELDS OF ^{11}C, ^{13}N AND ^{18}F PRODUCED BY IRRADIATION OF NITROGEN

Experimental values of the yields of ^{11}C, ^{13}N, and ^{18}F produced by irradiation of nitrogen with protons, deuterons, ^{3}He and ^{4}He ions are shown in Fig. 2. Data on cross-sections and isotopic yields for irradiation of nitrogen with

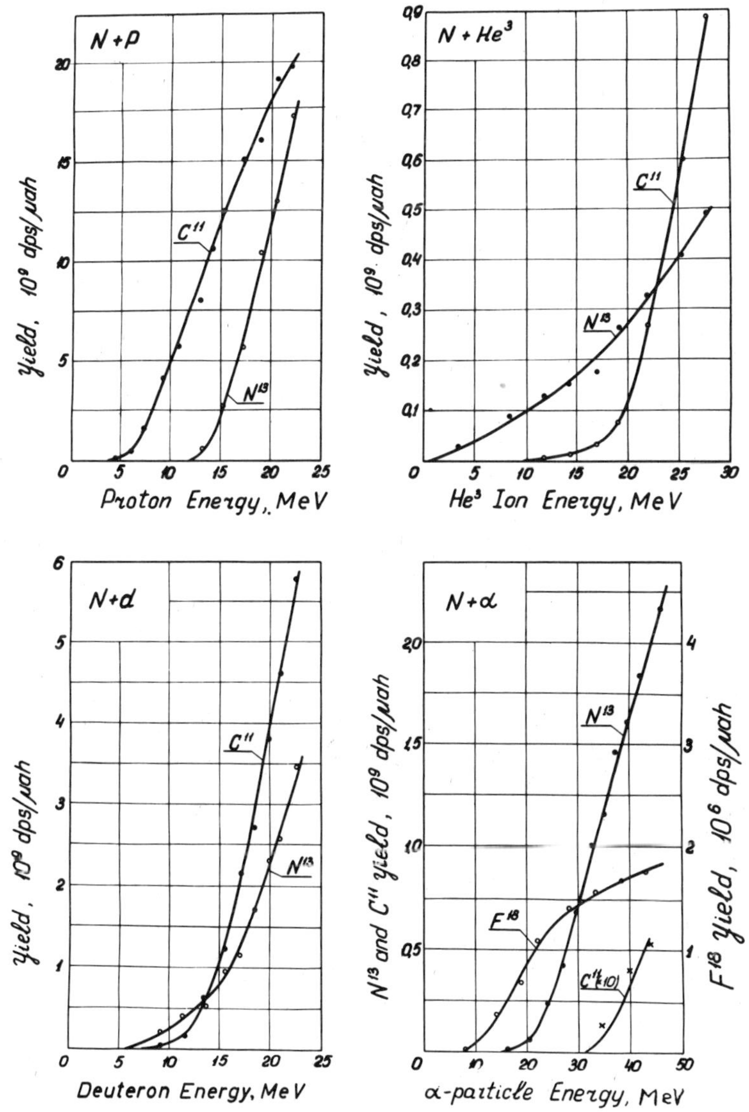

Fig. 2. Yields of ^{11}C, ^{13}N *and* ^{18}F *formed by irradiation of nitrogen with protons, deuterons,* ^{3}He *and* ^{4}He *ions*

charged particles of the energies indicated are scarce. Only two papers are known, in which only one reaction, $^{14}N(^{3}He, \alpha)^{13}N$, of the nine shown in Fig. 2 was studied, and in our opinion these data are erroneous. In ref. (5) the reaction cross-section was measured up to $E_{^3He} = 30$ MeV. Integration of this excitation function gave a yield of ^{13}N (at 30 MeV) five times greater than that measured in the present work. In ref. (9) the yield of ^{13}N was measured for the same reaction up to $E_{^3He} = 18$ MeV. This curve of ^{13}N yield is in good agreement with ours up to $E_{^3He} = 10$ MeV but then rises sharply. This is evidently explained by the fact that the activity of ^{11}C, which is formed at $E_{^3He} > 10$ MeV, was not considered in this case.

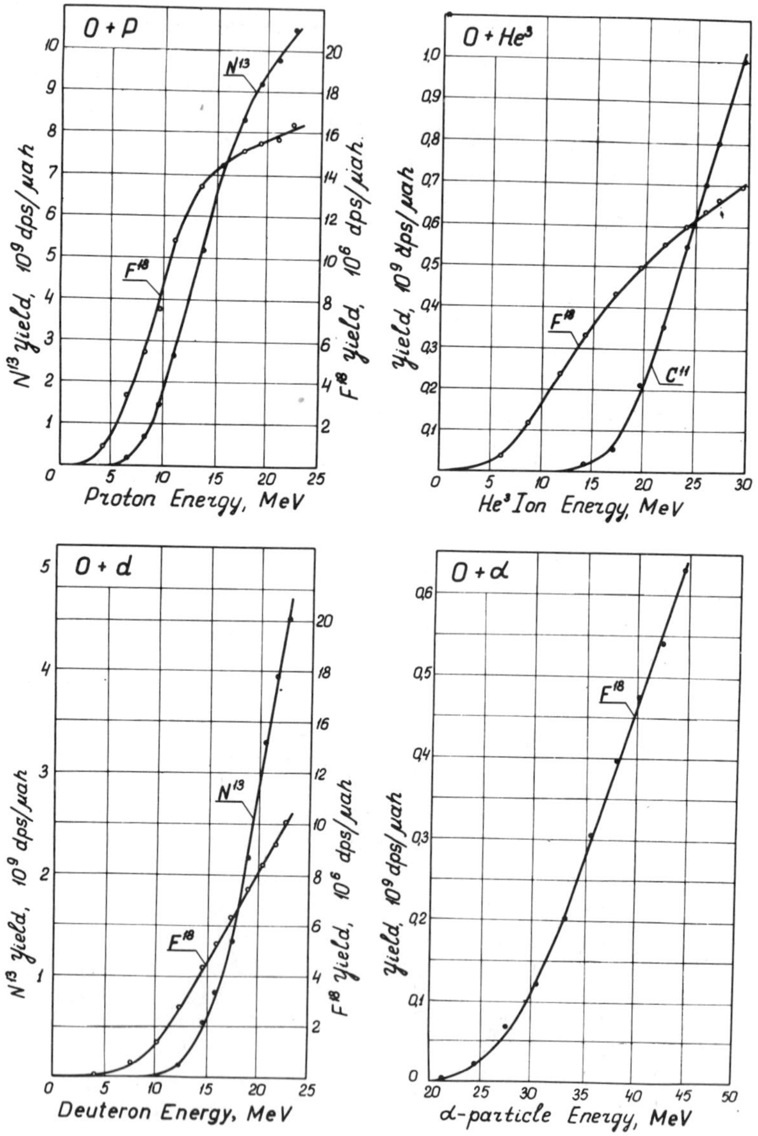

Fig. 3. Yields of ^{11}C, ^{13}N *and* ^{18}F *formed by irradiation of oxygen with protons, deuterons,* ^{3}He *and* ^{4}He *ions*

Consideration of the curves in Fig. 2 shows that all four particles can be used to determine nitrogen impurity. The highest yield is for the reaction $^{14}N(p, \alpha)^{11}C$, and hence it is more convenient to use proton irradiation for the determination of nitrogen. In this case rather a high yield of ^{11}C is obtained at low proton energy, which minimises activation of the material investigated. The use of ^{4}He ions with energy <22 MeV is of great interest, because in this case ^{18}F, which is convenient for radiochemical reasons, is produced only by nitrogen. However, in this case the low yield of ^{18}F results in very low sensitivity.

YIELDS OF ^{11}C, ^{13}N, ^{18}F PRODUCED BY IRRADIATION OF OXYGEN

Experimental data on the yields of ^{11}C, ^{13}N and ^{18}F produced by irradiation of oxygen with different charged particles are given in Fig. 3. Comparison of measured values of isotopic yields with other data is again only possible by integrating published excitation functions, as there are no direct data on the yields of these isotopes in the published literature. We have integrated the following published excitation functions: $^{16}O(\alpha\ pn)^{18}F$,[10] $^{16}O(^{3}He, p)^{18}F$,[11] $^{16}O(p, \alpha)^{13}N$.[12] The values of the yields calculated in this way differ from our data: in ref. (10) by + 30% (at 40 MeV), in (11) by –15% (at 30 MeV), in (12) by –5% (at 15 MeV).

The irradiation of oxygen with ^{3}He ions ($E > 20$ MeV) should result in the formation of ^{13}N by the reaction $^{16}O(^{3}He, \alpha\ pn)^{13}N$. However, due to the high activity of ^{11}C and ^{18}F, the activity of ^{13}N could not be determined by analysis of the decay curve. The data obtained only allow us to conclude that the ^{13}N yield in this case (at 30 MeV) is more than 100 times less than the ^{11}C yield. Consideration of the data in Fig. 3 shows that in principle all four particles can be used to determine oxygen impurities. When the measured isotope is isolated radiochemically from the material investigated ^{18}F is the most convenient nuclide. In this case high sensitivity will be achieved when the material is bombarded with ^{3}He or ^{4}He ions. Proton irradiation gives less sensitivity with ^{18}F, but because of the large activation thickness this irradiation may be preferable in some cases, since it allows us to analyse deep layers of a material which are not subjected to surface contamination. If oxygen impurities are determined by measuring ^{13}N, high sensitivity is achieved by proton irradiation.

YIELDS OF ^{11}C, ^{13}N, ^{18}F FORMED BY IRRADIATION OF LIGHT ELEMENTS (Be, B, F, Na, Mg, Al) WITH CORRESPONDING CHARGED PARTICLES

In practical applications of charged-particle activation analysis it often happens that the radioactive isotope by means of which the impurity of a particular element is determined may be formed from other elements, the presence of which in the matrix is unknown. In order to choose the best conditions for irradiation in this case it is necessary to have data on the isotope yield for all possible ways of formation. Thus, when carbon, nitrogen, and oxygen impurities are determined by measuring the radioactivity of ^{11}C, ^{13}N and ^{18}F it is necessary to have accurate data on the yields of these isotopes from all the neighbouring elements. For this purpose we have investigated all possible ways of ^{11}C, ^{13}N and ^{18}F formation and have measured their yields in reactions with protons, deuterons, ^{3}He and ^{4}He ions with particle energies of practical interest for activation analysis. Only the yield of ^{18}F from neon was not measured, because it was impossible to make a compact neon target. Experimental data on ^{11}C, ^{13}N and ^{18}F yields from light elements can also be used for determination of these element impurities in different materials. The yield of ^{11}C formed from beryllium by irradiation with ^{3}He and ^{4}He ions is given in Fig. 4 (the irradiation

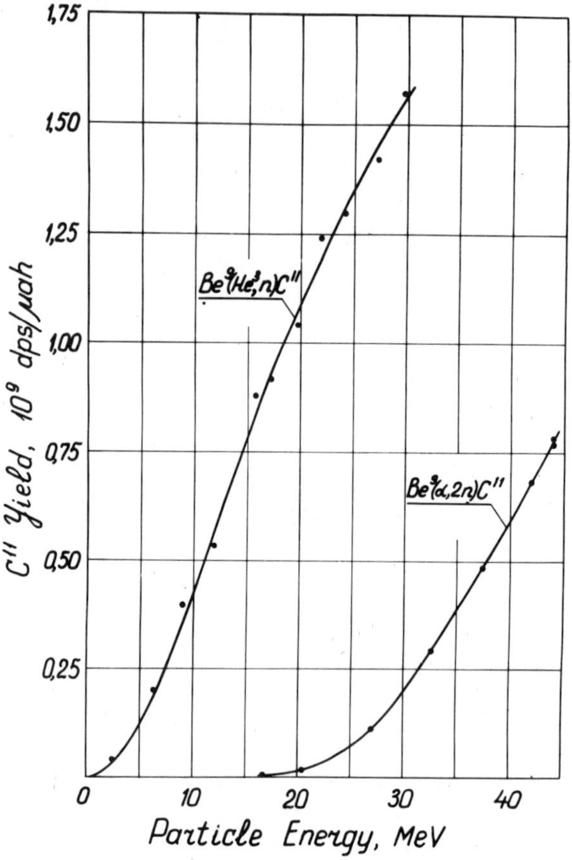

Fig. 4. Yield of ^{11}C *formed by irradiation of beryllium with* ^{3}He *and* ^{4}He *ions.*

of beryllium with protons and deuterons does not form the isotopes of interest). The curve of ^{11}C yield in the reaction $^{9}Be(^{3}He, n)\,^{11}C$ has also been measured[9] by irradiation of beryllium nitride with 18 MeV ^{3}He ions. The yield of ^{11}C for pure beryllium (at 18 MeV) is 10^{9} dps/μAh, which is in good agreement with our results. The excitation function of the $^{9}Be(\alpha, 2n)^{11}C$ reaction up to E_{α} = 38 MeV has also been measured,[7] and integration of this function gives the ^{11}C yield (at 38 MeV) as $0{\cdot}54 \times 10^{9}$ dps/μAh, which also is in good agreement with our results. Consideration of the curves in Fig. 4 shows that the ^{11}C yields in both cases are rather high, in particular on irradiation with ^{3}He ions.

The data on ^{11}C and ^{13}N yields formed by irradiation of boron with protons, deuterons, ^{3}He and ^{4}He ions are given in Fig. 5. The yields of ^{11}C and ^{13}N formed by irradiation of boron with 18 MeV ^{3}He ions have also been reported.[9]

The later data are in good agreement with ours. In two other papers[13, 14] the excitation function for $^{11}B(p, n)^{11}C$ has been obtained, and integration of this gives a value of the ^{11}C yield which coincides exactly with our results in one case;[13] the results from ref. (14) were 2·3 times greater than ours. Consideration of the curves in Fig. 5 shows that the best method of determining boron impurities is by proton irradiation. The curves of ^{18}F yields obtained by irradiation of fluorine with different particles are given in Fig. 6. Two papers give excitation functions for $^{19}F(^3He, d)^{18}F$[5] and $^{19}F(p, pn)^{18}F$[12]. Integration of these excitation functions gave values of yields coinciding with ours. In ref. (15) the curve of ^{18}F yield for irradiation of fluorine with 15 MeV deuterons was obtained; these data also coincide with ours. Consideration of the curves in Fig. 6 makes it clear that in determining oxygen impurities by measuring ^{18}F activity the presence of fluorine impurity may result in errors.

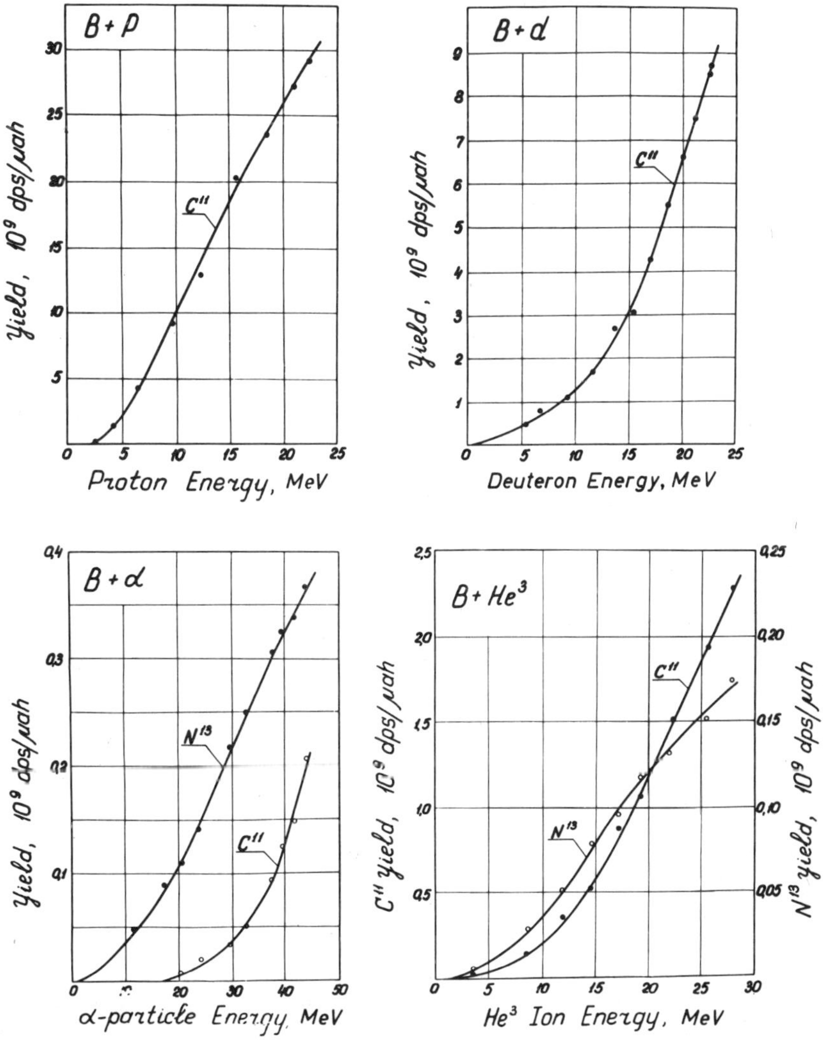

Fig. 5. Yields of ^{11}C *and* ^{13}N *formed by irradiation of boron with protons, deuterons,* ^3He *and* ^4He *ions*

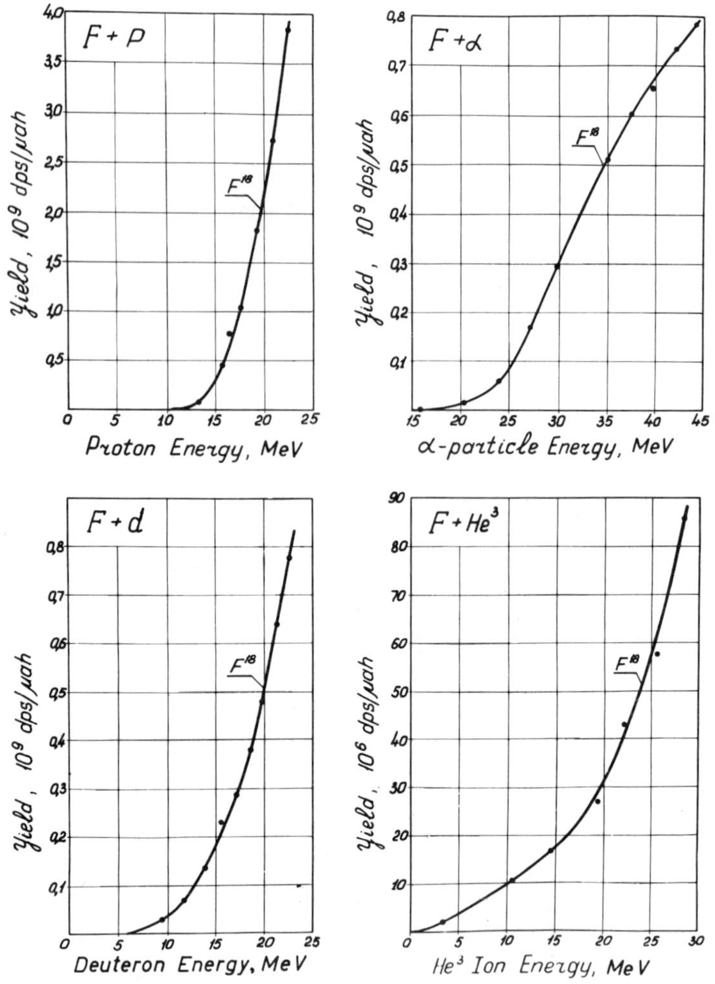

Fig. 6. Yield of ^{18}F *formed by irradiation of fluorine with protons, deuterons,* ^{3}He *and* ^{4}He *ions*

In Fig. 7 the curves of ^{18}F yields formed by irradiation of sodium with ^{3}He and ^{4}He ions are given. In this case ^{18}F is formed by means of ^{23}Na(^{3}He, 2α)^{18}F and ^{23}Na(α, 2αn)^{18}F reactions. The curves for the yield of ^{18}F formed by irradiation of magnesium and aluminium with ^{3}He ions are given in Fig. 8. ^{18}F is formed from magnesium by two reactions: ^{24}Mg(^{3}He, 2αp)^{18}F and ^{24}Mg(^{3}He, 2αn)^{18}Ne $\xrightarrow{\beta+}$ ^{18}F and from aluminium by the ^{27}Al(^{3}He, 3α)^{18}F reaction. The data on ^{18}F yields from sodium, magnesium, and aluminium are not available in the literature. Experimental data in Figs. 7 and 8 show that sodium, magnesium, and aluminium impurities may result in definite errors in determining oxygen impurities by measuring ^{18}F activity if the possibility of ^{18}F formation from these elements is not considered.

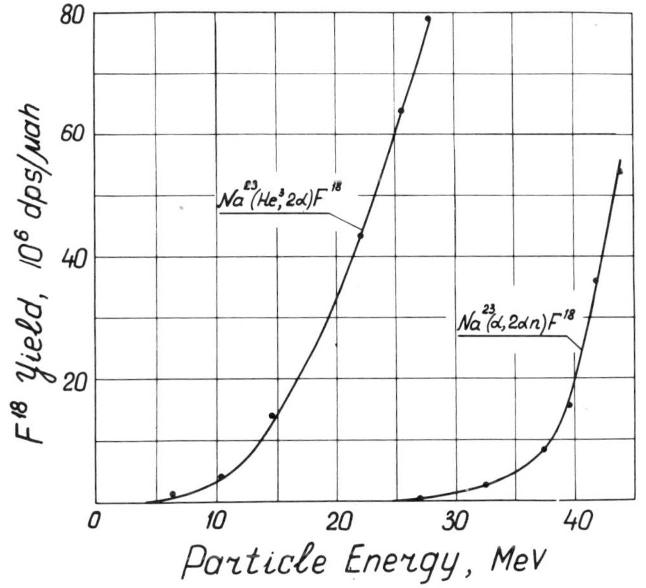

Fig. 7. Yield of ^{18}F *formed by irradiation of sodium with* ^{3}He *and* ^{4}He *ions*

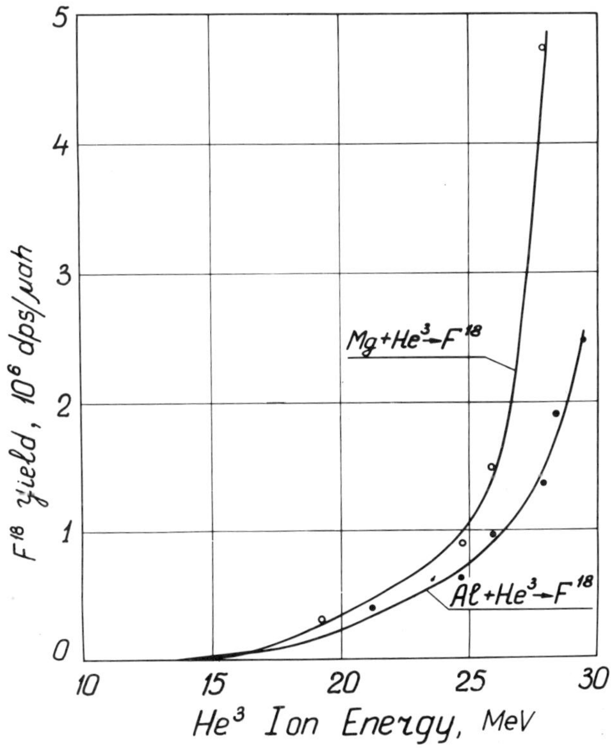

Fig. 8. Yield of ^{18}F *formed by irradiation of magnesium and aluminium with* ^{3}He *ions*

THE CHOICE OF IRRADIATION CONDITIONS FOR DETERMINATION OF CARBON, NITROGEN AND OXYGEN IMPURITIES

The determination of impurities can be done in two ways, with or without isolation of the observed nuclide from a matrix. Where the nuclide is isolated the matrix activity is not an obstacle in choosing irradiation conditions, and conditions are chosen to give the highest observed isotopic yield. It is this yield which determines the sensitivity of the analysis. If the nuclide is not isolated the matrix activity may interfere significantly. Irradiation conditions are then chosen for which the ratio of the observed isotope activity to the matrix activity is a maximum for the chosen procedure. It is this ratio that determines the sensitivity in this case. Of the three isotopes ^{11}C, ^{13}N and ^{18}F considered by us, only ^{18}F is convenient for radiochemical isolation due to its chemical properties and comparatively long half-life.

In Table 2 the data on ^{11}C, ^{13}N and ^{18}F yields, and the elements from which they can be formed, are given for particle energies which, in our opinion, are convenient for determination of carbon, nitrogen and oxygen impurities. The most convenient conditions for determining impurities of each of these three elements are underlined in the table.

Table 2. ISOTOPE YIELDS (10^9dps/μAh) FROM ELEMENTS FORMING ^{11}C, ^{13}N AND ^{18}F

Bombarding particles	^{11}C		^{13}N		^{18}F	
	Element	*Yield*	*Element*	*Yield*	*Element*	*Yield*
Protons, 10 MeV	B N	10 4·9	C O	0·22 1·8	O –	0·01 –
Deuterons, 8 MeV	B N	0·9 0·01	C N	5·6 0·1	O F	0·0004 0·01
^{3}He ions, 10 MeV	Be B C	0·43 0·17 0·9	C N –	0·3 0·1 –	O F Na	0·17 0·009 0·004
^{4}He ions, 15 MeV	–	–	B	0·07	N	0·0002

Carbon impurities are determined conveniently by deuteron irradiation at $\leqslant$8 MeV or by ^{3}He ions of ~10 MeV. Nitrogen is determined conveniently by ~10 MeV proton irradiation. High nitrogen impurity levels may be determined by ^{4}He ion irradiation at ~15 MeV. For determination of oxygen impurities it is convenient to use ~10 MeV protons and to determine them simultaneously via two isotopes (^{13}N and ^{18}F). This makes the results of the analysis more reliable. ^{3}He ions are also convenient for determination of oxygen impurities by measuring ^{18}F activity.

As is seen from Table 2, in most cases radioactive isotopes are formed simultaneously from two elements; if the contents of these two elements are unknown it is necessary to determine them by measuring the activity of one isotope formed from the two elements.[3]

In addition to the data in Table 2 an important value for practical purposes is the activation thickness, as defined by the charged-particle range. When the activation thickness is small, surface contamination may result in large errors. From this point of view protons have an advantage in comparison with other particles. For example, for the same energy of protons and ^{3}He ions the proton range is approximately nine times larger than that of ^{3}He ions. On the other hand, ^{3}He ions have another advantage arising from the low activation of matrix heavy elements due to the high Coulomb barrier. Thus the final choice of optimum irradiation conditions in each particular case must be made after consideration of the four main factors:

(1) the value of the observed isotopic yield from the element to be determined;
(2) the possibility of formation of this isotope from other elements;
(3) interference by the activity of isotopes formed from the matrix material;
(4) activation thickness.

REFERENCES

1. Krasnov, N. N., *Atomnaya energiya* **26**(3), 284, (1969).
2. Krasnov, N. N., Ognev, A. A., and Ponomaryev, A. A., *Pribory i tehnika eksperimenta* **4**, 22, (1965).
3. Maksimov, M. Z., *Zhurnal exsperimentalnoi i teoreticheskoi fiziki* **37**, 127, (1959).
4. Sternheimer, R., *Phys. Rev.* **115**, 137, (1959).
5. Brill, O. D., *Jadernaya fizika* **1**, 55, (1965).
6. Cochran, D. R. F. and Knight, I. D., *Phys. Rev.* **128**, 1281, (1962).
7. Brill, O. D. and Soomin, L. V., *Atomnaya energiya* **7**(4), 377, (1959).
8. Wilkinson, D. H., *Phys. Rev.* **100**, 32, (1955).
9. Hahn, R. H. and Ricci, E., *Nucl. Phys.* **A101**, 353, (1967).
10. Furukawa, M. and Tanaka, S., *J. Phys. Soc. Japan* **16**, 129, (1961).
11. Markowitz, S. S. and Mahony, J. D., *Anal. Chem.* **34**, 329, (1962).
12. Whitehead, A. B. and Foster, J. S., *Canad. J. Phys.* **36**, 1276, (1958).
13. Hintz, M. and Ramsey, N. F., *Phys. Rev.* **28**, 19, (1952).
14. Kalinin, S. I., *et al.*, *Atomnaya energiya* **2**(2), 171, (1957).
15. Bowen, Z. N. and Irvine, J. M., *Phys. Rev.* **127**, 1698, (1962).

DISCUSSION

P. Albert (C.N.R.S., Vitry-sur-Seine): I have two questions. Firstly, can you tell us if you carried out any special preparation of your NaCl before irradiation, to ensure that there was no oxygen in the NaCl plate? Secondly, for how long did you irradiate nitrogen, carbon and oxygen samples when quoting activities per μAh? Since the half-lives of ^{13}N and ^{11}C are 10 and 20 min. I do not understand the significance of μAh if you irradiate for more than a fraction of one half-life.

N. N. Krasnov (Obninsk): In reply to your first question, samples were heated in vacuum before irradiation. Secondly, we applied the formula

$$A = BI\left(\frac{1 - e^{-\lambda t}}{\lambda}\right)$$

where A = dps at the end of irradiation
B = dps per μAh
I = current in μA
t = irradiation time in h
λ = decay constant in h

Determination of vanadium and niobium by proton activation

J. L. Debrun, J. N. Barrandon, and Ph. Albert

C.N.R.S., Centre d'Études de Chimie Métallurgique,
15 rue Georges Urbain, 94-Vitry-sur-Seine, France

ABSTRACT

In most cases, vanadium and niobium cannot be determined at low concentrations by neutron activation,[1] for the half-lives of the radioisotopes are too short or the sensitivity is rather poor. We have therefore studied the possibility of proton activation. Our results show that it is possible to reach a high sensitivity for the determination of niobium, e.g. 3×10^{-2} ppm/μA in aluminium, for 1 h irradiation at 30 MeV. The nuclear reaction used is ^{93}Nb(p, n)^{93}Mo. To avoid any risk of interference by the reaction ^{94}Mo (p, pn)^{93}Mo, the energy must be limited to 10 MeV. Under these conditions (10 MeV and 1 h irradiation), the sensitivity is 4×10^{-1} ppm/μA.

For vanadium, our measurements show that $2{\cdot}4 \times 10^{-1}$ ppm/μA can be detected in aluminium by irradiating for 1 h at 30 MeV. The reaction studied is ^{51}V(p, n)^{51}Cr. The interference of chromium by ^{52}Cr(p, pn)^{51}Cr has been studied, and we show that the production of ^{51}Cr from ^{52}Cr can be neglected if $E_p < 15$ MeV. By limiting the proton energy to 15 MeV, the sensitivity does not decrease much; $3{\cdot}6 \times 10^{-1}$ ppm/μA can be detected with a 1 h irradiation.

By irradiating for several hours, a sensitivity of 10^{-1} ppm/μA can be easily reached for the determination of vanadium with 15 MeV protons and of niobium with 10 MeV protons, the half-lives of ^{51}Cr and ^{93}Mo being 27·8 d and 6·9 h.

INTRODUCTION

The use of thermal neutrons for the determination of vanadium and niobium leads to short-lived radioisotopes ($t_{1/2}$ = 3·75 and 6·3 min respectively). Under these conditions, it is possible to determine low concentrations, either if the analysis is non-destructive or if the chemical separations are very rapid; in other cases, the loss of time makes the determination of traces impossible.

With fast neutrons the sensitivities are poor, about 0·25 μg under the best conditions, and the interferences cannot be overcome. By activation with photons, vanadium determination is subject to the interference of titanium, which is serious at any energy;[2] for niobium there is no interference, but the intrinsic sensitivity is low.[2]

We have studied activation with protons, since it seemed possible to obtain high intrinsic sensitivities, and suitable radioisotopes (i.e. with long half-lives and no interference), by means of the reactions $^{51}V(p, n)^{51}Cr$ and $^{93}Nb(p, n)^{93}Mo$. Earlier workers[3] had already used proton activation with success.

DETERMINATION OF VANADIUM BY THE NUCLEAR REACTION $^{51}V(p, n)^{51}Cr$

Radioisotopes produced by irradiation of vanadium with 30 MeV protons. 30 MeV protons lead mainly to three radioisotopes, ^{47}Sc, ^{49}Cr, and ^{51}Cr. We have determined 'activation curves' for the reactions $^{51}V(p, \alpha p)^{47}Sc$, $^{51}V(p, 3n)^{49}Cr$ and $^{51}V(p, n)^{51}Cr$ by the 'stacked foil' method. The thickness of our foils was approximately 40 μ. For our calculations we used the data of Williamson[4] *et al,* giving the range-energy relationship for the charged particles. The measurement of the radioactivity of the three above radioisotopes is non-destructive, since the half-lives and production rates are quite different. γ-spectrometry of ^{49}Cr ($t_{½}$ = 42 min) is first carried out; when it has completely decayed it is possible to measure simultaneously the radioactivities of ^{47}Sc ($t_{½}$ = 3·4 d) and ^{51}Cr ($t_{½}$ = 27·8 d).

Fig. 1 shows that the experimental threshold of the reaction $^{51}V(p, 3n)^{49}Cr$ is 24·6 MeV; this value is very near to the calculated threshold of 24·18 MeV.

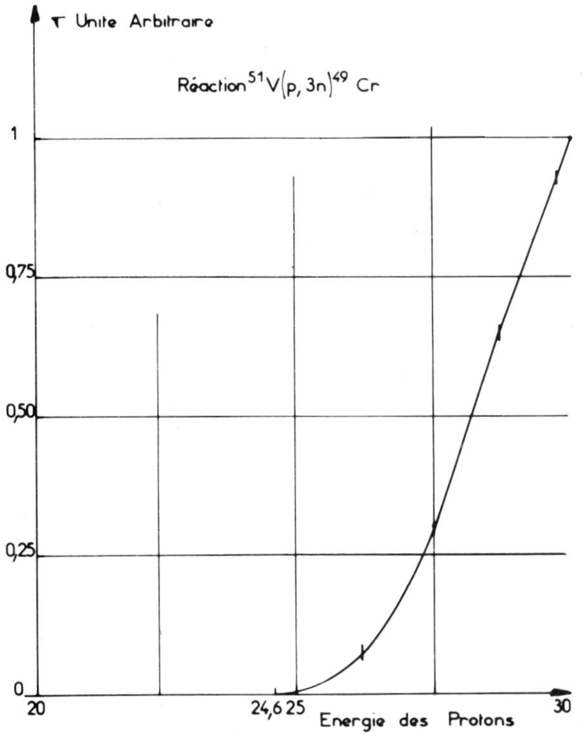

Fig. 1. Activation curve of the reaction $^{51}V(p, 3n)^{49}Cr$. *Mean curve from 70 experimental points*

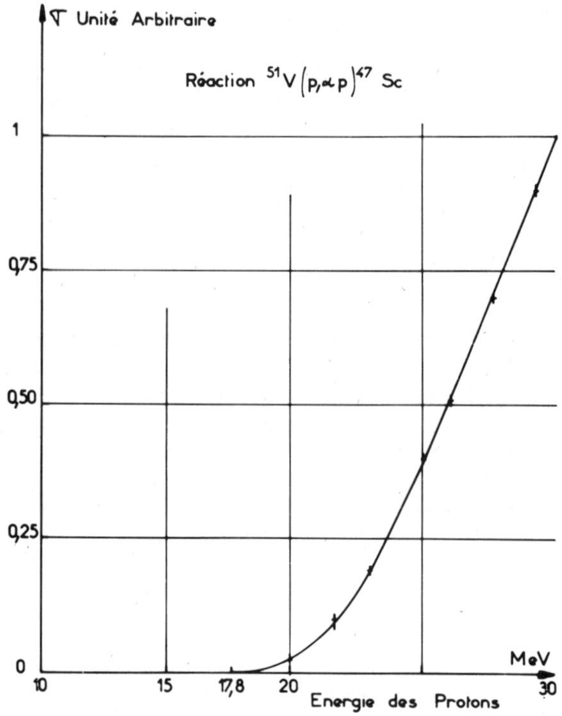

Fig. 2. Activation curve of the reaction ^{51}V(p, αp) ^{47}Sc. *Mean curve from 90 experimental points*

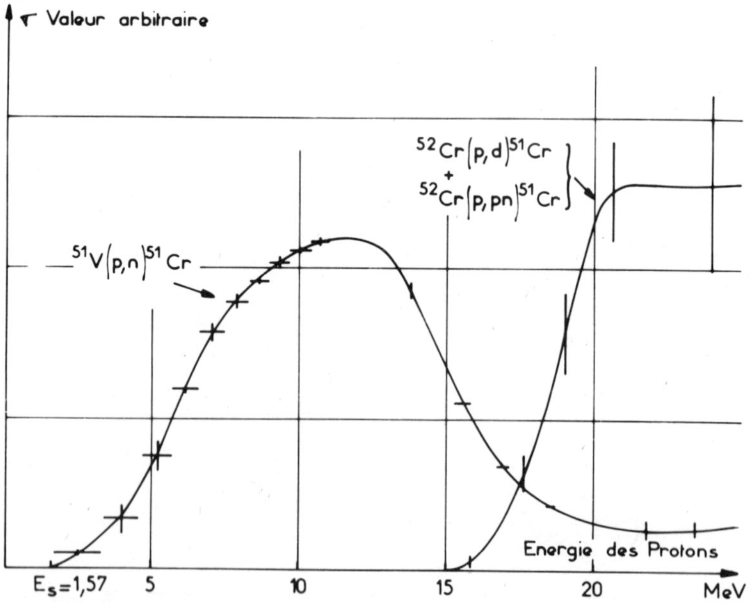

Fig. 3. Mean activation curves for the reaction leading to ^{51}Cr *from chromium, (120 experimental points), and from vanadium (180 experimental points), by irradiation with protons. The comparison is relative to isotopes* ^{51}V *and* ^{52}Cr

Fig. 2 indicates that the experimental threshold of the reaction $^{51}V(p, \alpha p)^{47}Sc$ is 17·8 MeV, which is very different from the theoretical value (10·47 MeV). From 10·47 to 17·8 MeV, the cross-section of the reaction is very low, probably because the two particles (α and p) must overcome the Coulomb barrier to escape from the nucleus.

Fig. 3 gives the excitation function of the reaction $^{51}V(p, n)^{51}Cr$; experimental and theoretical thresholds are identical ($\simeq$1·5 MeV). The cross-section is a maximum at ~12 MeV; it then decreases and becomes constant and 10 times lower beyond 20 MeV.

Considerations leading to the choice of the reaction $^{51}V(p, n)^{51}Cr$.
$^{51}V(p, n)^{51}Cr$ is the best of the three reactions cited above for determining vanadium at trace level, since

(1) its cross-section is high, as is shown below, and
(2) its threshold is low, so that the interferences can be overcome (see below).

The thresholds of the two reactions $^{51}V(p, \alpha p)^{47}Sc$ (i) and $^{51}V(p, 3n)^{49}Cr$ (ii) are high, much higher than those of the interfering reactions on chromium and titanium. Moreover, the radioactivity of ^{47}Sc and ^{49}Cr from reactions (i) and (ii) from their thresholds to 30 MeV is too low to allow the determination of low vanadium concentrations.

Sensitivity of γ-spectrometry for ^{51}Cr. The activity of ^{51}Cr ($t_{1/2}$ = 27·8 d, E_γ = 320 keV) is measured with a 3 in × 4 in NaI(Tl) crystal; the samples are placed on the crystal and the activities measured at the end of the irradiation (T_O). Under these conditions we measure 80 counts/minute/μA/ppm, if the irradiation is performed with 30 MeV protons for 1 h. This count-rate is relative to a vanadium standard. If we adopt Buchanan's[5] criterion and Ricci's[6] calculations, we can say that it is possible to determine $2{\cdot}4 \times 10^{-2}$ ppm of vanadium in aluminium by irradiating for 1 h with 10 μA of 30 MeV protons.

Study of interferences. Chromium may interfere by the nuclear reactions

$$^{52}Cr(p, d)^{51}Cr \quad \text{(iii) threshold: 10·01 MeV}$$
$$^{52}Cr(p, pn)^{51}Cr \quad \text{(iv) threshold: 12·28 MeV}$$

In order to measure the importance of this interference, we determined the activation curve for the sum of these two reactions. This determination, however, is not very precise, since thin chromium foils are not available and the experimental points were obtained after abrasion of a thick chromium plate. Fig. 3 shows that the production of ^{51}Cr may be neglected if E_p<15 MeV, i.e. about 5 MeV above the threshold of reaction (iii) and 3 MeV above the threshold of reaction (iv).

Fig. 4 shows the interference of chromium in terms of energy, for the same concentrations by weight of vanadium and chromium. For this calculation, we chose the values corresponding to the mean activation curves.

Conclusions. The half-life of ^{51}Cr (27·8 d) allows us to perform long chemical separations without reduction in sensitivity. This sensitivity, calculated for E_p = 15 MeV, i.e. when chromium interference can be neglected, is $3{\cdot}6 \times 10^{-1}$ ppm/μA for 1 h irradiation. The value of 1×10^{-1} ppm/μA can be easily reached by irradiating for several hours.

It appears that irradiation with 15 MeV protons is the only activation method

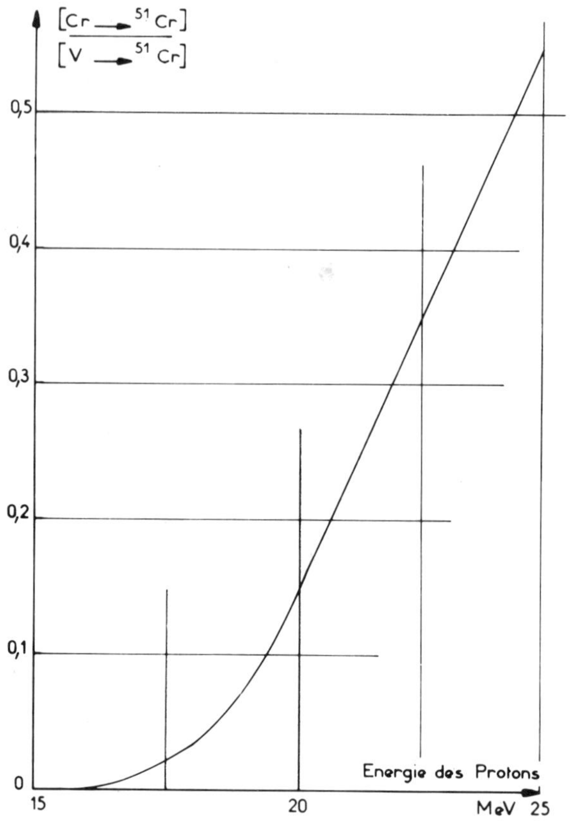

Fig. 4. Interference of chromium in the determination of vanadium by irradiation with protons. The calculation is relative to the mean curves of Fig. 3 and corresponds to equal weight concentration for vanadium and chromium

that can be used, in the general case, for the determination of low vanadium concentrations (~10^{-2} ppm).

DETERMINATION OF NIOBIUM BY THE NUCLEAR REACTION ^{93}Nb(p, n)^{93}Mo

Radioisotopes produced by irradiation of niobium with 30 MeV protons. The principal radioisotopes from the irradiation of niobium with 30 MeV protons are ^{91}Mo, ^{93}Mo, ^{92}Nb and ^{89}Zr. After decay of ^{91}Mo ($t_{1/2}$ = 15·5 min), the measurement of the radioactivity of ^{93}Mo ($t_{1/2}$ = 6·9 h) produced by the reaction ^{93}Nb(p, n)^{93}Mo is non-destructive and is carried out for each thin foil (≃30 μ) of the irradiated stack. Likewise, after decay of ^{93}Mo, it is possible to identify ^{89}Zr and ^{92}Nb produced by the reactions ^{93}Nb(p, αn)^{89}Zr and ^{93}Nb(p, d)^{92}Nb (v) or ^{93}Nb(p, pn)^{92}Nb (vi).

Fig. 5 shows the activation curve for the sum of the two reactions leading to ^{92}Nb; it can be seen that the experimental threshold is 10 MeV, while the theoretical values are 6·65 MeV for reaction (v) and 8·9 MeV for (vi).

Fig. 6 presents the activation curve of the reaction $^{93}Nb(p, n)^{93}Mo$, the calculated threshold of which is 1·27 MeV, the calculated Coulomb barrier being 8·3 MeV; it can be seen that in practice the production of ^{93}Mo starts at 3·8 MeV. The cross-section increases rapidly from 3·8 MeV, reaches a maximum at 13 MeV, and decreases to become constant above 22 MeV.

As for vanadium, we can determine from a study of the relative values of the

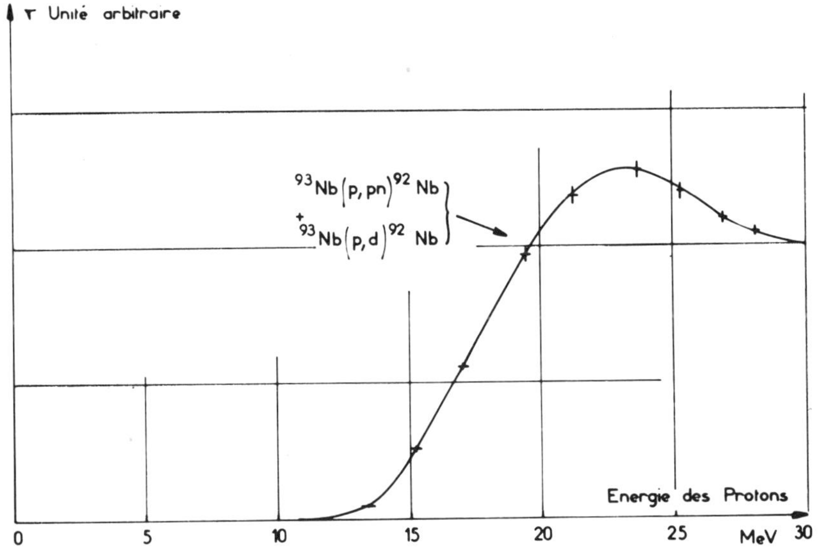

Fig. 5. Activation curve of the sum of the two reactions $^{93}Nb(p, pn)^{92}Nb$, $^{93}Nb(p, d)^{92}Nb$ *(130 experimental points)*

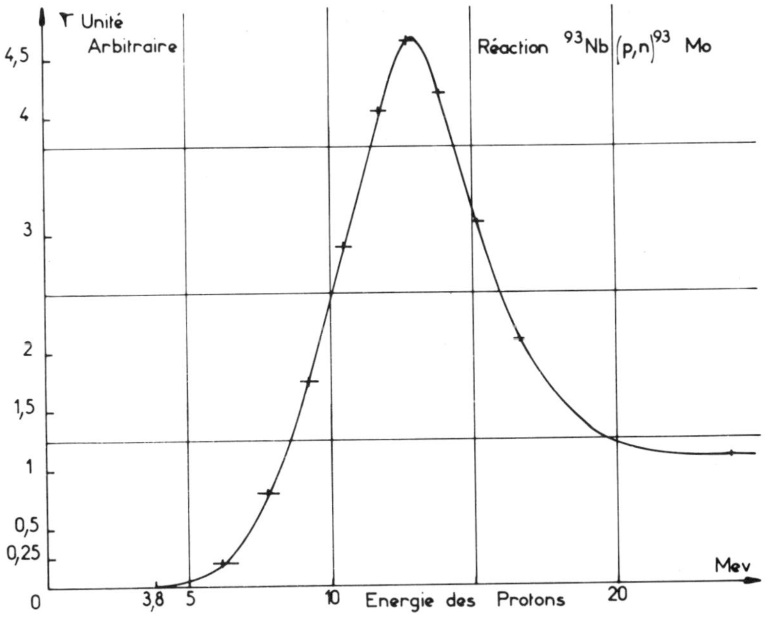

Fig. 6. Activation curve of the reaction $^{93}Nb(p, n)^{93}Mo$. *Mean curve from 170 experimental points*

thresholds and of the yields of the nuclear reactions that $^{93}Nb(p, n)^{93}Mo$ is the best reaction for the determination of low niobium concentrations.

Sensitivity obtained by γ-spectrometry of ^{93}Mo. The radioactive yield of ^{93}Mo for the photoelectric peak corresponding to 690 keV, measured under the same conditions as for ^{51}Cr, is equal to 450 cpm/ppm/μA, for a niobium standard irradiated for 1 h at 30 MeV. If we refer to aluminium instead of niobium, the activity becomes[6] 450 × 1·181/1·641 = 320 cpm/ppm/μA, since the respective ranges of 30 MeV protons in Al and Nb are[4] 1·181 and 1·641 g/cm^2.

This last count-rate corresponds to a sensitivity of 3×10^{-2} ppm of Nb in Al[5]. Assuming a possible proton intensity of 10 μA, this limit improves to 3×10^{-3} ppm.

Interferences. Molybdenum can interfere by $^{94}Mo(p, pn)^{93}Mo$ and $^{94}Mo(p, d)^{93}Mo$ reactions with thresholds at 9·84 and 7·59 MeV respectively. We did not measure the interference of molybdenum, for this would have taken too long to be carried out during the available irradiation time. Further measurements will be made to examine this possibility.

It is most probable however, that the reaction $^{94}Mo(p, d)^{93}Mo$ can be neglected from 7·59 to 9·84 MeV. We have noticed, by studying similar pairs of reactions, $^{35}Cl(p, d)^{34m}Cl$ and $^{35}Cl(p, pn)^{34m}Cl$, $^{52}Cr(p, d)^{51}Cr$ and $^{52}Cr(p, pn)^{51}Cr$, $^{93}Nb(p, d)^{92}Nb$ and $^{93}Nb(p, pn)\,^{92}Nb$, that the production of the radioisotope $^{A-1}_{Z}M$ from $^{A}_{Z}M$ is negligible if the energy is lower than the threshold of the (p, pn) reaction. The case of niobium is favourable because the isotopic abundance of ^{94}Mo is only 9·04%, while in the case of the determination of vanadium and sulphur the isotopic abundances of the interfering isotopes, ^{52}Cr and ^{35}Cl, are respectively 83·76% and 75·53%. Consequently, interference by molybdenum can be neglected if the proton energy is < 10 MeV.

Conclusions. If the proton energy is limited to 10 MeV in order to overcome interference by molybdenum, the limit for the determination of niobium in aluminium is 4×10^{-2} ppm for an irradiation of 1 h at 10 μA; this is 10 times higher than the limit corresponding to 30 MeV. If we irradiate for several hours, the other conditions remaining the same, a value of 10^{-2} ppm can be attained.

Thermal or fast neutrons do not offer so high a sensitivity; photons offer about the same sensitivity (10^{-2} μg for a 1 g sample) if we irradiate at high energy and for several hours. The interest in using protons lies in the possibility of determining very low concentrations of niobium in pure materials by irradiating at high energy. An irradiation of a few hours at 30 MeV and 10 μA provides a sensitivity of $\sim 10^{-3}$ ppm. When irradiating with 30 MeV protons, molybdenum interferes and we only determine an 'apparent concentration' for niobium. Nevertheless, it is interesting to have a method that can determine a niobium concentration of 10^{-3} ppm, even if it is an 'apparent concentration'.

REFERENCES

1. (a) Yule, H. P., *Anal. Chem.* **37**, 129, (1965); (b) Yule, H. P., *et al, Nuclear Instruments and Methods* **33**, 277, (1965).
2. Debrun, J. L. and Albert, Ph., *Bull. Soc. Chim. France* **3**, 1020, (1969).
3. (a) Schweikert, E. and Albert, Ph., *C. R. Acad. Sci. Paris* **262**, 87, (1966); (b) Schweikert, E. and Albert, Ph., ibid. **262**, 342, (1966).
4. Williamson, C. F., Boujot, J. P., and Picard, J., Rapport C.E.A. R 3042 (1966).
5. Buchanan, J. D., *Atompraxis* **8**, 272, (1962).
6. Ricci, E. and Hahn, R. L., *Anal. Chem.* **37**, 742, (1965).

DISCUSSION

Mrs. M. C. Palcos (Buenos Aires): Have you completed any determinations of niobium or vanadium? If so, how do you standardise the activities which you measure?

J. L. Debrun: We have not yet completed an analysis for niobium or vanadium, but we would standardise by our usual methods of either placing a thin foil of the standard element in front of the sample, or by placing a monitor foil of another element in front of both sample and standard.

SESSION 8

Excitation Functions

Chairman: Dr. T. B. Pierce

A novel method for the measurement of excitation functions for charged-particle reactions on gases

W. L. Rautenbach, J. Steyn, D. Richards and H. J. Smith

Nuclear Physics and Radioactivity Division, National Physical Research Laboratory, P.O. Box 395, Pretoria, Republic of South Africa

ABSTRACT

Excitation functions are presented for the production of ^{18}F by irradiation of ^{16}O with 32 MeV α-particles and 35 MeV ^{3}He ions at the Pretoria cyclotron. The maximum measured cross-section for the production of ^{18}F by ^{3}He irradiation of oxygen was 380 mbarn at 7·7 MeV. The excitation curve for the production of ^{18}F by irradiation of oxygen with α-particles reached 120 mbarn at 31·5 MeV, and was still rising at this energy. The ^{18}F activity was determined absolutely by $4\pi-\beta\gamma$ coincidence measurements which were carried out directly on the aluminium catcher foils.

The difficulties of using the 'stacked layer' technique on gaseous elements were overcome by introducing the gas between a number of accurately-spaced thin metal foils. The foils also served to degrade the beam energy. Provided that the gas layers were thin enough, all the radioactive nuclei produced in the gas by a charged-particle beam passing through the stack were collected by recoil onto the foil situated in the forward direction.

INTRODUCTION

The energy dependence of the cross-section for a specific nuclear reaction is generally known as the excitation function for that reaction. Experimentally-determined excitation functions are useful for activation analysis and radioisotope production, and also supply information on the mechanism of the nuclear reaction.

Excitation functions for charged-particle reactions leading to radioactive products are generally measured using the 'stacked foil technique': a suitable number of homogeneous foils of the element under consideration is stacked one on another and irradiated with the charged-particle beam of known energy and known integrated flux. If the thickness of the foils and the specific energy loss of the charged particles are known, the degraded beam energy at each foil can be calculated. Absolute measurement of the activity produced in the target foils is then necessary to obtain the excitation curve. In practice the recoil of radioactive atoms out of the target foils poses a problem, which is solved however by using suitable catcher foils.

The application of the stacked foil technique depends on the availability of homogeneous targets of suitable thickness. This requirement is troublesome for elements which are non-metallic, and especially so when the element is gaseous under normal conditions. In the past the general procedure has been to incorporate the gas in a suitable compound in order to prepare thin targets. It is, however, usually difficult to prepare a target of homogeneous thickness in this manner, with the result that in many cases the accuracy of the measured excitation curve is limited by the quality of the target foils. The preparation of suitable target foils is also especially difficult for noble gases.

We have solved this problem by using thin gaseous targets in a modified stacked foil assembly. The targets to be irradiated consist of gas introduced between accurately-spaced parallel thin metal foils. If the pressure, temperature and thickness of the gas layers are known, the thickness of the gas target can be calculated in mg/cm^2. Provided the targets are thin enough, the recoiling radioactive nuclei produced by the nuclear reaction in the gas penetrate the gas layer and are quantitatively collected on the catcher foils bordering on the gas target. By using a number of such gas targets in series with suitable catcher and degrader foils, excitation functions can be measured for gases without the complications of difficult target preparation.

The application of this method to the measurement of the excitation functions for the production of ^{18}F by the irradiation of oxygen with α-particles and ^{3}He nuclei is described in the following sections.

EXPERIMENTAL

Gas target thickness. The success of this type of measurement hinges on the collection of all of the radioactive nuclei on the catcher foils. This limits the gas target thickness to a maximum value, which can be calculated if the energy, direction, and range of the recoil nuclei are known. These quantities are generally not known very accurately, the energy and direction being also dependent on the reaction mechanism. However, an estimate of the maximum target thickness can be obtained by assuming that the reaction takes place by the formation of a compound nucleus to which the full linear momentum of the incident particle is transferred. The subsequent evaporation of one or more low-energy particles will modify this momentum only slightly. The heavy product of the reaction will thus move in the direction of the beam with a kinetic energy not very different from that of the compound nucleus, which is given by:

$$E_c = mE_m/M_c \qquad (1)$$

where E_c and M_c are the kinetic energy and the mass, respectively, of the compound nucleus, and E_m and m the corresponding quantities for the incident particle. The values of E_m will range from the threshold of the reaction to the maximum energy of the incident beam.

In the case of the irradiation of oxygen with α-particles, the following three reactions result in the production of ^{18}F:

$^{16}O(\alpha, d)^{18}F$ with a Q-value of $-16{\cdot}331$ MeV,[1]
$^{16}O(\alpha, pn)^{18}F$ with a Q-value of $-18{\cdot}556$ MeV,[1]
$^{16}O(\alpha, 2n)^{18}Ne$ with a Q-value of $-23{\cdot}560$ MeV;[1]

^{18}Ne, with a half-life of 1·5 s, decays to ^{18}F. According to Eqn (1) the recoil energies are between 4 and 6 MeV for a maximum beam energy of 32 MeV. The ranges of the ^{18}F recoils in oxygen can be calculated from Lindhard's theory[2] and are between 763 and 958 $\mu g/cm^2$.

For the production of ^{18}F by the irradiation of oxygen with ^{3}He particles, the following reactions are dominant:

$^{16}O(^{3}He, p)^{18}F$ with a Q-value of $+2{\cdot}201$ MeV, and
$^{16}O(^{3}He, n)^{18}Ne$ with a Q-value of $-2{\cdot}980$ MeV,

where again ^{18}Ne decays to ^{18}F. The threshold for the first reaction will be low, because it is determined only by the penetration of the Coulomb barrier by the ^{3}He particle. The recoil energy will therfore vary from ~150 keV at 1 MeV incident energy of the ^{3}He particle to ~5 MeV for a 35 MeV particle. At 150 keV energy the calculated range of ^{18}F in oxygen is 64 $\mu g/cm^2$.

Target thicknesses obtained by these considerations are maximum values, however, since the contribution by direct reactions will lead to lower momentum transfer and lower recoil ranges. It is therefore necessary to measure the excitation curves at different target thicknesses (gas pressure) to determine whether a change in target thickness has an effect on the collection of the recoils.

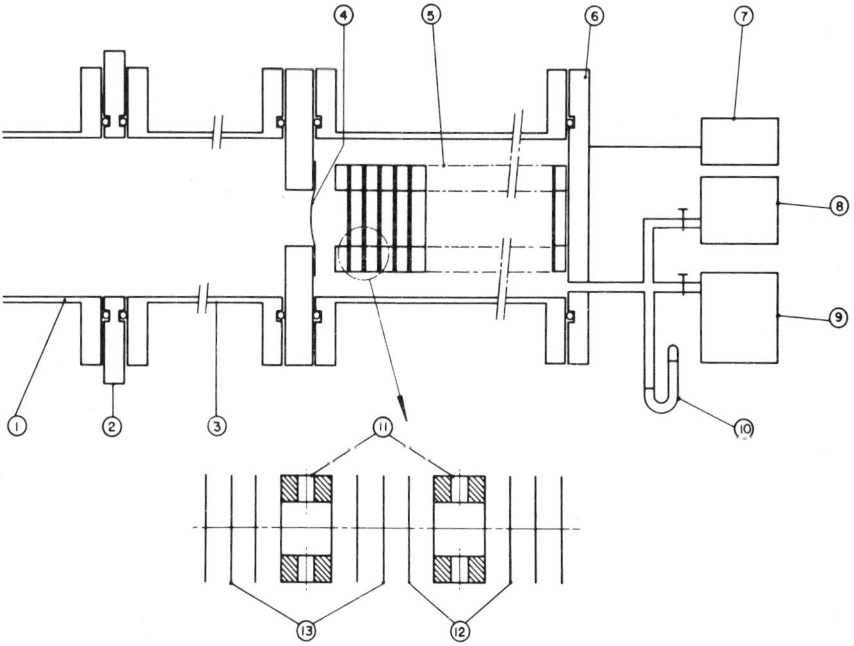

Fig. 1. Gas target assembly with (1) Beam tube from cyclotron, (2) Insulator for Faraday cup, (3) Entrance tube of Faraday cup, (4) Vacuum window of thin Havar foil, (5) Target assembly, (6) Rear plate of target assembly, (7) Precision current integrator, (8) Fore vacuum pump, (9) Oxygen supply, (10) Mercury manometer, (11) Brass spacer rings, (12) Aluminium catcher foils, (13) Aluminium degrader foils

Target assembly. The target assembly, as installed in a direct beam line of the Pretoria cyclotron, is shown in Fig. 1.

The oxygen targets are defined by aluminium catcher foils (~3 mg/cm^2) on the two sides of a brass spacing ring of thickness 5 mm, as shown in the exploded part of Fig. 1. The gas targets are separated from one another by one or more aluminium degrader foils of the same thickness as the catcher foils. The neighbouring degrader foils are then used also to determine the background activity produced by the beam in the catcher foils.

The target assembly is isolated from the vacuum in the beam pipe by a Havar[3] foil of thickness 4·8 mg/cm^2. The target volume is pumped down by a fore-vacuum pump to ~0·2 torr, filled with oxygen and pumped down again before final filling to the appropriate pressure, which is measured on a mercury manometer. The target and the pipe in front of it are insulated from earth, thus forming a Faraday cup which is connected to an accurate current integrator.[4]

After lining up the target assembly with the cyclotron beam and adjusting the quadrupole system between cyclotron and target to yield a beam spot of roughly 6 × 12 mm^2 at the rear of the target assembly, a collimator of 5 mm diam. is used 265 mm upstream from the insulator to ensure that the beam does not strike either the holder of the Havar foil or the brass rings in the target assembly. The collimator and the target assembly are cooled by compressed air. The temperature in the target assembly is measured before and after irradiation and generally varies by not more than ~1°C.

Irradiations. Irradiation conditions are summarised in Table 1.

Table 1. IRRADIATION CONDITIONS

Irradiation No.	*Projectile*	*Incident energy* (MeV)	*Average beam intensity* (μA)	*Duration of irradiation* (min)	*Pressure* (torr)	*Temperature* (°C)	*Equivalent gas target thickness* (μg/cm²)
1	^4He	32·6 ± 0·1	0·106	8·0	401·3	24·5	346
2	^4He	32·6 ± 0·1	0·104	8·0	403·0	25·4	347
3	^4He	32·6 ± 0·1	0·115	5·0	602·5	25·5	517
4	^4He	32·6 ± 0·1	0·139	5·0	10^{-1}	–	–
5	^3He	35·2 ± 0·1	0·076	8·58	198·0	26·4	170
6	^3He	35·2 ± 0·1	0·147	10·0	200·5	26·7	172
7	^3He	35·2 ± 0·1	0·163	10·0	100·5	26·3	86
8	^3He	35·2 ± 0·1	0·076	10·0	10^{-1}	–	–

Beam energy measurements. The energy of the incident beam is measured by scattering it from a thin gold foil onto a Si (Li) detector which is calibrated with the α-particle groups of Th(B + C + C″). The scattered beam is degraded to a suitable energy by means of an aluminium absorber of known thickness. The energy degradation in the target assembly is calculated from the expressions of Williamson *et al.*[5]

Measurement of ^{18}F *activity.* After irradiation the catcher and degrader foils are counted in a fixed position in a NaI(Tl) well crystal in counting intervals of 100 s. The count rates are corrected for background and dead time. Selected foils are counted repeatedly to obtain a measure of the half-lives of the activities on the foils. For the α-particle irradiation the only interfering short-lived activities are ^{28}Al (2·3 min), ^{29}Al (6·6 min) and ^{30}P (2·6 min). About 70 min after irradiation, essentially only ^{18}F is present. Under these conditions decay curves yielded values of the half-life of ^{18}F very near to the accepted value of 109·7 min. After ^{18}F has decayed, slight residual amounts of ^{56}Co and ^{57}Co, probably arising from activation of some impurity in the aluminium foils, were identified by γ-spectroscopy.

The amounts of ^{18}F present on a few selected foils are measured absolutely by using the catcher foils directly as sources in a $4\pi-\beta\gamma$ coincidence-counting system. The counter for the β-particles is a proportional gas-flow counter, and a 3 X 3 in NaI(Tl) crystal is used in γ-channel to register the annihilation radiation in coincidence with the positrons. This method of determining the counting efficiency of a 4π counter for positrons has previously been discussed more fully.[6] Since the β-counting efficiency is very high (96%), it is not necessary to correct for the fact that the annihilation radiation comes from an extended source. The standardised foils are then used to calibrate the well counter.

The same procedure is employed to measure ^{18}F produced by ^{3}He irradiation. In this case, however, several complications arise. An appreciable amount of a positron activity with a half-life of ~10 min is produced in the oxygen, while ^{24}Na is formed in the foils at the high-energy end of the stack. This problem is solved by irradiating stacks of aluminium foils in vacuum in order to determine the amount of ^{24}Na produced at each foil position. In addition, the amounts of ^{18}F produced in surface layers of aluminium oxide on the degrader foils adjacent to the catcher foils are also measured. After subtracting the background due to these activities, decay curves for representative catcher foils are used to correct for the contribution of the 10 min activity, which varies strongly with the time elapsed between irradiation and counting. Very little contribution by activities other than those from ^{18}F are found on the foils at the low-energy end of the target assembly, and some of these foils are therefore selected for absolute standardisation.

RESULTS AND DISCUSSION

It is generally found that the activity of the upstream catcher foil is practically the same as that of the degrading foils immediately in front of it. This confirms the assumption deduced from the kinematics of the reactions that all the recoil ^{18}F atoms move in the forward direction.

The excitation curves are shown in Figs. 2 and 3.

For the irradiation with α-particles, systematic differences exist for the two different gas pressures, proving that the collection of ^{18}F is complete even at the highest pressure and lowest energy. A similar conclusion can be reached for the ^{3}He irradiations down to a ^{3}He energy of ~6 MeV. At lower energies not enough experimental points were obtained to determine whether there exists a significant difference at the two oxygen pressures used.

In order to obtain more results at lower energies, the experiment should be repeated at lower incident energy and with thinner degrader foils.

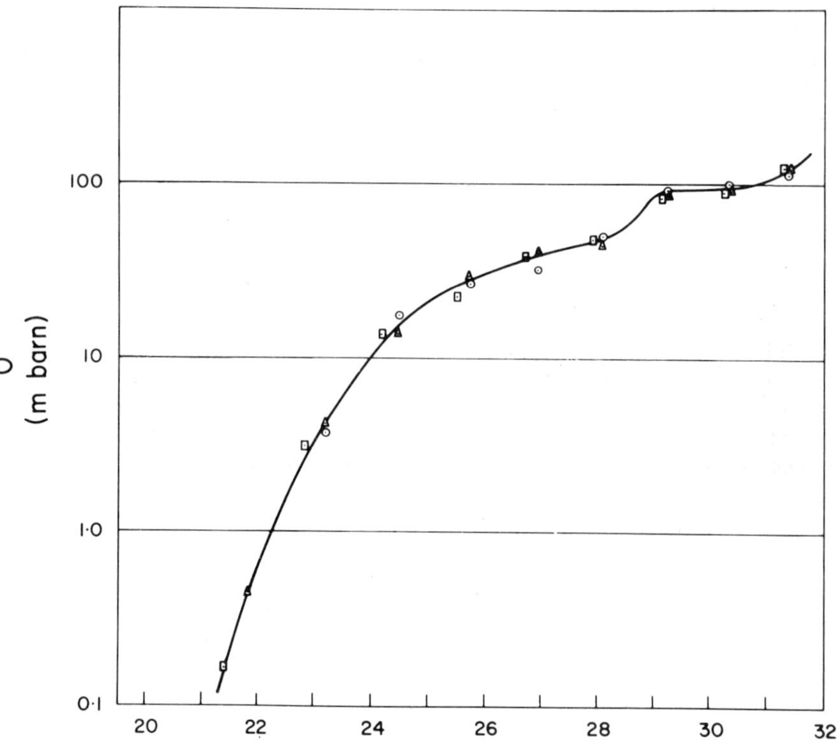

Fig. 2. The measured excitation function for the production of ^{18}F by the irradiation of oxygen with α-particles

⊡ *oxygen pressure 603 torr* ⊙ *oxygen pressure 403 torr* △ *oxygen pressure 401 torr*

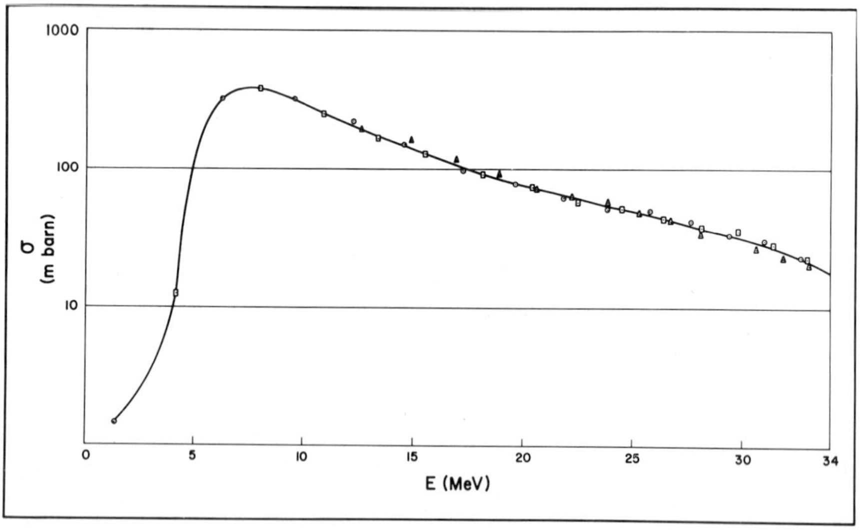

Fig. 3. The measured excitation function for the production of ^{18}F by the irradiation of oxygen with ^{3}He ions

⊡ *oxygen pressure 100 torr* ⊙ *oxygen pressure 200 torr* △ *oxygen pressure 198 torr*

The scatter of the experimental points in Figs. 2 and 3 can be attributed to the fact that the catcher foils sometimes tend to wrinkle slightly when the target assembly is pumped down. This gives rise to variation in the thickness of the oxygen layer.

For the production of fluorine by α-particle irradiation of oxygen the experimental results at 30 MeV agree within about 20% with the previous determination by Furukawa and Tanaka,[7] who irradiated oxygen in the form of Na_2CO_3. The results of these authors, however, do not show the prominent rise in cross-section by a factor of nearly two between 28 and 29 MeV. Since the threshold for the (α, 2n) reaction is 29·4 MeV, the rise in cross-section cannot be attributed to this reaction. It is probably caused by the (α, np) reaction, which has a threshold at 24·6 MeV, but which will only contribute appreciably to the cross-section several MeV above threshold because of the Coulomb barrier for the proton. The cross-section below 24·6 MeV must be attributed to the (α, d) reaction with a threshold of 20·4 MeV.

The experimental results show that this is a feasible method of determining excitation curves for gases. It can in principle be extended to elements for which gaseous or volatile compounds can be prepared. It may also be possible to use the method for expensive materials such as separated isotopes, since the gas can be recovered and used repeatedly.

REFERENCES

1. Everling, F., Koenig, L. A., Mattauch, J. H. E., and Wapstra, A. H., 1960 Nuclear Data Tables, Part 1, 36, U.S. Government Printing Office, Washington 25 D.C. (1961).
2. Lindhard, J., Scharff, M.. and Schiøtt, H. E., *Matematisk-fysiske Meddelelser* 33, 14, (1963).
3. Manufactured by Hamilton Precision Metals, Reading, Pa. 19603, U.S.A.
4. Röhm, H. F., Verwey, C. J., Steyn, J., and Rautenbach, W. L., to be published in *J. of Inorg. and Nuc. Chem.* (1969).
5. Williamson, C. F., Boujot, J.-P., and Picard, J., Tables of range and stopping power of chemical elements for charged particles of energy 0·05 to 500 MeV, Rapport CEA-R 3042, Paris (1966).
6. Steyn, J. and de Jesus, A. S. M., CSIR Special Report Fis 15, Pretoria, CSIR (1966).
7. Furukawa, M. and Tanaka, S., *J. Phys. Soc. Japan* 16, 129, (1961).

DISCUSSION

J. L. Need (New England Nuclear Corp.): Did you perform bombardments at different currents? There has been a continuing argument among those using gas targets of deuterium or tritium for neutron production as to whether the measurement of temperature and pressure describes the real conditions in the gas within the beam.

W. Rautenbach: ^{3}He bombardments were performed at currents differing by a factor of two, and no difference was noticed.

D. J. Silvester (M.R.C., Hammersmith): Have you compared your excitation function curves with those published by Nozaki *et al?*

W. Rautenbach: At an α-particle energy of 30 MeV our result differs by ~20% from that published by Furukawa and Tanaka in 1961. We have not compared it with more recent results.

M. A. Chaudhri (M.R.C., Hammersmith): Could you please tell us the overall accuracy of your experimental method in measuring the excitation function.

W. Rautenbach: The current integration was accurate to ±1% and the measurement of the activity to about ±3%. The scatter in the experimental points is probably due to foils which do not remain completely flat.

M. Lefort (Orsay): The shoulder in the excitation function for ^4He on oxygen in the production of ^{18}F has not been found in experiments which were performed some years ago on solid targets of SiO_2 and other oxygen-rich materials. Did you try to think of an explanation connected with the method of collecting ^{18}F?

W. Rautenbach: Since the shoulder is reproduced for oxygen pressures differing by 50%, it is difficult to see how differences in collection-efficiency can be responsible.

P. Albert (C.N.R.S., Vitry): What are the volume and pressure of gas between two catcher-foils?

W. Rautenbach: The distance between the catcher-foils is 5 mm; the oxygen pressure for ^4He irradiation was 400 and 600 torr respectively, and for ^3He irradiation 100 and 200 torr.

M. A. Chaudhri (M.R.C., Hammersmith): I am puzzled by your mention of the Coulomb barrier in connection with emission of the proton. Do you mean to say that the Coulomb barrier for the deuteron is less than that for the proton?

W. Rautenbach: The kinematic threshold for the reaction $^{16}O(\alpha, np)^{18}F$ is 24·6 MeV. Since the proton has to surmount the Coulomb barrier before this reaction can contribute appreciably to the cross-section, it may be possible that the (α, np) reaction is responsible for the shoulder, since the kinematic threshold of the (α, 2n) reaction is 29·4 MeV.

Evaluation of unknown excitation functions for (α, xn) and (d, xn) reactions

H. Münzel and J. Lange

Institute of Radiochemistry, Gesellschaft für Kernforschung m.b.H. 75 Karlsruhe, West Germany

ABSTRACT

In order to evaluate the best irradiation conditions for the cyclotron production of radionuclides the excitation function for the reaction has to be known. Unfortunately, excitation functions for (α, xn), (α, pxn), (d, xn), and (α, pxn) reactions are known only for few target-projectile combinations. In most cases therefore, we have to make a rather arbitrary guess about optimal target thickness and projectile energy. We have therefore investigated the possibility of estimating unknown excitation functions by interpolation. It can be shown that the influence of the charge and mass number of the target nucleus on characteristic data such as position, height, and shape of the known excitation functions can be explained by qualitative theoretical arguments.

INTRODUCTION

One of the main purposes of a cyclotron is the production of radioactive materials for scientific research or technical applications. Because the operation of such a machine is rather expensive, the necessary irradiation time should be kept as short as possible. We have attempted, therefore, to find a procedure for estimating the optimum irradiation conditions.

The activity A produced in a target depends on particle flux F, cross-section σ, half-life T, irradiation time t and the number N of target atoms per cm^2. If the target thickness is larger than the range of the projectiles with energy E_p, then the activity produced can be calculated according to Eqn (1).

$$A = kF\left(1 - e^{-\frac{0{\cdot}7\,t}{T}}\right) \int_{o}^{E_p} \frac{\sigma}{(\mathrm{d}E/\mathrm{d}x)}\,\mathrm{d}E \qquad (1)$$

k is a numerical constant and $(\mathrm{d}E/\mathrm{d}x)$ the specific energy loss of the projectiles in the target material. Values of $(\mathrm{d}E/\mathrm{d}x)$ for light projectiles in elements or in compounds are either well known or can be estimated with fairly good accuracy.

To evaluate the integral in Eqn (1) it is also necessary to know the excitation function of the reaction. Unfortunately, only a very limited number of excitation functions have been experimentally determined. Therefore, quite often one cannot perform the numerical integration and the activity that will be produced under certain irradiation conditions cannot be calculated. We have now tried to develop a method for estimating unknown excitation functions.

METHOD

As long as the excitation energy is not too high, it may be assumed that the main part of the reactions proceed via compound nucleus formation.[1] If shell effects are neglected, then a smooth dependence of the formation cross-sections, as well as of the decay properties of compound nuclei, with excitation energy E^*, mass number A and charge number Z may be expected. Because the variation of A in target nuclides of a single element is relatively small, we can, within the accuracy of the following considerations, neglect A or Z as variables. We therefore assume that the shape and position of the excitation function depends on E^* and Z only.

To characterise the position and shape of an excitation function we chose the following three properties:

(1) Excitation energy E^*_M at the maximum of the excitation function. It was assumed that E^* is equal to the sum of the projectile energy E_p and the Q-value of the reaction.

(2) Maximum cross-section, σ_M.

(3) Half width of the excitation function, H_M.

These characteristic data were extracted from nearly 300 published excitation

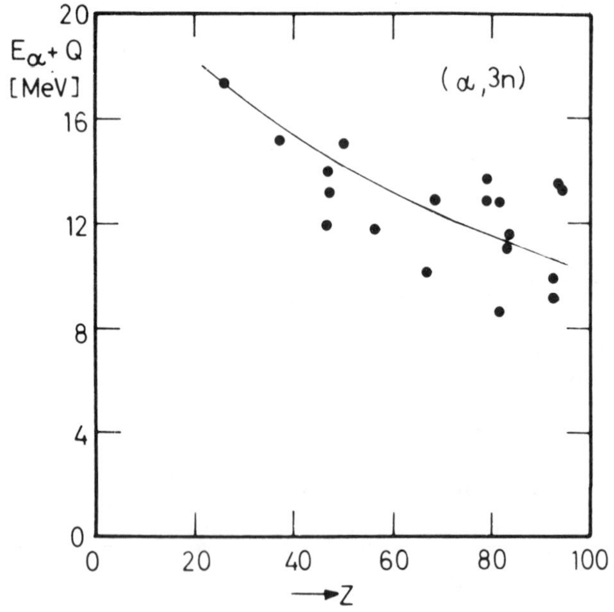

Fig. 1. Dependence of the excitation energy E^*_M ($=E_\alpha + Q$) *at the maximum of the excitation function for* (α, 3n) *reactions on the charge number Z of the target nuclides. The curve is a visual fit to experimental data*

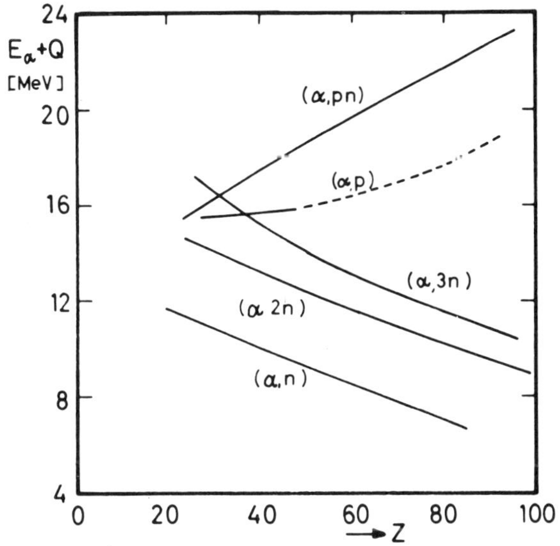

Fig. 2. Dependence of E_M^* ($=E_\alpha + Q$) *on* Z *for several* (α, xn) *and* (α, pxn) *reactions. The dashed line is an extrapolated curve*

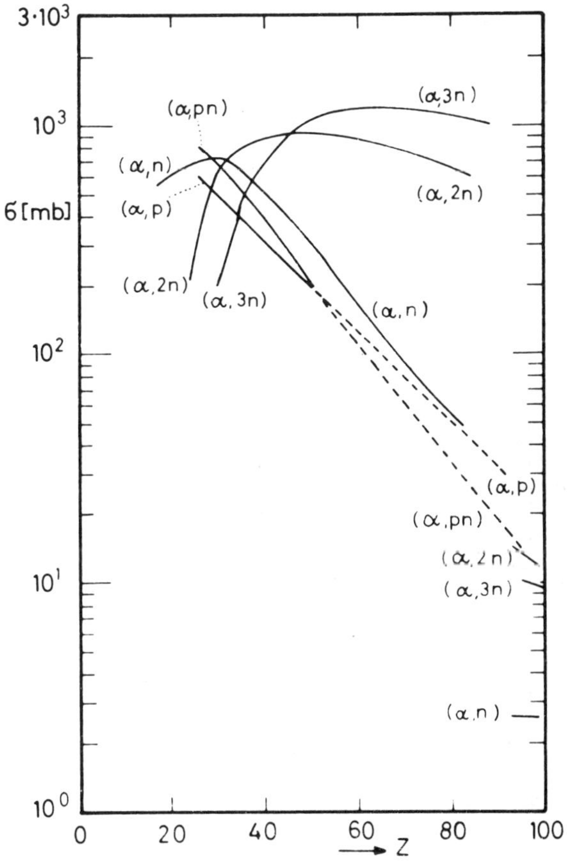

Fig. 3. Maximum cross-section vs Z *for several* (α, xn) *and* (α, pxn) *reactions. The dashed lines are extrapolated curves*

functions. Fig. 1 shows a typical dependence of E_M^* upon Z, the charge of the target nucleus. In spite of the appreciable scatter, the experimental points show a general tendency to decrease with increasing Z. The curve drawn is a visual fit to the data. In Fig. 2 several such curves are shown for different (α, pxn) and (α, xn) reactions. We see that the slope of the curves for all the (α, xn) reactions is similar, and that E_M^* decreases with increasing Z. This trend is to be expected, because the average binding energy as well as the average kinetic energy of the emitted neutrons decreases with increasing Z. For (α, p) and (α, pxn) reactions the slope has the opposite sign. This is due to the increase of the Coulomb barrier.

The curves in Fig. 3, which were again obtained by visual fit to experimental data, show the dependence of the maximum cross-section σ_M on Z for several α-particle reactions. The decrease in σ_M for (α, p) and (α, pn) reactions with increasing Z is due to the Coulomb barrier, which influences the emission probability of the protons. This in turn explains the increase of σ_M for the (α, xn) reactions.

The unusual variation in σ_M for the (d, n) reactions is also due to the influence of the Coulomb barrier. For target nuclides with large Z the Coulomb barrier is already so high that the corresponding excitation energy of the compound nucleus is sufficient to emit more than one neutron. The strong decrease of σ_M for very large Z is due to the increase in the fission cross-section.

Using these general trends for the characteristic data (E_M^*, σ_M, H_M), we constructed standard excitation functions and compared them with experimental values. In Figs. 4 and 5 two examples are given for such a comparison. The broken line shows the standard excitation function, whereas the other curves represent experimental data. The agreement in the case of the (α, 2n) reactions with target nuclides with $61 \leqslant Z \leqslant 83$ (Fig. 4) is quite good; for the (α, n) reactions (Fig. 5) and $20 \leqslant Z \leqslant 40$ it is still satisfactory. Generally we expect an error of less than a factor of two.

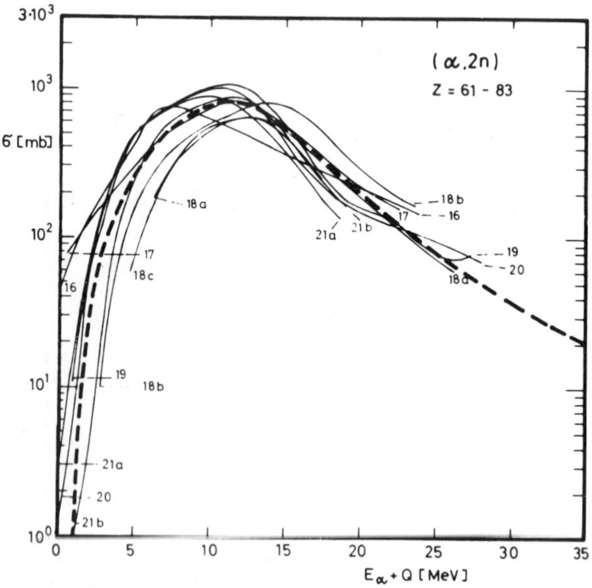

Fig. 4. Excitation functions for (α, 2n) *reactions with target nuclides with* $61 \leqslant Z \leqslant 83$. *The dashed line shows the estimated standard excitation function*

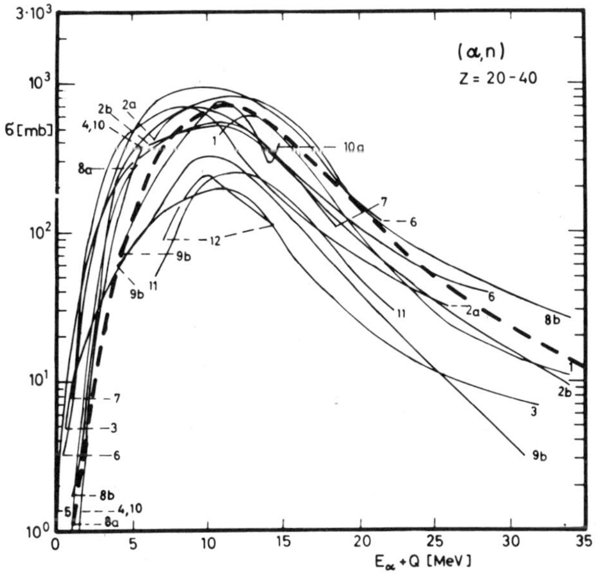

Fig. 5. Excitation functions for (α, n) *reactions with target nuclides with* 20 ⩽ Z ⩽ 40. *The dashed line shows the estimated standard excitation function*

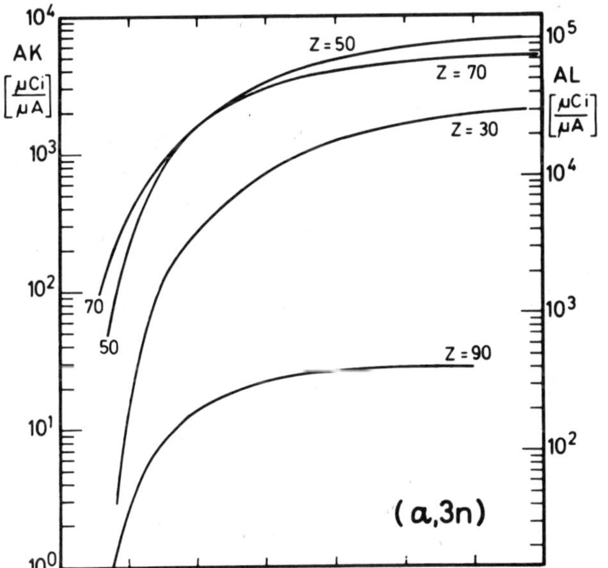

Fig. 6. Thick target yields for (α, 3n) *reactions for target nuclides with* Z *equal to 30, 50, 70, and 90. The left-hand scale gives the activity* AK *for an irradiation time* t = 0·1 T, *whereas the right-hand scale shows the saturation activity* AL (t ≫ T)

In the same way standard excitation functions for (d, xn), (d, p), and (p, 3n) reactions were constructed. The agreement of these curves with experimental data is also quite satisfactory.

Using the standard excitation functions and the range-energy relationships calculated by Williamson *et al*[2] we evaluated Eqn (1) by numerical integration. Figs. 6 and 7 show two typical examples of the calculations. The left-hand ordinate gives the activity *AK* for an irradiation time *t* that is 1/10 of the half-life *T* of the nuclide produced, whereas the right-hand ordinate gives the saturation activity *AL* ($t \gg T$).

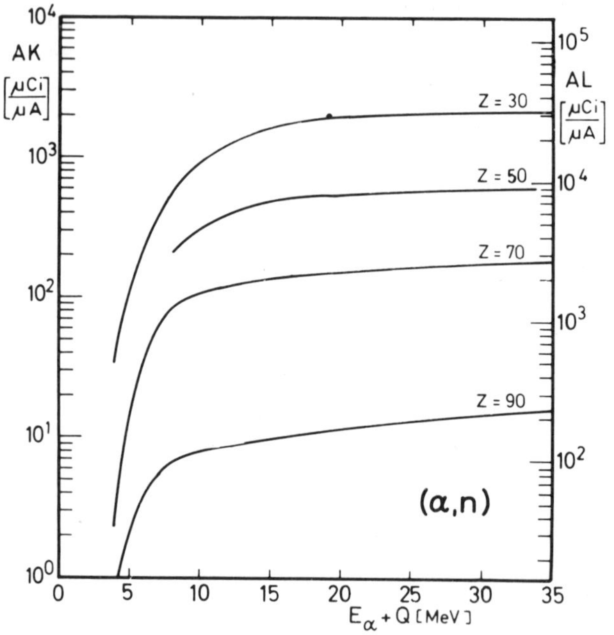

Fig. 7. Thick target yields for (α, n) *reactions for target nuclides with* Z *equal to 30, 50, 70, and 90. The left-hand scale gives the activity* AK *for an irradiation time* t = 0·1 T, *whereas the right-hand scale shows the saturation activity* AL ($t \gg T$)

APPLICATION

With the aid of these curves, we can estimate relatively accurately the optimum irradiation conditions for the production of radionuclides. In addition the amount of impurities produced during the irradiation may be estimated.
An example of the application of our standard excitation functions is shown in Fig. 8. A. Karpeles, in our laboratory, was interested in the production of ^{68}Ge by irradiating natural Zn with α-particles. The dashed curves in Fig. 8 show the estimated contributions of the different reactions of the Zn isotopes to the ^{68}Ge formation cross-section. The sum of these contributions agrees very well the experimental data determined by Karpeles.[3]

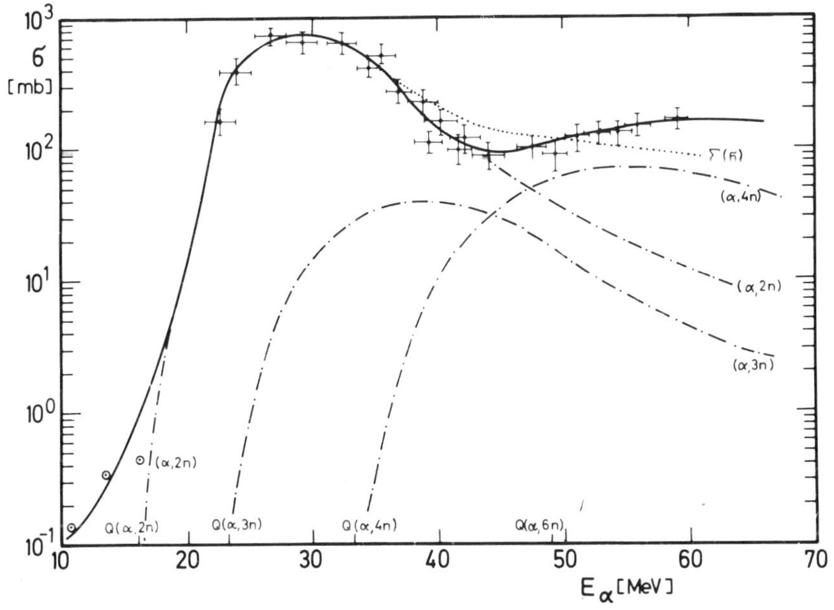

Fig. 8. Cross-section for the production of ^{68}Ge *by α-irradiation of natural* Zn[3]. *The full line is a visual fit to the experimental points, whereas the dashed line is the sum of the estimated contributions (– · – · – · –) of the possible reactions leading to* ^{68}Ge

REFERENCES

1. Ewart, A., Valentine, C., and Blann, M., *Nuclear Phys.* **69**, 625 (1965); Hazan, J. P. and Blann, M., *Phys. Rev.* **137**, B 1202, (1965).
2. Williamson, C. F., Boujot, J., and Picard, J., CEA–R 3042 (1966).
3. Karpeles, A., *'Radiochimica Acta* in press.

DISCUSSION:

H. A. C. McKay (A.E.R.E.): I am a little surprised that you were able to proceed without distinguishing between even-even, even-odd, etc., nuclei, especially in the case of proton irradiations.

H. Münzel: These effects are partly allowed for in using the individual *Q*-values for the reactions.

J. Vervier (Centre de Physique Nucléaire, Louvain): Which analytical function is used to represent the empirical cross-section curves you use?

H. Münzel: No analytical function was used. The curves are a visual fit to the general shape.

B. D. Pate: (Simon Fraser University): Did you consider using, in addition to *Z*, the *Z*/*N* value as a parameter for interpolation? My reason for asking is that for targets with a given value of *Z*, but where *Z*/*N* values change (e.g. ^{92}Mo ⟶ ^{100}Mo) values of σ can change by a factor as large as 5, due to binding-energy changes. This may be the source of the scatter by a factor of about 10 which appears in some of your data.

H. Münzel: We have tried several additional parameters in describing the data, without much success. This may be due mainly to the lack of, or inaccuracy in, the excitation function. We are now trying to calculate the magnitude of the expected trends for comparison with the experimental data.

M. A. Chaudhri (M.R.C., Hammersmith): I would like to ask why you have neglected all the data concerning nuclei with $Z < 20$.

H. Münzel: We are afraid that the compound nucleus hypothesis will break down with such light target nuclei.

H. A. C. McKay (A.E.R.E.): In view of the work reported, how far is it still worth while making excitation function measurements? It appears that it might be possible to predict excitation functions for $Z > 20$ almost as accurately as it is at present possible to measure them.

H. Münzel: There is no doubt that one should continue to determine excitation functions experimentally. The estimated excitation functions should only be considered as a rule of thumb.

Round-Table Discussion

Dùring the conference a discussion was organised on the above topic, in the hope that some measure of agreement might be reached on this important topic. Although this aim was (not surprisingly) not realised, the subject was usefully ventilated and proposals were made for continuing discussion.

The session was opened by an introductory paper given by Dr. K. Svoboda (Řež), which is given below together with a summary of the discussion prepared by Dr. D. J. Silvester (M.R.C., Hammersmith).

Terms used in the production of radioisotopes by charged-particle bombardment

K. Svoboda

Nuclear Research Institute, Czechoslovak Academy of Sciences, Řež, near Prague, Czechoslovakia

The yield of a nuclear reaction is characterised by the nuclear cross-section σ, usually given in barns or millibarns. When producing radioisotopes in a nuclear reactor by irradiation with thermal neutrons, which is the most common method, the total yield of a radioactive product may be calculated by means of very simple, well-known formulae, since the energy of the bombarding particles, and therefore the cross-section for the reaction, does not change when the particle passes through the sample. A significant practical aspect is the fact that the samples are usually irradiated in hermetically-sealed capsules, so that losses of the radioactive product during the irradiation process itself do not play a role.

The use of charged particles to produce radioisotopes gives trouble in both these respects. The energy of the bombarding particles decreases very rapidly on passing through the irradiated sample, causing large changes in the cross-section in successive layers of the sample, and the formulae become more sophisticated. As the energy of charged particles would be substantially decreased on passing through the wall of a capsule, most of the irradiations are carried out with samples which are covered only with a thin foil, or very often without any protection, under the high vacuum of the accelerator. Since the high-energy bombardment often causes overheating of the irradiated sample there may be significant loss of the product during the irradiation due to evaporation, possibly combined also with recoil effects.

Besides the above-mentioned factors which are intrinsic to the irradiation method, two other factors should be mentioned which characterise the present state of the problem. The first is that at the present time there are still very few data available giving cross-sections for charged-particle reactions ('excitation functions'—ref. 1). The second is the fact that accurate measurement of the intensity and energy of the bombarding particles is not very easy to achieve, so that significant errors may have occurred in some laboratories. Both factors may explain some discrepancies in earlier results, but fortunately they do not seem to cause great difficulty nowadays.

Faced with these difficulties, isotope producers have been forced to find a term to characterise the output of the radioactive product during a charged particle bombardment. The most widely-used term has been the so-called 'thick-target yield'.[2–7] It is difficult to find out who in fact introduced this term, and nowhere can an accurate definition be found. It is supposed to be the activity produced in one hour's irradiation, and is usually quoted in μCi/μAh or mCi/mAh. The tick-target yield Y is given by

$$Y = \frac{A}{It} \tag{1}$$

where A is the total activity (in μCi or mCi)
I is the bombarding particle beam intensity (in μA or mA)
t is the irradiation period (in h)

This definition supposes tacitly that the activity produced is proportional to the irradiation time, and therefore that the thick target yield can be easily evaluated by dividing the total activity obtained (quoted in mCi or μCi) by the product of irradiation time (in h) and irradiation beam intensity (in μA or mA). Taken to absurd limits, it follows that the same total activity should be obtained when irradiating with 10 μA for 10 h as by irradiation with 100 μA for 1 h.

It is obvious that this statement has no general validity. It is well known[8] that the total activity of a radioactive product depends on the irradiation time, and is given by:

$$A = RI\ [1 - \exp(-\lambda t)] \tag{2}$$

where besides the symbols already introduced R is the production rate. For charged-particle bombardments, let us quote it for convenience in μCi/μA or mCi/mA. R corresponds to the saturation activity (for bombardments with $t \gg 1/\lambda$). Provided that the irradiation time is short relative to the half-life of the radioactive product ($\tau = 0{\cdot}693/\lambda \gg t$) we can use the approximation

$$A = RI\ [1 - (1 - \lambda t + \frac{\lambda^2 t^2}{2!} - \frac{\lambda^3 t^3}{3!} + \ldots)] \approx RI\lambda t \tag{3}$$

Only under this condition can Eqn (1) be used with sufficient accuracy, where

$$Y = R\lambda \tag{4}$$

When the ratio of irradiation time to half-life is $<0{\cdot}10$ then the error introduced by using Eqn (1) does not exceed 4%. Until recently the species produced usually had half-lives of at least a few days and the irradiation times did not exceed a few hours, so the thick-target yield was quite a good practical value.

However, due to the availability nowadays of small, relatively inexpensive cyclotrons installed in the vicinity of the user, the production and utilisation of short-lived radioisotopes is increasing very rapidly. It does not seem reasonable to use the term 'thick-target yield' (as defined above) for short-lived radioisotopes, where the irradiation times are of the same order of magnitude as the half-life. This can be demonstrated as follows. Consider the production of an isotope of 2 h half-life.

Case I: Irradn. period 1 h, beam intensity 100 μA.

$$A = 100\,R\ [1 - \exp(-0{\cdot}5)] = 100\,R \times 0{\cdot}29 = 29\,R$$

Case II: Irradn. period 10 h, beam intensity 10 μA.

$$A = 10\,R\ [1 - \exp(-5)] = 10R \times 0{\cdot}969 = 9{\cdot}7\,R$$

If we use Eqn (1) to evaluate Y in both cases we should obtain in the first case a value three times greater than in the second case. Without stating clearly for what irradiation time the given thick-target yield Y was derived, the value is not defined sufficiently for short-lived radioisotopes. However, we find in the reviews referred to above values of Y for very short-lived as well as long-lived radioisotopes (e.g. thick-target yields can be found for ^{11}C (20 min), ^{13}N (10 min), ^{18}F (109 min), ^{27}Mg (9·5 min), ^{38}Cl (38 min)).[2–7]

The incomplete definition of the thick-target yield is probably the reason why other terms have been proposed, for example 'saturation activity per 1 μA'.[9] This term is identical with R. Practically the same term was used also by other authors,[1] but under another name, 'thick-target yield for the condition of saturation activity' (or, in short, 'long' thick-target yield). The same authors introduced another term 'thick-target yield for an irradiation time of 0·1 half-life' (or 'short' thick-target yield). There have also been other proposals, e.g. thick-target yield in mCi/kWh,[10] or 'specific activity for 100 μA and a thin target of 1 mg/cm^2'. This was estimated for irradiation times corresponding to saturation activity or for irradiation times of 8 h for longer-lived radioisotopes.[11]

Without criticising in detail the various terms which have been proposed previously it seems that R—the production rate (quoted in μCi/μA or mCi/mA) or 'saturation activity of a thick target for 1 μA' or 'long thick-target-yield' might be regarded as the most general term. The reasons for recommending this are as follows:

(1) it is independent of irradiation time,
(2) it is independent of the half-life of the radioisotope produced, and
(3) it can be easily evaluated on the basis of the known cross-sections and given energy (cf. Appendix 1).

The following aspects also require definition:

(1) its definition,
(2) its symbol and name, and
(3) conditions under which the estimated value can be regarded as 'true'.

Something has been already said about the first two items, even if only implicity. Let us now consider the last one.

When estimating the production rate experimentally significant errors may arise. These seem to be important when the irradiation is carried out at high ion-beam intensities or under inefficient cooling conditions. This can be demonstrated by reference to some experimental measurements of 'thick-target yields' for long-lived radioisotopes, where the relation between R and Y is satisfactorily defined by the expression (4).

Gleason and co-workers[12] have pointed out that the thick-target yield often varies substantially from one production run to another, and that thick-target yields determined from irradiations inside the accelerating chamber—where higher ion-beam intensities are usually employed and greater focusing of the beam takes place—are mostly lower than thick-target yields obtained during irradiations in the deflected beam, where lower beam intensities are used and the beam strikes the target over a larger area. They explain this effect by evaporation and/or diffusion of the product. Their results are reproduced in Table 1.

Table 1. THICK-TARGET YIELDS OF SEVERAL RADIOISOTOPES FOR PRODUCTION RUNS AND EXTERNAL BOMBARDMENT[12]

Isotope	*Target*	*Proton energy* (MeV)	*Thick-target yield*				
			External target mCi/mAh	*Internal target*			
				Maximum value		*Average value*	
				mCi/mAh	ε (%)	mCi/mAh	ε (%)
^{7}Be	Li	17·1	254	214	84·3	140	55·2
^{22}Na	Mg	20·8	0·64	0·50	79·6	0·43	67·2
^{51}Cr	V	17·1	595	226	37·9	226	37·9
^{55}Fe	Mn	17·1	14·2	13·0	91·5	12·7	89·4
^{56}Co	Fe	17·1	76	–	–	70	92·2
^{57}Co	Ni	20·8	31·6	28·0	88·6	24·6	77·8
^{65}Zn	Cu	17·1	14·2	9·8	69·0	8·5	59·8
^{109}Cd	Ag (plated)	17·1	5·53	1·1	19·9	1·1	19·9
^{109}Cd	Ag (solid)	17·1	5·53	3·4	61·4	3·4	61·4

Table 2. DISCREPANCIES IN SOME ESTIMATES OF THICK-TARGET YIELDS GIVEN BY DIFFERENT AUTHORS

Isotope	*Target*	*Reaction*	*Thick-target yield* (mCi/mAh) *at given particle energy*					
			14 MeV	Ref.	15 MeV	Ref.	19 MeV	Ref.
^{7}Be	Li	d, 2n	–	–	48	5	2	6, 7
^{51}Cr	V	d, 2n	–	–	–	–	8 0·2	6 7
^{52}Mn	Cr	d, 2n	–	–	81 144	3 5	8	6, 7
^{55}Fe	Mn	d, 2n	–	–	0·67 3·7	3 5	0·02	6, 7
^{57}Co	Fe	d, n	–	–	9·5	5	5	6, 7
^{74}As	Ge	d, 2n	–	–	2·0 31	3 5	–	–
^{85}Sr	Rb	d, 2n	0·13	6	16	5	0·6	7
^{88}Y	Sr	d, 2n	0·1	7	38	3	1·0	7

Unnoticed evaporation, as well as insufficiently-accurate determination of the beam current intensity, may explain large differences in thick-target yield estimates by different authors (Table 2).

There has been, however, no previous evidence of any correlation between the decrease in the estimated thick-target yield and the factors influencing it. In our laboratory we have tried to study the effect of the deuteron beam intensity on the thick-target yield for targets constructed or prepared in various ways. Some of our results with ^{85}Sr during its preparation by deuteron bombardment of RbCl have been presented by Hrubý.[13] In addition I should mention briefly results with ^{52}Mn produced by deuteron bombardment of Cr[14] (Fig. 1).

We define ϵ as the ratio of individual estimates of the thick-target yield to the thick-target yield estimated in the external beam at low beam intensity. This

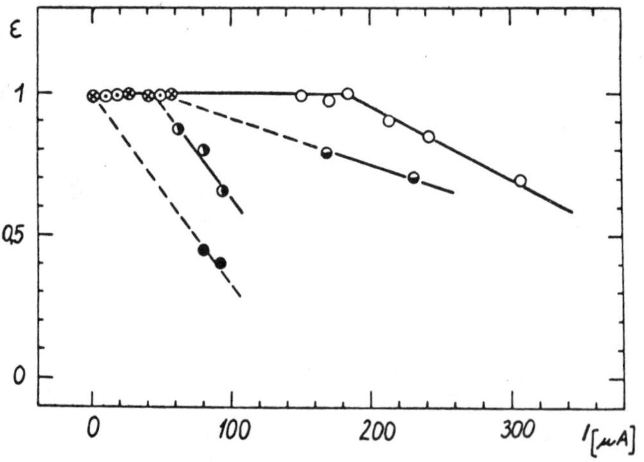

Fig. 1. Dependence of the ratio ε on the deuteron beam intensity for various irradiation conditions. Radionuclide: Mn 52. *Cylindrical target:* ○ *annealed,* ◒ *not annealed. Loop-shaped target;* ◑ *annealed,* ● *not annealed. (Annealing at 300°C for at least 6 h.) Flat target, irradiated in external beam, irradiated area:* ⊗ *2·8 cm², ⊙ 12·5 cm²*

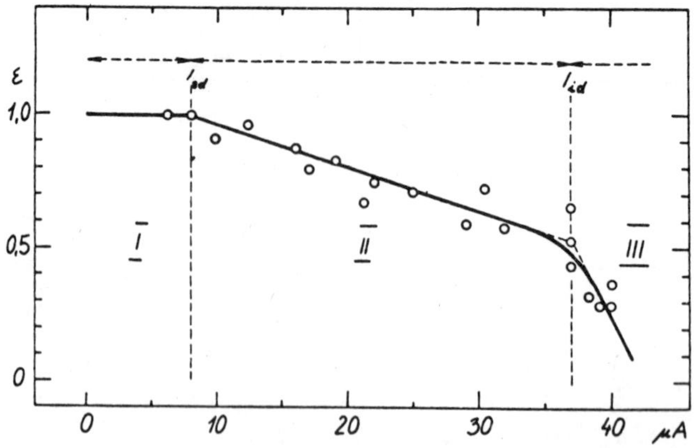

Fig. 2. Dependence of the ratio ε on the deuteron beam intensity. Radionuclide: Sr 85

value is in very good agreement with that of Kafalas and Irvine.[15] Fig. 2 shows similar results for the production of ^{85}Sr.[13] Even though the results are rather incomplete they give some idea about the probable features of the decrease in thick-target yield caused by increasing the beam current intensity for different irradiation conditions.

The examples given show that the activity A produced by charged-particle bombardment cannot be simply calculated according to Eqn (2) but that an additional coefficient should be introduced which takes into account the loss of activity due to evaporation during irradiation. Let us call this coefficient 'irradiation effectiveness', ϵ, and modify Eqn (2) accordingly:

$$A = \epsilon RI\,[1 - \exp(-\lambda t)] \tag{5}$$

The coefficient ϵ was in fact previously introduced by Dmitriev and Krasnov,[10] who consider it to be a function of the energy dissipated in the target during irradiation. Unfortunately, they do not give any experimental data showing time-dependence or the dependence of the coefficient on energy input.

To simplify the problem, let us suppose that we can assume ϵ to be constant during any one experimental run. It is governed by such irradiation conditions as target shape, spreading of beam, target preparation, cooling, particle energy, etc. Under the same conditions the coefficient ϵ is a function of the beam current, as can be seen from Figs. 1 and 2. The coefficient for low beam intensities is always unity, and only above a certain beam intensity can a decrease of ϵ be observed.

It does not seem reasonable at the moment to propose a 'general target efficiency equation', giving ϵ as a function of I for different irradiation conditions, as the experimental data are too scarce to make any general conclusion possible. However, from the results given in Figs. 1 and 2 some general features of the 'efficiency curve' may be drawn. From zero to a given value of I, called I_{sd} (beam current at the start of ϵ decrease), ϵ remains constant at unity. Thus:

$$\text{Region 1: When } 0 < I < I_{sd}, \quad \epsilon = 1 \tag{6a}$$

Beyond this region, ϵ decreases practically linearly with I, i.e.

$$\text{Region 2: When } I_{sd} < I < I_{id}, \quad \epsilon = 1 - a(I - I_{sd}) \tag{6b}$$

where a is the decremental decrease.

On increasing the beam current further, a third region is reached:

Region 3: When $I > I_{id}$ (ion-beam current at the start of intensified decrease)

$$\epsilon = 1 - a(I_{id} - I_{sd}) - b(I - I_{id}) \tag{6c}$$

where b is the decrement for the intensified decrease ($a < b$).

It is hoped that in addition to the function $\epsilon(I)$, experimental and theoretical attention will also be paid in future to the significance of the 'constants' (a, b, I_{sd}, I_{id}) for the given irradiation conditions.

How does the coefficient ϵ influence the estimation of the production rate R? It can readily be seen that Eqn (5) is identical with Eqn (2) when $\epsilon = 1$.

However, how may we check whether we are in this region ($I < I_{sd}$)? The solution of this problem is available. To estimate the production rate R, at least two experimental runs should be carried out differing in beam current at least by a factor of 2. If the activity produced in both experiments is proportional to the beam current I multiplied by the time factor $(1 - \exp(-\lambda t))$, I should be regarded as $I < I_{sd}$ and $\epsilon = 1$. In this case Eqn (2) is applicable and the estimated value corresponds to the true value of R. If this is not so, other experiments with lower beam currents should be carried out until the region is reached where ϵ is constant. If the region of constant ϵ is still not reached, R should be calculated according to Eqn (2) and plotted against I. The value of R extrapolated to $I = 0$ should be regarded as a true value.

In the appendices some of the general relationships between cross-section, production rate, etc., are given for the convenience of participants in the discussion, as well as proposals about terms and units for possible approval.

APPENDIX 1

Relation between the production rate R and the cross-section of the nuclear reaction σ

$$R = \frac{K}{zN} \int_0^S \sigma(s)\,\mathrm{d}s = \frac{K}{zN} \int_{E_m}^0 \sigma(E)\,(\mathrm{d}s/\mathrm{d}E)\,\mathrm{d}E$$

where R is the production rate in μCi/μA
- z is the number of elementary charges of the bombarding particles
- σ is the nuclear cross-section (in millibarns)
- s is the depth of the target layer under consideration (in mg/cm^2)
- S is the range of the bombarding particles (in mg/cm^2)
- E is the actual energy of the bombarding particles in the given depth of target s under consideration (in MeV)
- E_m is the energy of the bombarding particles at the target surface (in MeV)
- N is the number of target nuclei
- K is a constant which takes into account that the bombarding particle beam is given in μA (1 μA of protons or deuterons equals $6{\cdot}25 \times 10^{12}$ particles per second), that the amount of atoms in one gram-atom is $6{\cdot}02 \times 10^{23}$, that the cross-section is expressed in millibarns ($10^{-3} \times 10^{-24}$ cm^2) and that the activity is expressed in μCi ($3{\cdot}7 \times 10^4$ disintegrations/s), i.e.

$$K = \frac{6{\cdot}25 \times 10^{12} \times 6{\cdot}02 \times 10^{23} \times 10^{-27}}{3{\cdot}7 \times 10^4} = 1{\cdot}02 \times 10^5$$

In many cases it is convenient to calculate the value of R from experimentally-estimated excitation functions by means of graphical or numerical integration. In this case

$$R = \frac{K}{zN} \Sigma_i \frac{S_{i+1} - S_i}{E_{i+1} - E_i} \int_{E_i}^{E_{i+1}} \sigma(E)\mathrm{d}E = \frac{K}{zN} \Sigma_j \Delta s_j\, \sigma_j(E)$$

The increment of target thickness Δs_j should be chosen so that the value of σ would not change significantly and can be replaced by its average value.

When we irradiate a target which does not consist solely of the nuclides yielding the required nuclear reaction, but which contains also other isotopes and which may be a chemical compound, the production rate is given by

$$(R)_{a,\,ch} = R\frac{Nma}{M}F$$

where $(R)_{a,\,ch}$ means the production rate in the target consisting of the chemical compound of type *ch*, with the isotopic abundance of the true target nuclide *a*

n is the number of target element atoms in the target compound molecule

M is the molecular weight of the target compound

F is a factor which takes into account the difference of the stopping powers of the pure element and of the chemical compound used.

This factor can be expressed as for example in ref. (1):

$$F = \frac{S_{ch}}{S} = \frac{1}{x_1 + x_2\frac{S_1}{S_2} + x_3\frac{S_2}{S_3} + \ldots}$$

where S_1, S_2, S_3 are the ranges of the bombarding particles in the individual elements forming the given chemical compound

x_1, x_2, x_3 are the weight proportions of the individual elements in the target compound

Activity produced in a thin target

$$A = \frac{K}{zN} I\sigma s\,(1 - \exp(-\lambda t))$$

A thin target should fulfil the condition $\sigma \approx$ const. for its whole thickness; in any case $s < 10$ mg/cm^2

Krasnov and Dmitriev's activity equation

$$A = B\int_0^t \epsilon I\,\mathrm{d}t$$

where A is the total activity (in mCi)

B is the thick-target yield (in mCi/kWh)

ϵ is the irradiation efficiency

I is the bombarding beam power dissipated in the target (in kW)

t is the irradiation time (in h)

APPENDIX 2

Proposals for discussion

1. The production rate R should be regarded as the general unit for characterisation of the quantity of radionuclide produced by charged-particle bombardment in a thick target. It is defined as follows:

   R, the production rate (in μCi/μA), is the activity (in μCi) of the radionuclide produced in a thick target by charged particles of a given energy E_m corresponding to an ion beam current of 1 μA, to an irradiation time such that saturation activity is reached, and to irradiation conditions such that no loss of product occurs during bombardment ($\epsilon = 1$).

   R without subscript should signify the value for a target consisting of pure target nuclides. For other targets (e.g. polyisotopic mixtures, natural elements, or chemical compounds) appropriate subscripts should be added, e.g. R_{Rb}, R_{RbCl}, etc.

2. The thick-target yield Y (in μCi/μAh, mCi/mAh, etc.) should be regarded as a derived, practical value. It should be used for radionuclides produced only when the irradiation time is less than one-tenth of the half-life of the radionuclide under consideration. In any case it should not be used for radionuclides with half-lives less than one day. The thick-target yield value should be regarded rather as a technical than as an accurate physical value. When publishing any value of Y, it will be essential to state whether or not it was checked that the irradiations were carried out under the condition $\epsilon = 1$.

REFERENCES

1. Lange, J. and Münzel, M., KFK 767, Kernforschungszentrum Karlsruhe (1968).
2. Aten, A. H. W. Jr. and Halberstadt, J., *Philips techn. Rundschau* **15**, 357, (1954).
3. Irvine, J. W., *Nucleonics* **3**, 2, 5, (1968).
4. Martin, J. A., Livingston, R. S., and Murray, R. L., *Nucleonics* **13**, 3, 28, (1955).
5. Gruverman, I. J. and Kruger, P., *Int. J. Appl. Rad. Isot.* **5**, 21, (1959).
6. Garrison, W. M. and Hamilton, J. G., *Chem. Rev.* **49**, 237, (1951).
7. Nesmejanov, A. N., Lapickij, A. V., and Rudenko, N. P., Production of Radioisotopes, Goschimizdat, Moscow (1954) (in Russian).
8. Friedlander, G., Kennedy, J., and Miller, J. M., Nuclear and Radiochemistry, J. Wiley, New York (1964).
9. Svoboda, K., Hrubý, J., and Hradil, M., *Isotopenpraxis* **3**, 313, (1967).
10. Dmitrijev, P. P. and Krasnov, N. N., *Chem. zvesti* **21**, 611, (1967).
11. Lorenz, W. J., Sheer, K. E., Lindner, H. W., and Chaudri, M. A., *Atompraxis* **13**, 521, (1967).
12. Gleason, G. I., Gruveman, I. J., and Need, J. L., *Int. J. Appl. Rad. Isot.* **13**, 223, (1962).
13. Hrubý, J., contribution at this conference.
14. Hrubý, J. and Svoboda, K., unpublished results.
15. Kafalas, P. and Irvine, J. W., *Phys. Rev.* **104**, 703, (1956).

SUMMARY OF DISCUSSION

(Rapporteur D. J. Silvester, M.R.C., Hammersmith)

The interest aroused by the presentation of Svoboda's introductory paper can be judged by the fact that over half of the 32 conference participants who came to hear it contributed to the discussion which followed. This ensured that the subject was given a lively, if largely inconclusive, airing, but made the task of the rapporteur an unenviable one. What follows is therefore but the briefest summary of the many points of view that were expressed.

Initial discussion centred around the relative merits of the terms R (production rate) and Y (thick-target yield) as defined by Svoboda, and each term had its protagonists. Those whose prime interest appeared to be activation analysis favoured the use of R, whilst those who were chiefly concerned with isotope production preferred Y, though perhaps with some minor modification. Others with divided interests expressed the hope that a single term could be found that would satisfy everyone's requirements, thereby avoiding the confusion that would probably arise if two terms were in general use.

The case for wider use of the term R was based on its close relationship to the nuclear cross-section, σ, and the relative ease with which it may be calculated from published excitation functions. Against this, some believed that there would be greater risk of misinterpretation of tabulated values of R than of Y by those who had not clearly understood their definitions.

Y was felt by many to be a convenient, practical term, although the restriction regarding the relative lengths of bombardment and half-life made it unattractive as defined at present. Furthermore, since a consideration of competing nuclear reactions sometimes makes it preferable that a target should not be 'thick' in the nuclear sense, would it still be reasonable to talk of a 'thick-target yield' in such a case?

Several participants spoke of the numerous parameters that would have to be specified when publishing values of either term in order to define them fully. These were conveniently listed, in a memorandum circulated by R. S. Tilbury, as follows:

(1) The bombarding particle and its ionic charge, e.g. $^{4}He^{2+}$, $^{40}Ar^{10+}$.
(2) The energy of the incident particles, e.g. 30 MeV.
(3) The target nuclide, e.g. ^{50}Cr.
(4) The chemical composition of the target, e.g. H_2O, $CaSO_4$.
(5) The target thickness, e.g. g of compound/cm^2.
(6) The product radionuclide.

To this list, Svoboda's 'irradiation effectiveness' coefficient, ϵ, must of course be added.

Whilst all seemed to agree that it would be useful to define terms such as R and Y and to encourage their general use, there seemed to be little sign of agreement about the units to be employed in the definitions. Some would have been content to follow Svoboda, who proposes units of μCi/μA for R, and μCi/μAh for Y. Others, however, would prefer R, for instance, to be defined as 'the number of disintegrations per second of the product nuclide produced at saturation by bombardment with a beam of 10^{10} particles per second'. Needless to say, a number of variations on both these proposals were offered.

A little time was also spent in discussing the accuracy with which the necessary measurements of absolute activity and beam current can be made. It was agreed that the former could only be as precise as a knowledge of the decay scheme for the nuclide of interest would allow, and it was pointed out that a beam current measurement might sometimes depend upon the nature of the target material—for example, the extent of its surface oxidation.

At the conclusion of the discussion, which lasted about an hour, it was proposed that the advice of bodies such as the I.C.R.U. (International Commission on Radiation Units and Measurements) and the I.A.E.A. should be sought, regarding the best way to proceed towards the formulation of an internationally-acceptable term defined in internationally-acceptable units.

Approval greeted the further proposal that those who held firm opinions on this issue should be invited to put them in writing, so that they might be circulated, and given detailed consideration, prior to some future meeting. It was suggested that such memoranda be sent to K. Svoboda and D. J. Silvester, who agreed to find some means of canvassing the views of those who were known to be interested in the subject of the discussion, but who had been unable to attend the present conference. On this note of general agreement the discussion ended.

List of Participants

Dr. P. Albert, Centre d'Etudes de Chimie Metallurgique, Vitry, France

Dr. R. G. Allas, U.S. Naval Research Laboratory, Washington, D.C., U.S.A.

Dr. C. B. Amphlett, Chemistry Division, A.E.R.E., Harwell

Dr. C. P. Baker, Brookhaven National Laboratory, Upton, New York, U.S.A.

Professor J. Bardwick, Cyclotron Laboratory, University of Michigan, Ann Arbor, Michigan, U.S.A.

Mr. R. A. Barton, Chemistry Division, A.E.R.E., Harwell

Mr. G. J. Batra, M.R.C. Cyclotron Unit, Hammersmith Hospital, London

Dr. R. J. Berry, Radiobiology Laboratory, Churchill Hospital, Headington, Oxford

Professor H. G. Blosser, Cyclotron Laboratory, Michigan State University, East Lansing, Michigan 48823, U.S.A.

Dr. J. W. Blue, Physics and Chemistry Division, National Aeronautics and Space Administration, Lewis Research Center, Cleveland, Ohio, U.S.A.

Dr. B. J. Bowles, Chemistry Division, A.E.R.E., Harwell

Dr. J. T. Brennan, Donner Center, Dept. of Radiology, Hospital of the University of Pennsylvania, Philadelphia, U.S.A.

Dr. J. J. Broerse, Radiobiological Institute TNO, Lange Kleiweg, Rijswijk Z.H., The Netherlands

Mr. P. D. Buckingham, M.R.C. Cyclotron Unit, Hammersmith Hospital, London

Dr. W. G. Burns, Chemistry Division, A.E.R.E., Harwell

D. J. A. Cairns, Solid State Division, A.E.R.E., Harwell

Professor W. L. Caldwell, Division of Radiotherapy and Radiation Research, Vanderbilt University, School of Medicine, Nashville, Tennessee, U.S.A.

Dr. B. Candoni, Instituto di Fisica, University of Milano, Italy

Mr. M. A. Chaudhri, M.R.C. Cyclotron Unit, Hammersmith Hospital, London

Mr. N. S. Chen, Physics Department, University of Birmingham, Birmingham 15

Mr. J. C. Clark, M.R.C. Cyclotron Unit, Hammersmith Hospital, London

Dr. D. Comar, C.E.A., Département de Biologie, Service Hospitalier Frédéric Joliot, Orsay, France

Dr. C. Corno, Istituto di Microbiologia, Milan, Italy

Mr. I. F. Croall, Chemistry Division, A.E.R.E., Harwell

Mr. D. R. A. Cuff, Analytical Sciences Division, A.E.R.E., Harwell

Dr. R. B. Cundall, Department of Chemistry, University of Nottingham

Mr. J. G. Cuninghame, Chemistry Division, A.E.R.E., Harwell

Mr. J. F. B. Dealler, Physics Department, University of Birmingham

Dr. J. L. Debrun, c/o Professor R. E. Jervis, Dept. of Chemical Engineering, University of Toronto, Ontario, Canada

Mr. Yu. N. Denisov, Joint Institute of Nuclear Research, Dubna, Moscow, U.S.S.R.

Mr. V. P. Dmitrievskii, Joint Institute of Nuclear Research, Dubna, Moscow, U.S.S.R.

Mr. B. Donnadille, Thomson-CSF, Orsay, France

Mr. J. W. Doornebos, N.V. Philips-Duphar, Cyclotron and Isotope Laboratories, Petten, The Netherlands

Mr. M. Dubus, Section des Accélérateurs, Centre d'Etudes Nucléaires de Grenoble, Cedex No. 85, 38 Grenoble-Gare, France

Mr. M. J. Duck, Chemistry Division, A.E.R.E., Harwell

Dr. A. A. Van Dulmen, N.V. Philips-Duphar, Cyclotron and Isotope Laboratories, Petten, The Netherlands

Dr. J. Ehrlich, Gesellschaft für Kernforschung m.b.H., Karlsruhe, Postfach 3640, West Germany

Mr. C. Engelmann, Centre d'Etudes Nucléaires de Saclay, S.R.M.P.C., Gif-sur-Yvette, France

Mr. K. V. Ettinger, Department of Physics, University of Birmingham, Birmingham 15

Dr. J. C. Evans, Radiation Therapy Division, Duke University Medical Center, Durham, North Carolina, U.S.A.

Mr. E. P. Ferreira, Cyclotron Laboratory, C.N.E.A., Buenos Aires, Argentina

Mr. M. F. Finlan, The Radiochemical Centre, White Lion Road, Amersham, Buckinghamshire

Mr. R. M. Fry, Australian Atomic Energy Commission, Canberra House, London, W.C.2

Dr. W. van Genderen, Instituut voor Kernphysisch Onderzoek, Amsterdam, The Netherlands

Mr. J. Giroux, Institut de Physique Nucléaire, Lyon, France

Mr. A. A. Glazov, Joint Institute of Nuclear Research, Dubna, Moscow, U.S.S.R.

Mrs. K. M. Glover, Chemistry Division, A.E.R.E., Harwell

Mr. A. B. M. Golam Mostafa, Physics Department, University of Birmingham, Birmingham 15

Mr. L. J. Goodman, Radiological Research Laboratory, Department of Radiology, Columbia University, New York, U.S.A.

Dr. C. L. Graham, Physics Department, University of Birmingham, Birmingham 15

Dr. A. M. Gurewitsch, Consultant–Applied Sciences and Technology, General Electric Co. U.S.A., European Office, Zurich, Switzerland

Mr. H. J. J. de Haan, N.V. Philips-Duphar, Cyclotron and Isotope Laboratories, Petten, The Netherlands

Professor H. L. Hagedoorn, Cyclotron Laboratory, Technische Hogeschool, Eindhoven, The Netherlands

Mr. J. W. Hale, Chemistry Division, A.E.R.E., Harwell

Professor C. R. Hansen, Jeffersen Medical College, Philadelphia, Pennsylvania, U.S.A.

Dr. H. Hazewindus, Philips Research Laboratories, Geldrop, The Netherlands

Dr. F. Helus, Royal Postgraduate Medical School, Hammersmith Hospital, London

Mr. J. A. Hetherington, 124 Newtorn Road, Carlisle, Cumberland

Mr. B. Hoop, Massachusetts General Hospital, Boston, Massachusetts, U.S.A.

Mr. K. J. Howard, Chemistry Division, A.E.R.E., Harwell

Mr. M. L. Howlett, c/o Inorganic Department, The Radiochemical Centre, White Lion Road, Amersham, Buckinghamshire

Mr. J. Hrubý, Nuclear Research Institute, Rez near Prague, Czechoslovakia

Dr. R. Jahr, Physikal. Tech. Bundesanst., Braunschweig, West Germany

Dr. O. N. Jarvis, Nuclear Physics Division, A.E.R.E., Harwell

Dr. I. L. Jenkins, Applied Chemistry Division, A.E.R.E., Harwell

Professor W. P. Johnson, Department of Physics, University of Maryland, College Park, Maryland, U.S.A.

Mr. E. J. Jones, Chemistry Division, A.E.R.E., Harwell

Mr. T. R. Jones, Division of Research, U.S. Atomic Energy Commission, Washington, D.C., U.S.A.

Dr. T. Karasawa, Cyclotron Laboratory, Institute of Physical and Chemical Research (RIKEN), Yamato machi, Saitama Pref., Japan

Professor C. Kellershohn, C.E.A. Département de Biologie, Service Hospitalier Frédéric Joliot, Orsay, France

Mr. K. T. Khoe, Argonne National Laboratory, 9700 South Cass Avenue, Argonne, Illinois, U.S.A.

Dr. R. T. King, Oak Ridge National Laboratory, Oak Ridge, Tennessee, U.S.A.

Dr. W. S. Klutz, Radiosotope Unit, Roger Williams General Hospital, Providence, Rhode Island, U.S.A.

Mr. R. A. Koolhof, Philips Research Laboratories, Geldrop, The Netherlands

Mr. A. A. von Kranenburg, N.V. Philips' Gloeilampenfabrieken, Hig Pit Comm. Dept. Cyclotrons, Eindhoven, The Netherlands

Mr. N. N. Krasnov, Power Physics Institute, Obninsk, U.S.S.R.

Mr. O. Krauss, Deutsches Krebsforschungszentrum Heidelberg, Institut für Nuklearmedizin, Heidelberg, West Germany

Mr. P. N. Kuin, N.V. Philips Research Laboratories, Eindhoven, The Netherlands

Dr. D. N. Kundu, Saha Institute of Nuclear Physics, Calcutta, India

Dr. R. Langkau, I. Institut für Experimentalphysik, Hamburg, West Germany

Professor M. Lefort, Laboratoire Chimie Nucléaire, Orsay, France

Mr. A. Leonard, C.E.N.-S.C.K., Mol, Belgium

Dr. L. Lindner, Chemistry Department, Brookhaven National Laboratory, Upton, New York, U.S.A.

Mr. R. N. Litunovskii, Electrophysical Apparatus Research Institute, Leningrad, U.S.S.R.

Dr. H. D. Maccabee, M.R.C. Radiobiology Unit, Harwell, Didcot, Berkshire

Mr. R. W. McIlroy, Chemistry Division, A.E.R.E., Harwell

Dr. H. A. C. McKay, Applied Chemistry Division, A.E.R.E., Harwell

Dr. W. Maier-Borst, Deutsches Krebsforschungszentrum Heidelberg, Institut für Nuklearmedizin, Heidelberg, West Germany

Dr. M. L. Mallory, Oak Ridge National Laboratory, Oak Ridge, Tennessee, U.S.A.

Dr. A. M. Marko, Biology and Health Physics Division, Atomic Energy of Canada Limited, Chalk River, Ontario, Canada

Mr. W. R. Marsh, Chemistry Division, A.E.R.E., Harwell

Dr. S. May, Centre d'Etudes Nucléaires de Saclay, Gif-sur-Yvette, France

Mr. B. Morris, Chemistry Division, A.E.R.E., Harwell

Dr. R. W. Müller, AEG-Telefunken, Nuclear Energy Division, E 33, D-8752, Grosswelzheim, West Germany

Dr. H. Münzel, Gesellschaft für Kernforschung m.b.H., Institut für Radiochemie, Karlsruhe, West Germany

Professor W. Myers, Ohio State University Hospitals, Columbus, Ohio, U.S.A.

Mr. T. Nagai, I.A.E.A., Vienna, Austria

Dr. G. J. Neary, M.R.C. Radiobiology Unit, Harwell, Didcot, Berkshire

Dr. J. L. Need, New England Nuclear Corporation, Boston, Massachusetts, U.S.A.

Dr. R. L. Nelson, Chemistry Division, A.E.R.E., Harwell

Dr. R. S. Nelson, Solid State Division, A.E.R.E., Harwell

Mr. J. M. van Nieuwland, Philips Research Laboratories, Geldrop, The Netherlands

Dr. H. Ostertag, Deutsches Krebsforschungszentrum Heidelberg, Institut für Nuklearmedizin, West Germany

Mrs. M. C. Palcos, Comision Nacional de Energia Atomica, Buenos Aires, Argentina

Dr. D. A. Parkes, Chemistry Division, A.E.R.E., Harwell

Professor W. C. Parkinson, Cyclotron Laboratory, North Campus, University of Michigan, Ann Arbor, Michigan, U.S.A.

Professor B. D. Pate, Department of Chemistry, Simon Fraser University, Burnaby, British Columbia, Canada

Dr. W. Paul, Dept. of Pathological Chemistry, University of Toronto, Toronto 5, Canada

Mr. P. F. Peck, Chemistry Division, A.E.R.E., Harwell

Dr. T. B. Pierce, Analytical Sciences Division, A.E.R.E., Harwell

Dr. J. J. Pinajian, N.V. Philips-Duphar, Cyclotron and Isotope Laboratories, Petten, The Netherlands

Mr. L. Quaglia, University of Liege, Laboratory Van de Graaff, Liege, Belgium

Dr. W. Rautenbach, Council for Scientific and Industrial Research, Pretoria, South Africa

Mr. C. R. V. Reed, Chemistry Division, A.E.R.E., Harwell

Dr. B. Rose, Nuclear Physics Division, A.E.R.E., Harwell

Mr. J. L. Roux, Institut de Physique Nucléaire, Villeurbanne, France

Mr. P. C. Röding, c/o AEG-Telefunken, Nuclear Energy Division, E 33, D-8752 Grosswelzheim, West Germany

Dr. M. J. Sametband, Cyclotron Laboratory, C.N.E.A., Buenos Aires, Argentina

Dr. U. Schryber, SIN-Schweiz. Institut für Nuklearforschung, Zurich, Switzerland

Mr. F. Schulz, Kernforschungszentrum, Zyklotron Laboratorium, Karlsruhe, West Germany

Dr. D. J. Silvester, M.R.C. Cyclotron Unit, Hammersmith Hospital, London

Dr. H. Sinn, Deutsches Krebsforschungszentrum, Institut für Nuklearmedizin, Heidelberg, West Germany

Mr. E. M. Smith, School of Medicine, University of Miami, Miami, Florida, U.S.A.

Mr. E. E. Smolko, Comision Nacional de Energia Atomica, Buenos Aires, Argentina.
attached: Astbury Dept. of Biophysics, University of Leeds

Dr. A. H. Snell, Oak Ridge National Laboratory, Oak Ridge, Tennessee, U.S.A.

Dr. J. V. Sodd, Nuclear Medicine Laboratory, Bureau of Radiological Health, Cincinnati General Hospital, Cincinnati, Ohio, U.S.A.

Mr. D. Stinchcombe, Chemistry Division, A.E.R.E., Harwell

Professor A Strzalkowski, Institute of Nuclear Physics, Cracow, Poland

Dr. G. B. Strini, Istituto di Fisica dell' Universita, Via Celoria, Milan, Italy

Dr. F. J. Stubbs, Chemistry Division, A.E.R.E., Harwell

Dr. H. Stüssi, SIN-Schweiz. Institut für Nuklearforschung, Zurich, Switzerland

Mr. A. Svanheden, Instrument AB, Scanditronix, Acc.avd., Uppsala, Sweden

Dr. K. Svoboda, Nuclear Research Institute, Řež near Prague, Czechoslovakia

Professor G. Tagliaferri, Institute of Physics, University of Milan, Milan, Italy

Dr. J. H. Tait, Theoretical Physics Division, A.E.R.E., Harwell

Mr. J. F. W. Tertoolen, N.V. Philips, Central Laboratory for Phys. Chem., Building BL, Eindhoven, The Netherlands

Mr. R. T. Theus, U.S. Naval Research Laboratory, Washington, D.C., U.S.A.

Mr. M. Thistlethwaite, Chemistry Division, A.E.R.E., Harwell

Mr. J. P. Thomas, Institut de Physique Nucléaire, Lyon, France

Dr. U. Thyssen, Deutsches Krebsforschungszentrum, Institut für Nuklearmedizin, Heidelberg, West Germany

Dr. R. S. Tilbury, Sloan-Kettering Institute, New York, N.Y., U.S.A.

Professor J. Tousset, Institut de Physique Nucléaire, Villeurbanne, France

Mr. M. D. Tran, Institut de Physique Nucléaire, Villeurbanne, France

Mr. N. G. Trott, Institute of Cancer Research, London

Dr. W. Vaalburg, Isotopen Laboratorium, Acad. Ziekenhuis, Oostersingel 59, Groningen, The Netherlands

Mr. N. I. Venikov, Kurchatov Institute of Atomic Energy, Moscow, U.S.S.R.

Professor J. Vervier, Centre de Physique Nucléaire, Heverle-Louvain, Belgium

Mr. D. D. Vonberg, M.R.C. Cyclotron Unit, Hammersmith Hospital, London

Dr. D. M. Wagnerová, Institute of Inorganic Chemistry, Czechoslovak Academy of Sciences, Praha, Czechoslovakia

Mr. A. G. Wain, Applied Chemistry Division, A.E.R.E., Harwell

Mr. G. N. Walton, Chemistry Division, A.E.R.E., Harwell

Mr. S. L. Waters, M.R.C. Cyclotron Unit, Hammersmith Hospital, London

Mr. L. J. White, Engineering Support Division, A.E.R,E., Harwell

Dr. J. E. Whitley, Scottish Research Reactor Centre, East Kilbride, Near Glasgow, Scotland

Dr. W. Wild, Chemistry Division, A.E.R.E., Harwell

Mr. L. Wilkinson, Engineering Support Division, A.E.R.E., Harwell

Mrs. L. A. Winsborrow, Department of Physics, King's College, London, W.C.2

Mr. J. A. Winter, Chemistry Division, A.E.R.E., Harwell

Dr. G. K. Wolf, Institut für Heisse Chemie, Kernforschungszentrum Karlsruhe, Karlsruhe, West Germany

Mr. C. W. Wormley, Department of Physics, University of Birmingham, Edgbaston, Birmingham

Mr. J. H. Worth, Metallurgy Division, A.E.R.E., Harwell

Mr. J. Wright, Applied Chemistry Division, A.E.R.E., Harwell

Professor L. Yaffe, Department of Chemistry, McGill University, Montreal, Quebec, Canada